Praise for Bryan Bu

"Burrough offers a Hollywood-worthy st[...] [...]ative journalism and dramatic family history. . . [...] [...]ped characters that seems boxed up and read[...] [...]*Time*

"In this riveting history, Burrough charts the decades-long rush that made Texas oil into a political and economic powerhouse through the lives of the four great barons. . . . Burrough brings each of his outsized subjects brilliantly to life, pitching their individual epics against a grand narrative of rise and decline."
—*The New Yorker*

"What's not to enjoy about a book full of monstrous egos, unimaginable sums of money, and the punishment of greed and shortsightedness by the march of events? . . . A ripping . . . read from start to finish." —*The Economist*

"Lively . . . impeccably rendered."
—Mimi Schwartz, *The New York Times Book Review*

"*The Big Rich*, a four hundred-page opus on the oil-powered rise of the Texas elite, has so many characters and entertaining subplots it reads like a petroleum-based *Lord of the Rings*. This is, of course, a compliment. . . . In Burrough's captivating story, done with the same keen eye on excess as his corporate classic *Barbarians at the Gate*, it's clear these men cast a shadow so wide they contributed more to our economic, national, and political identities than almost any other titans of industry." —*GQ*

"A vivid tale . . . Bryan Burrough has produced an elegantly interwoven chronicle of the lives of mid-twentieth-century independent oil barons and their families. . . . A smart, readable book with lessons for today's executives." —*BusinessWeek*

"A Lone Star epic . . . A galloping history of the wildcatters whose drive, ranches and gaudy mansions inspired both the *Beverly Hillbillies* and their malevolent neighbor, J. R. Ewing. . . . Burrough introduces his protagonists with a novelist's eye for detail." —*Bloomberg News*

"Winning . . . well researched and briskly told. Burrough has produced an indispensable guide to the knotty fascination that Texas spurs in the imagination."
—*BookForum*

"Gossipy, engrossing . . . Burrough knows a good story when he hears one."
—*Austin American-Statesman*

"It would be hard to ask for a literally or figuratively rich cast of characters than those in *The Big Rich*. . . . Nicely detailed and suspenseful."
—Harry Hurt III, *The New York Times* (Business Section)

"Eminently readable."   —*Texas Monthly*

"Bryan Burrough has long been one of this nation's best storytellers, but he has outdone himself with his tour de force, *The Big Rich*. Set amid the rough and tumble of the Texas oil fields and stretching to the halls of political power in Washington, this epic tale reveals the hidden undercurrents of modern American history that flowed from four families of unimaginable wealth and recklessness. With an unerring eye for detail, Burrough dissects their lives and histories, starting with the patriarchs—struggling, poorly educated men who might have remained forever unknown if not for their success at pulling black ooze from the ground. *The Big Rich* lays bare their arrogance and aspirations, their principles and hypocrisy, their daring and foolishness, taking readers deep inside a world of affluence that has remained secret for far too long. It is, quite simply, a triumph."
                —Kurt Eichenwald, author of *The Informant!* and *Conspiracy of Fools*

"Capitalism at its most colorful oozes across the pages of this engrossing study of independent oil men. . . . This is a portrait of capitalism as white-knuckle risk taking, yielding fruitful discoveries for the fathers, but only sterile speculation for the sons—a story that resonates with today's economic upheaval."
                                          —*Publishers Weekly* (starred review)

"The most improbable people of all must live in Texas and, in the good old days, they hunted for oil, found it, sold it, made fortunes and eventually blew most of it. In *The Big Rich*, Bryan Burrough, former *Wall Street Journal* reporter and now a *Vanity Fair* special correspondent, tells a wonderful tale of the four biggest Texas millionaires."   —*St. Louis Post-Dispatch*

"*The Big Rich* is simply a 'must-read.'"   —*The Capitol Annex*

"It's hard to imagine a greater literary marriage than that of the oil barons of Texas and Bryan Burrough. On the one hand, you have a collection of gargantuan personalities who in the 1920s struck it rich and then, in the decades that followed, used their wealth to transform American business, culture, and politics. On the other, you have an author—and native Texan—who writes, as he always does, with enormous insight and panache. *The Big Rich* has all the hallmarks of a classic American saga."   —David Margolick, author of *Beyond Glory: Joe Louis vs. Max Schmeling, and a World on the Brink*

"Burrough . . . invokes a tale of bitter competition, family feuds, booms, and bankruptcies that more than lives up to the legends."   —*Booklist* (starred review)

"An entertaining look at the larger-than-life histories of the incomprehensibly rich and powerful."   —*Library Journal*

"Full of schadenfreude and speculation—and solid, timely history too."
                                                               —*Kirkus Reviews*

PENGUIN BOOKS

THE BIG RICH

Bryan Burrough is the author or coauthor of seven books, four of them *New York Times* bestsellers, including the Wall Street classic *Barbarians at the Gate: The Fall of RJR Nabisco* and, most recently, *Forget the Alamo: The Rise and Fall of an American Myth*. A longtime correspondent at *Vanity Fair*, he lives in Austin.

THE
BIG RICH

The Rise and Fall of the
Greatest Texas Oil Fortunes

BRYAN
BURROUGH

PENGUIN BOOKS

PENGUIN BOOKS

Published by the Penguin Group
Penguin Group (USA) Inc., 375 Hudson Street, New York, New York 10014, U.S.A.
Penguin Group (Canada), 90 Eglinton Avenue East, Suite 700, Toronto,
Ontario, Canada M4P 2Y3 (a division of Pearson Penguin Canada Inc.)
Penguin Books Ltd, 80 Strand, London WC2R 0RL, England
Penguin Ireland, 25 St Stephen's Green, Dublin 2, Ireland (a division of Penguin Books Ltd)
Penguin Group (Australia), 250 Camberwell Road, Camberwell,
Victoria 3124, Australia (a division of Pearson Australia Group Pty Ltd)
Penguin Books India Pvt Ltd, 11 Community Centre, Panchsheel Park, New Delhi – 110 017, India
Penguin Group (NZ), 67 Apollo Drive, Rosedale, North Shore 0632,
New Zealand (a division of Pearson New Zealand Ltd)
Penguin Books (South Africa) (Pty) Ltd, 24 Sturdee Avenue,
Rosebank, Johannesburg 2196, South Africa

Penguin Books Ltd, Registered Offices:
80 Strand, London WC2R 0RL, England

First published in the United States of America by The Penguin Press,
a member of Penguin Group (USA) Inc. 2009
Published in Penguin Books 2010

ScoutAutomatedPrintCode

Grateful acknowledgment is made for permission to reprint an excerpt from
"Breakfast in America" by Roger Hodgson and Richard Davies. © 1979 Almo Music Corp.
and Delicate Music. All rights administered by Almo Music Corp. (ASCAP).
Used by permission. All rights reserved. www.rogerhodgson.com

THE LIBRARY OF CONGRESS HAS CATALOGED THE HARDCOVER EDITION AS FOLLOWS:
Burrough, Bryan, 1961–
The big rich : the rise and fall of the greatest Texas oil fortunes / Bryan Burrough.
p.   cm.
Includes bibliographical references and index.
ISBN 978-1-59420-199-8 (hc.)
ISBN 978-0-14-311682-0 (pbk.)
1. Petroleum industry and trade—Texas—Biography.   I. Title.
HD9567.T3B873 2008
338.2'7280922764—dc22   2008027043

Printed in the United States of America
Designed by Amanda Dewey

To Marla, and to Griffin and Dane

CONTENTS

INTRODUCTION

I t's hard to tell people about Texas. It is. It's hard to explain what it means to be a Texan. To anyone who grew up in the North, it probably means nothing. The idea of a state "identity," or that a state's citizens might adopt it as part of their own self-image, seems a quaint, almost antebellum notion. Folks in Iowa don't strut around introducing themselves as Iowans, at least none I know.

But if you grew up in Texas, as I did, it becomes a part of you, as if you're a member of a club. It's a product of the state's enduring, and to my mind endearing, parochialism, a genetic tie to the days when Texas was a stand-alone nation born of its own fight for independence, which produced its own set of national myths. Ohio doesn't have an Alamo. I'm not sure Ohioans, as wonderful as they are, have a distinct culture. As a child I was always vaguely ashamed I wasn't born in Texas. I'll never forget the day a boy in my fifth-grade class actually called me a carpetbagger. How on earth would he even know what that was?

The myths about Texas die so hard, mostly because Texans love them so. So much of it is wrapped up in oil. Non-Texans probably think it's all-pervasive; it's not. The fact is, growing up in Central Texas during the 1970s, I never met an actual oilman. There were a few pump jacks out in the fields around our little town, but we never gave a thought to who owned them. It wasn't until I was sixteen, the weekend I served as an escort at Waco's Cotton Palace debutante ball, that I was introduced to the class of Texans known as the Big Rich: boys from Highland Park and River Oaks in white dinner jackets and gleaming hair, willowy Hockaday girls with enormous eyes and glistening jewels. Ogling them from within my rumpled rented tux, they seemed like royalty.

And they were, Texas royalty at least. They had flasks in their pockets and talked of boarding schools and weekends in Las Vegas and the wine in Paris and jetting to London and my head just spun and spun and spun. It wasn't for another five years that, as a cub reporter for the *Wall Street Journal* in Dallas and later in Houston, I began to read—and write a little—about the Big Rich. Though I didn't realize it at the time, those were the years, the early and mid-1980s, when their era was ending. The fathers of those boys from River Oaks were going bankrupt; their buildings were being sold or torn down. The state was completing a decades-long maturation, and the new Texas, chockablock with northern-owned corporations and Yankee executives and their shimmering office parks, was fast becoming something different and somehow artificial, a Texas-flavored Ohio. Something was being lost.

I haven't lived in Texas for twenty years, but in some ways I've never left. My parents are still there. I visit often. Still, when my editor suggested some kind of book on Texas oil, I was surprised how quickly a structure sprang to mind. It took barely thirty seconds, in fact. It would be not about the oil industry per se but about the great Texas oil families, the ones who generated all those myths. The Hunts. The Basses. The Murchisons. The Cullens. I thought of them as the Big Four, though it wasn't until I began my research that I found they had been called exactly that, although not since the 1950s.

This book is built on three years of research, during which I plunged into dozens of Texas and out-of-state archives, interviewed surviving members of the Big Four families, and read more than two hundred books and thousands of newspaper and magazine articles. Some of the choicest information I found in county courthouses, in the mammoth, musty suitcase-sized ledger books where Texas clerks for decades scrawled out the minutiae of land records, oil leases, and lawsuits. The published literature is hefty but uneven, and hasn't been refreshed since the great bankruptcies of the 1980s.

There have been four books published on the Hunts, two on the Murchisons, one on Roy Cullen, and nothing but bits of journalism on the Basses and their paterfamilias, Sid Richardson. The best of these are Harry Hurt III's 1981 history of the Hunts, *Texas Rich*, and Jane Wolfe's 1989 *The Murchisons*. Both books are definitive; I've done my best to add new material to their stories despite having far fewer pages to use, a product of telling all four families' tales at once. Also of great help were books written by the foremost historian of the Texas oil industry, Roger Olien.

These and other histories served as a starting point to explore the rise and fall of the greatest Texas oilmen, many of whom are fast being forgotten. There's never been anything lasting written about Houston's flamboyant Glenn McCarthy, though there should be; it's hard to find anyone under sixty who remembers a Texas legend so famous in his day he adorned the cover of *Time*. There's been even less written about the secretive Sid Richardson, once the richest man in America. My research led into areas other historians have downplayed or ignored, notably the Big Four's involvement in national politics. Texas Oil's contribution to America's rightward shift in recent decades became a major theme for me. I mean, the Bushes had to come from somewhere.

This book is what you might call an engineered history; that is, I've superimposed a narrative framework onto disparate events that may be familiar to Texans of a certain age. There are concepts introduced here with which some academic historians may disagree, such as the idea that Texas oil wealth was "discovered" by the national press in 1948. The story of the Big Four's introduction to, and ignominious departure from, Washington politics during the 1950s has likewise never been told; few realized what was happening at the time, and even fewer wrote about it afterward. In fact, the very idea that the highwater years of the Big Four constituted an "era" from 1920 until about 1986 is likely to be challenged in a state where, when I was growing up at least, every student took a year of Texas history in the seventh grade.

The joys of writing this book were multitude. There's nothing I love more than cruising the Texas back country. It's beautiful land, calming, serene. Before working on this book, I thought I knew the state well. Then one morning in 2005 I was in far West Texas heading out of Midland on Rural Route 91, out toward the state's most remote corner, Winkler County, in the crook of the elbow where southeastern New Mexico tucks into the Texas border. There's nothing but barbed wire and blue sky for miles around, or so I thought, until I crested a rise, just east of Kermit, and entered a tableland whose view was so breathtaking I had to pull to the side.

There, as far as the eye could see, were oil wells. Hundreds, maybe thousands, of robotic pump jacks, their metal heads bobbing up and down like metronomes, laid out from my windshield to every horizon. It was like something out of Edgar Rice Burroughs, a lost plateau, filled not with dinosaurs but with the steel and wire and sweat of American industry. Men had been out here for years, I realized, mapping the land, drilling holes in the earth, and, I imagined, returning home to Dallas and Houston and Fort Worth with millions of dollars in their pockets. This was a Texas, an America, I had never seen, and I suddenly needed to know what became of these men and their fortunes. Their stories, it turned out, were everything I had imagined and more.

I hope you enjoy *The Big Rich*. It certainly was a pleasure to write. Everything you read here is true; any errors are mine and mine alone. If you have any question or comments, feel free to e-mail me at bburrough@comcast.net. I live with my wife and two sons in suburban New Jersey now, a half hour west of the Holland Tunnel. For you native Texans out there, I hope you won't hold that against me.

BRYAN BURROUGH
SUMMIT, NEW JERSEY

ONE

"There's Something Down There..."

I.

On Friday morning, January 10, 1901, the people of Beaumont, in southeast Texas, woke beneath their blankets to a chill dawn, the winds of a rare blue Norther whistling past the buckboards and wagons bumping along the unpaved downtown streets. Splayed along the banks of the Neches River, about fifteen miles inland from the Gulf of Mexico, Beaumont was a lumberman's town, ringed by sawmills that split and cut the giant tree trunks railroad cars brought in from East Texas.

Four miles southwest of town, out on the barren coastal moors pockmarked with marshes and sinkholes, rose a lonely mound. Cattle nibbled grass all around it. The locals called it the Big Hill, but it was hardly a hill at all, more a bump on the earth, barely fifteen feet high. The prairie grasses, waving now in the north wind, gave way at its crest to a whitish spray of alkali, making it look like a grandfather's balding head. The older children sometimes tried to scare the little ones by saying the Big Hill was haunted. There were legends the pirate Jean Lafitte had buried treasure there. Sometimes at night you could see strange dancing lights. The Big Hill certainly smelled satanic, owing to the sulfur springs that spat and bubbled all around its weary base.

Atop the mound, silhouetted against the gray morning sky, stood the skeleton of an oil derrick, a latticework triangle of wood, sixty-four feet high. At its base that morning two men could be seen working, a square-jawed young man named Curt Hamill and his helper, Peck Byrd. At around nine, a buckboard pulled up below. Their boss, Curt's brother Al, trudged up the hill with a new drill bit. The old one had ground to pieces the day before when they struck solid rock.

The Hamills, honest, strongly built men known for working themselves to exhaustion, were water-well drillers who had switched to oil three years earlier, nursing the odd trickle of crude out of the ground at Corsicana, south of Dallas. Called to Beaumont, they had been working this new well for seven weeks now, reaching eleven hundred feet, and they had already found the rainbow sheen of oil in the muck they drew from below. They told no one but the owner, who instructed them to drill deeper. Yet even that modest showing of oil would surprise folks in Beaumont. Most everyone around town thought they were daft to be looking for oil here anyway. Everyone knew the only serious oil under American soil was back east, in Pennsylvania, or up north, in Kansas.

When Al brought up the new bit, he and Curt attached it to the drill stem, then began lowering steel pipe into the hole. Curt clambered high on the derrick, onto the swivel boards, where he worked the elevator wires as the pipe slowly disappeared into the earth. An hour later they had fed thirty-five joints of pipe into the hole, reaching seven hundred feet or so, when suddenly the pipe began to shake violently. A moment later, the rig began to shudder and quake.

As the Hamills watched, dumbstruck, reddish-brown drilling mud began to bubble up from the hole, slowly at first, then faster, till the entire drilling platform was awash. With the rush of mud came a hissing sound, then a gurgling, as if some giant beast below was spitting up. Beneath them the Big Hill began to rumble. The flow of drilling mud began to jump and leap, creating a fountain right there in front of them. Then, suddenly, the fountain exploded, an eruption of mud that shot straight up into the morning sky high above the derrick.

Al and Peck Byrd dived to the side, rolling beneath a barbed-wire fence. The geyser of mud drenched Curt, high on the swivel boards, momentarily blinding him. He leaped for the ladder and skittered to the ground like a circus acrobat, though years later he would say he had no memory of doing so. Al yelled for him to run, but Curt paused. Now rocks began to shoot out of the hole, firing into the sky and falling around them. With them came the rotten-egg smell of natural gas. The traveling block, caught in the geyser, began to lift off the ground, and Curt saw that if it hit the top of the derrick, it could destroy everything. Bravely wading through the rain of mud and rocks, Curt wobbled across the derrick floor and kicked out the clutch, stopping the machinery. Then he dived out, rolling down toward Al.

Just then drill pipe began shooting out of the hole, tons of it, rocketing up through the crown block, destroying the top of the derrick. The pipe surged high in the air, then broke into sections, tumbling back to the ground as the men covered their heads. After a moment Curt took his shirttail and cleared his eyes. It was then he thought of the boiler. It was still on fire. If flames lit the gas escaping from the hole, well, Curt didn't want to think of that. All three men crawled to their feet and ran for the boiler pit, where they took buckets of water and threw them into the firebox, dousing the flame. Then they ran. From safe spots on the hillside below, they watched as the geyser of mud slowly fell, foot by foot, then stopped.

Silence.

Slowly the three men crept back up the hill. The derrick still stood, just barely, but all their machinery was a wreck. To one side the drill stem, heaved from the hole during the tumult, protruded from the hillside like a thrown spear. Six inches of drilling mud covered the derrick floor. Discarded pipe lay everywhere. Al stared at it. It was rented pipe, and he knew they would have to pay for it.

Drillers hated blowouts like this; they could ruin months of hard work. Al had just grabbed a shovel, starting the cleanup, when suddenly a six-inch plug of mud exploded from the hole like a cannon shot. Then, once again, silence enveloped the Big Hill. The men glanced at one another. Al edged toward the hole. Taking care, as if perched on the edge of a balcony, he looked down into it. It took a moment for his eyes to focus, but then he saw it: Something was down there. Something moving evenly up the hole, then receding back into the darkness. As Al watched, it flowed up again, then down again, the movement regular now. For the longest few seconds, the thing down in the shadows once again heaved toward him, then eased back. To Al Hamill, it was as if he was peering into the very heart of the Earth, and it was breathing.

What happened next would change the course of history.

II.

There is a legend in America, about Texas, about the fabulously wealthy oil-men there who turned gushers of sweet black crude into raw political power, who cruised their personal jets over ranches measured in Rhode Islands, who sipped bourbon-and-branch on their private islands as they plotted and schemed to corner entire international markets. In popular culture the Texas oilman tends to come in two guises, the overbearing, dim-witted high roller with a blonde on either arm, and his evil twin, the oilman of Oliver Stone and Mother Jones, the black-Stetsoned villain whose millions pull the levers of power in Washington. He can be young and conflicted and obscenely rich, as in James Dean's portrayal of the wildcatter Jett Rink in *Giant*, or smooth and conniving and obscenely rich, like J. R. Ewing of television's *Dallas*, but he is almost always crass and loud and a tad mysterious, a classic American other.

There is truth behind the legend, a surprising amount in fact. There really were poor Texas boys who discovered gushing oil wells and became overnight billionaires, patriarchs of squabbling families who owned private islands and colossal mansions and championship football teams, who slept with movie stars and jousted with presidents and tried to corner an interna-tional market or two. Back before television, before *Lifestyles of the Rich and Famous*, they were the original Beverly Hillbillies, counting their millions around the cement pond as they ogled themselves on the cover of *Time*. They helped make Texas Oil an economic and political powerhouse, a world whose contributions, large and small—from Enron and two George Bushes to the Super Bowl—shaped the America we know today.

This is their story, told through the lives of the four Texas families—and a few of their peers—who rose the highest and, in some cases, fell the hardest. Each of their patriarchs began in obscurity, and all, through a historic quirk of fate, laid the foundations of their fortunes in a single four-year span. They married, bore and lost children, and became the country's first shirtsleeve billionaires, transforming America's idea of what it meant to be rich even as they played host to kings and queens and accumulated every toy of their age: mansions and ranches and castles, airplanes and yachts and limousines, skyscrapers, hotels, and cabinet members. As they reshuffled the deck of America's most powerful families, eventually helping to propel three of their favorite sons into the Oval Office, the country was enthralled by them, then suspicious, then, after a fateful fusillade of gunfire, came to view them as nefarious caricatures.

In time their salad days dissolved into a sordid litany of debauchery, family feuds, scandals, and murder, until collapsing in a tangle of rancorous bankruptcies. Some survived, others didn't. A few count their millions today. As the movies say about almost every story set in Texas, theirs is a big, sprawling American epic, marked by exhilarating highs and crashing lows, and it all began, sort of, with a queer character named Patillo Higgins and that odd hump of dirt they called the Big Hill. History would know it as Spindletop.

III.

For outsiders, it can be tough to wrap one's mind around Texas. The first challenge is its sheer size: 801 miles from north to south, 777 miles across; if plopped down on the Eastern Seaboard it would subsume all of Maine, Vermont, New Hampshire, Massachusetts, Connecticut, Rhode Island, New York, Pennsylvania, and Ohio, with room left over for North Carolina. Its enormity encompasses every North American landscape short of glaciers: pine forests in East Texas, deserts and sand dunes in West Texas, marshland and moors along the coast, mountains as high as eight thousand feet, and enough flat scrubby plains, stretching from Dallas to El Paso, to sate a Mongol horde.

The second challenge is the Texas mind-set: proud, stubborn, and independent, perhaps unsurprising for a state that in 1901 was only fifty-six years removed from its previous status as an independent country. In the intervening years Texans had defeated the last Comanches and pushed the line of settlement out onto the plains. The state's suffocating parochialism was a by-product of its isolation; while Europeans poured into Eastern cities, Texas experienced almost no immigration in the late 1800s. Old ways hardened. Change was resisted. Outside interests, especially commercial interests, were met with suspicion; a wave of progressivism swept the state in the 1890s, during which Texas attorneys general filed several antitrust suits against the monopolistic Standard Oil, even though the company had practically no presence in the state. Self-sufficient and inward-looking, people thought of themselves as Texans first, Americans second. By 1901, "the early 19th-century American values were in no way eroded in Texas," the historian T. R. Fehrenbach has written. "There was no reason why they should have been. During a century of explosive conquest and settlement, the land changed very little, and the people not at all."

Turn-of-the-century Texas was half Old South, half Old West, a rural frontier society that, for all its ballyhooed pride, had very little to preen about. As the rest of America industrialized, Texans still lived off the land. In the eastern half of the state cotton, still picked from the bush by Negro and poor white sharecroppers, was the cash crop. Vast ranches, one or two as large as Delaware, spread across West Texas and the Panhandle, where cowboys herded cattle and sheep onto new railroads for sale to meat-packers up north. The only industry to speak of was lumber, hewn in sawmills dotting the East Texas backwoods. Not until 1901 did the state boast a multimillion-dollar corporation, when the Houston lumberman John Henry Kirby, the fabled "Prince of the Pines," formed the giant Kirby Lumber Company.

It was into this backward, agrarian world that Patillo "Bud" Higgins, the Johnny Appleseed of Texas Oil, emerged in the 1890s. Higgins was the dreamer every new era demands, a one-armed Beaumont eccentric who in a day when Texas was known for cattle and cowboys and not much else envisioned giant pools of oil beneath its dirt and giant cities that would sprout above them. Born a gunsmith's son in 1863, confident, rail-thin, and peripatetic, Higgins had been a juvenile delinquent, losing his left arm during a drunken evening in which he and a pair of teenage pals wildly fired pistols into a Negro neighborhood. Confronted by a local deputy, Higgins shot and killed him, while the deputy's return fire hit his arm, which later had to be amputated. A jury called it self-defense.

In his twenties Higgins became what is known today as a born-again Christian, emerging as a church deacon and throwing himself into a variety of money-making schemes. He worked on the railroad and carted dust at a sawmill before trying his hand as a cabinetmaker, fishmonger, and real estate agent. In 1886, noticing that all of Beaumont's bricks were imported from Houston and New Orleans, he opened a brickyard, the area's first. Three years later, determined to find ways to make his kiln more efficient, he embarked on a tour of northern brickyards, examining facilities in Indiana, Ohio, and Pennsylvania. In Dayton and Indianapolis he saw how a kiln burning over an oil fire burned more evenly than over a wood fire. Intrigued, he headed on to the oil fields of western Pennsylvania to see what this oil business was all about.

He found an industry in its infancy, barely thirty years old. A man named Edwin Drake had found the first oil there in 1859; most of it was refined into kerosene and various lubricants. In the ensuing years much of the industry had come under the control of John Rockefeller's giant Standard Oil, which as Higgins made his rounds was under mounting attack for its monopolistic practices. What he learned in Pennsylvania fired Higgins's imagination. Telltale signs of oil found there had come on the surface of the land, in seeps and sulfur springs any man could see. His father had been a Confederate soldier, and Higgins remembered stories he told of rebels around Beaumont lubricating their guns with some sort of oily substance they found south of town, at the Big Hill. He had played and bathed in the sulfur springs at its base as a child, and taken his Sunday school students on picnics nearby. Higgins returned to Texas certain there was oil under the Big Hill.

Believing the oil beneath Spindletop would transform Texas, Higgins worked out a Utopian scheme of development, complete with pipelines, refineries, iron smelters, a deepwater port, and a metropolis he named Gladys City, after his favorite Sunday school student, Gladys Bingham. No one around Beaumont took any of this seriously, but in 1892 Higgins did manage to entice a member of his church, a lumberman named George W. Carroll, to cosign the five-thousand-dollar note he needed to lease a thousand-acre parcel that covered about half of Spindletop hill. Once Carroll became involved, several adjoining landowners signed over their lease rights. They founded a company, Gladys City Manufacturing.

A water-well driller was hired to sink a hole; he gave up upon hitting quicksand. A second attempt did no better. Higgins was already low on cash when a nationwide depression hit in 1893, scattering his investors. Still, he refused to give up. Hoping an official endorsement of Spindletop's oil potential might lure new investors, Higgins invited a state geologist to assess his land; to his dismay, the geologist not only declared his venture hopeless, he published his opinion in the Beaumont newspaper, advising the town to avoid "frittering her money away upon the idle dreams or insane notions of irresponsible parties in the vain outlook for either oil or useful gas."

It was at that point, as the specter of a local bailout evaporated, that Higgins got lucky—or so he thought. The discovery of minute amounts of oil at Corsicana, south of Dallas, had drawn a number of curious eastern oilmen to Texas, and one group, Savage Brothers of West Virginia, approached the Gladys City board about acquiring its lease rights in return for a royalty on any oil it might find. Higgins, fearful of losing control, objected, but other board members, desperate to recoup their money, overruled him. Savage Brothers, however, got nowhere. Like the previous two efforts, its driller used a cable-tool rig, which created a hole by repeatedly pounding downward, a strategy that time and again proved unable to defeat quicksand. When Higgins and his board members fell to squabbling, he quit the company; George Carroll sued and Higgins was eventually forced to sell back his stock. By 1898, while still determined to pursue his dream of oil on Spindletop, Higgins had returned to the real estate business.

Still, he couldn't forget the Big Hill. In one final stab at seeing it drilled, Higgins placed an advertisement in a national magazine. He got precisely one answer, from a down-on-his-luck character named Anthony Lucas, a onetime captain in the Austro-Hungarian navy who had become fascinated with the possibilities of a geologic feature known as a salt dome, literally domes of salt that poked up, pimplelike, in mounds all along the American Gulf Coast. Lucas thought salt domes often harbored caches of sulfur or sometimes oil. After meeting with Higgins, he judged the Big Hill a classic salt dome.

Lucas drilled a well and found a thin shean of oil in the dirt he drew to the surface, but the hole soon collapsed. Neither Lucas nor Higgins had any more money to start again. Now it was Lucas who took up the banner of the Big Hill. He spent months canvassing the large oil companies back east, even gaining an audience at Standard Oil. But the professionals said he was crazy; any fool knew salt domes held only one thing—salt. Finally, in Pittsburgh, Lucas found a willing ear in James Guffey and James Galey, two of Pennsylvania's best-known wildcatters. They took over the acreage Lucas had amassed, cutting him in for a share. It was Guffey and Galey who hired the Hamill brothers and dispatched them to drill the Big Hill. It was there four months later that the well appeared to explode and Al Hamill thought he saw the Earth breathe.

IV.

The Hamills had just started to pull debris off the derrick floor when a sudden roar emanated from the hole. Once again, mud bubbled to the surface. Once again came a blast of natural gas. Then, as the Hamills and Peck Byrd scrambled for cover, a geyser of greenish-black liquid erupted from the ground. As the crew stared in amazement, it grew and grew, until it reached its apex a hundred feet into the morning sky. Stultified, they sent Byrd running to fetch Captain Lucas, who appeared at a dead run a half hour later. "Al! Al!" Lucas shouted, pointing at the black spout. "What is it? What is it?"

"Oil, captain!" Hamill yelled. "Oil! Every drop of it!"

Lucas had never seen anything like it. No human had. The Lucas No. 1 well changed the world forever. That first well produced at a greater rate than all other American oil wells in existence—*combined*. In a matter of days, in fact, the pastures around Spindletop would be producing more than the rest of the world's oil wells—*combined*. Of those first six Texas oil wells, three produced at a higher rate than the entire country of Russia, then the world's top producer.

Spindletop not only created the modern American oil industry, it changed the way the world used oil. Its dirty little secret was that the oil found around Beaumont was of such poor quality it could not be refined into kerosene. But it made fine fuel oil—and that's what changed everything. So much black crude flowed from Beaumont that oil prices dropped to three cents a barrel—a cup of water cost five cents—making it economical for railroads and steamship companies to convert from coal to oil. The Santa Fe Railroad went from one oil-fired locomotive in 1901 to 227 in 1905.[1] Others followed, as did the British, American, and German imperial navies. Everything that today runs on oil and its by-products, from automobiles to jet fighters to furnaces, barbecue grills, and lawn mowers—all of it began at Spindletop.

The rush to drill near Spindletop triggered the greatest speculative boom since the California Gold Rush. It took ten long days just to cap the Hamills' first well, but within days more derricks began going up all around it—214 wells in all by the following summer. For a time the discovery of oil transformed Beaumont into a classic American boomtown, an orgy of mud and blood, oilmen, prostitutes, and thieves of every stripe, anyone and everyone who was willing to work for a quick buck. It was the beginning of Texas Oil.

But if Spindletop created an oil industry for Texas, little of it ended up controlled by Texans. The big money at Spindletop was initially split between groups of powerful Texas businessmen and seasoned oilmen from back east. One Texas faction was an alliance of Austin politicians and Gulf Coast attorneys led by the former governor, Jim Hogg, who acquired a valuable lease on Spindletop hill for the bargain price of $180,000 in July 1901, six months after the first gusher. Other Texas syndicates formed around East Texas's leading lumbermen, including John Henry Kirby. Dry-goods merchants from Dallas, cattlemen from Fort Worth—just about any Texas businessman with cash threw it at Spindletop.

The Texans, unfortunately, knew next to nothing about oil. Time and again they were outmaneuvered by eastern oilmen with experience in Pennsylvania and other fields. The pivot on which everything turned, at least initially, was James Guffey, who struck "the deal of the century" when he sold much of Spindletop's oil to a company he had never heard of—and whose executives needed a map to locate Beaumont—Royal Dutch Shell, Europe's largest oil producer; the deal made Shell an international colossus. Guffey was backed by and later sold out to the Mellon family of Pittsburgh, who roared into the Gulf Coast fields with a new company it named, appropriately, Gulf Oil, which became one of America's greatest oil companies. The Pew family of Philadelphia, founders of Sun Oil, swept into Spindletop in a blizzard of activity, laying pipelines, buying storage facilities and so much oil it had to build a new refinery at Marcus Hook, Pennsylvania. Another of the companies weaned at Spindletop was the Texas Company, later known as Texaco.

In the face of such competition, many of the Texas groups wilted away, leaving the Spindletop fields largely in eastern hands. Still, the boom created the first Texas oilmen, and as the prospect for new fields around Beaumont ebbed, they fanned out in search of new salt domes to drill. They found them quickly, bringing in miniature Spindletops in an arc of new fields scattered around Houston, at Batson and Sour Lake and, in 1903, at Humble, eighteen miles northeast of downtown. And then...nothing. In 1904 and 1905 and into 1906, every discernible salt dome on the Texas coast was poked like a patient, but no new Spindletops turned up. The boom began to wane. When, in 1906, oil was discovered in Oklahoma, hundreds of men began streaming north.

In their wake, Texas politicians confronted a troubling new reality. The state, in effect, found itself in the same position as Patillo Higgins: it had found oil, lost control over it to eastern businessmen, and had no guarantee the same thing wouldn't happen the next time a gusher was struck. While groups led by native Texans still controlled a share of the new Gulf Coast reserves, almost all the strange new infrastructure of Texas Oil—the storage tanks, the pipelines, the refineries—was controlled by eastern interests. Ominously, easterners were in a position to dictate the terms and to some extent the price of the Texans' oil; with that kind of power, it was only a matter of time before the hated Yankees used it to squeeze Texans out of the oil bonanza altogether.

The state's salvation, it turned out, lay in the fine print of its antitrust laws, which forbade the integration of oil companies, that is, companies that stored, transported, and refined oil weren't allowed to produce it. There were exceptions galore, but between 1905 and 1910, politicians in Austin moved to ensure that native Texans retained their toehold in oil, defeating two measures that would have allowed integration amid the din of anti-eastern rhetoric. Integration, one legislator decried, would "enable the great capitalists of the North to seize complete control of the Texas oil industry." Legislators put teeth in their words when the hated Standard Oil tried to creep into Texas, secretly funding a flamboyant empresario named George A. Burt, who built the world's largest oil refinery, outside Beaumont. When Texas oil prices rose in 1906, Burt tried to drive them down by threatening to import cheaper oil from Oklahoma. The Texas attorney general, Robert Vance Davidson, responded by strafing Burt with a series of lawsuits and fines, eventually forcing Standard to dismantle its refinery and rebuild it in Baton Rouge, Louisiana. The message was clear: don't mess with Texas.

The upshot, by 1910 or so, was that control of Texas Oil remained split. Eastern interests—Shell, Sun, Gulf, and others—handled the purchasing, transporting, refining, and retailing of oil and its by-products, while Texans remained active in exploration and production. The official division lasted until 1917, when state legislators, realizing that out-of-state funding was crucial to the industry's growth, finally passed a law allowing the large companies to integrate. The intervening years, however, saw the creation of a native Texas presence in the risky business of oil-finding that would endure for decades. Hundreds of small companies, many one-man operations, popped up to look for oil. During the 1910s at least, they did a dismal job of finding any. Year after year passed with no new Spindletop. Many companies disappeared. Scores of oilmen returned to their jobs as farmers, clerks, or lawyers. Thanks to their brass-knuckled political leaders, Texans had earned the right to look for their own oil.

Now they just had to find some.

V.

Spindletop may have changed the world, but it didn't change Texas that much—not at first. What the Spindletop boom provided native Texans and new arrivals was not so much a vault of black treasure as a classroom where the oil business could be *learned*.

A few caught on quickly. Among them was a young Iowa heir named Howard Hughes, who fled a lead mine in Southwest Missouri for Spindletop within weeks of that first gusher. Captivated by oil, Hughes and a partner, Walter Bedford Sharp, who had drilled one of Patillo Higgins's early wells, began sinking wells in fields opening in northern Louisiana. Frustrated by their drillers' inability to penetrate solid rock, Hughes and Sharp developed a drill that could. Patented in 1908, the Hughes rock bit became an industry standard, and by the time the Hughes family returned to Houston in 1909, Hughes Sr. was fast becoming a wealthy man. In time the company he founded, Hughes Tool, would make his son, the legendary Howard Hughes Jr., the wealthiest Texas oilman of all, though in name only. When he reached adulthood in the 1920s, Howard Hughes fled Texas for Hollywood and never returned.

In those early years the best-known and most profitable of Texas-run oil companies was Humble Oil & Refining, formed in 1917 by the mass merger of an all-star team of Houston and Beaumont oilmen, many of whom began their careers at Spindletop. In later years its founders, men like Walter Fondren, Robert Lee Blaffer, and future Texas governor Ross Sterling, became what passed for "old money" in Houston. Savvy, aggressive finders and purchasers of oil reserves, the Humble men sold half the company to Standard Oil in 1919 for seventeen million dollars, the largest transaction in the Texas oil industry's brief history. Humble, which later became better known as Exxon, would remain a power in the Texas oil fields for decades to come. Its counterpart in Dallas was the Magnolia Oil Company, formed from the remnants of the company Standard Oil left behind when it fled the state. Standard retained a sizable minority stake it slowly increased in Magnolia, whose claim to fame was the giant red-neon Pegasus it erected over its downtown headquarters, a Dallas landmark visible from fifty miles away.

The oilmen who ran Humble and Magnolia and scores of smaller rivals survived the lean years of the 1910s by dividing the risk in a very risky business. In doing so, they also divided the upside; as wealthy as they were by local standards, no one in that first generation of Texas oilmen—with the asterisked exception of Howard Hughes—created anything like a true American fortune, nor left any lasting footprints on history. Their success, however, raised a tantalizing question: What if there really was another Spindletop out there, and what if it were discovered not by a large company but by a single Texan, working alone? Just how wealthy would Patillo Higgins have become had he been able to keep Spindletop to himself?

One well, one fortune. It was the stuff of myth, the El Dorado of Texas Oil, and as a new decade dawned, a horde of young second-generation oilmen would begin trying to find it.

TWO

The Creekologist

Some day we'll have a big white house to live in.
—HUGH ROY CULLEN AT AGE TWELVE

I.

By 1920, two decades after Spindletop, the discovery of oil hadn't changed Texas much. Of the forty-eight states, it ranked just fourth in oil production. During the 1910s many oil field workers actually fled the state to work in fields opening in Oklahoma. "Why do you waste your time in Texas?" one asked a soon-to-be famous geologist named Wallace Pratt. "Why don't you come up to Oklahoma where all the oil is?" What oil was pumped from Texas fields was still largely controlled by eastern interests; the biggest companies that flowered at Spindletop were now headquartered back east, Gulf in Pittsburgh, Sun in Philadelphia. Even the Texas Company, after a squabble between its Houston-based executives and eastern investors, was now based in New York. Academics viewed the state of Texas as an economic colony of the East, a view that would endure for years after it was no longer true.

All this began to change in the months after the armistice that ended World War I in November 1918. The war years had been good to middle America; between 1914 and 1918, the average wage rose 63 percent. The United States was still an agricultural country and farmers did even better; during the war, wheat and corn prices doubled. Cotton prices tripled. Wartime restrictions, reinforced by the frugality of uncertain times, obliged many Americans to stow their new money in mattresses and savings accounts—until the war ended.

Then, in 1919, an orgy of consumer spending inaugurated the Jazz Age. All across America, wartime parsimony gave way to an explosion of household purchasing as families finally opened their wallets to buy the new technologies the new American oil had wrought. Farmers bought tens of thousands of tractors. Housewives ordered new stoves and ranges. Corporations ordered new industrial boilers. No item, though, sold faster than automobiles. In 1900 there were eight thousand cars in the United States. By 1916 there were three and a half million. By 1921 there were 10.5 million. And every single one needed gasoline. Between 1900 and 1920, American energy demand quintupled.

The surge in demand produced postwar gasoline shortages, as American refineries strained to produce enough fuel to satisfy the country's new appetites. Major oil companies doubled and redoubled their exploration budgets in the scramble to find new oil, hiring geologists and scouts by the scores and sending them scurrying down every paved road in the Southwest, and quite a few that weren't. Nearly thirty-four thousand American oil wells were drilled in 1920, more than twice the number just five years earlier. Booms came (and went) in Kansas, Arkansas, Kentucky, Louisiana, and California.

Suddenly, though, it wasn't just oilmen who were looking for oil. The Jazz Age ushered in a period of "get-rich" mania, producing a classic American frenzy of speculative ventures. Everywhere slick promoters, backed by modern advertising, pushed ordinary citizens to plunge their savings into a range of new and often risky investments: real estate in Florida and California, gold and silver mines in Nevada, all manner of Wall Street offerings—and oil. Magazines thrummed with stories of the fortunes being made in Oklahoma, Kansas, and Texas, a few of which were actually true. Southwestern oil fields were "making millionaires at...a dizzy rate," *Scientific American* noted in 1917, mentioning Oklahomans such as Tom Slick and Harry Sinclair, who founded Sinclair Oil. "Men who three or four years ago were in the down-and-out class are millionaires many times over today." Oil fields, the *Saturday Evening Post* told readers in 1918, are "where fairy tales come true," where "new kings of oil appear upon the stage, and fortunes by the hundreds—yes, even by the thousands...all spring up like magic overnight."[1]

Spurring much of this publicity was Texas's first genuine set of gushers since Spindletop. They erupted around the drowsy country town of Ranger, west of Fort Worth, in October 1917. The Ranger wells were a fluke, discovered by engineers working for the Texas Pacific Coal Company, who drilled a hole looking for coal but found traces of oil instead. Town fathers offered the company twenty-five thousand acres in return for the drilling of four wells. The first produced only natural gas, but the second and third were epic gushers, geysers of a type of greenish black crude that produced high levels of gasoline. The rush was on.

Ranger and a series of smaller finds drew thousands of newcomers into the Texas oil fields. The vast majority became laborers, working as "roughnecks" on drilling rigs, hauling pipe or working in the broiling heat laying the pipelines that began to snake across the state. Many, however, sought to find oil themselves, an activity that until now had mostly been the province of large companies. These men were called "independent" oilmen or "wildcatters," for their propensity to drill unexplored areas, known as wildcat wells. The first independents were businessmen who made their money in other realms—cotton or cattle or dry goods—and looked upon oil as a sidelight.

Many, however, were farmers and ranchers who were *sure*, absolutely *sure*, there was oil beneath their back forty. All across Texas, men who sank water wells for a living were besieged with offers to try their hand at discovering oil. A few actually found some. The 1918 boom around the North Texas town of Burkburnett began when a beleaguered cotton farmer, S. K. Fowler, pressed by creditors to liquidate his spread, decided to drill a well before giving up. He raised twelve thousand dollars from a group of townspeople, hired a contractor and, to his amazement, struck oil, the discovery well pouring out a strong twenty-two hundred barrels a day.

The barrier to entry, as economists would put it, was low. In those early years drilling a shallow well to sixteen hundred feet could cost as little as ten thousand dollars, the kind of cash a group of local businesspeople could assemble from savings. Anyone, it seemed, could become an independent oilman, and by the early 1920s wildcatters born during the booms at Ranger and Burkburnett constituted a burgeoning new middle class of Texas commerce. Making a living as an independent, however, was a more complex task than drilling a single well.

For many wildcatters, success was defined by one's relationship with the large oil companies, including Gulf and the Texas Company; Magnolia

Petroleum of Dallas, later gobbled up by Amoco; and especially the savvy
operators at Houston's Humble Oil, half-owned by the Rockefellers' Standard
Oil of New Jersey. The majors were the Greek gods of Texas Oil, corporations of enormous power who could and often did dictate the destinies of
mortal wildcatters. Men such as Humble's chief geologist, Wallace Pratt, or
L. P. Garrett, Gulf's man in Houston, were the Zeuses and Apollos of Texas
Oil, their leasing and oil-purchase policies the lightning bolts that, when
flung across Texas, could enrich or ruin almost any independent.

Everyone understood the game. Few independents had the money to
build the infrastructure necessary to refine and sell gasoline to the public;
in the 1920s only the majors did that. By and large, an independent who
found oil was looking to sell his production as quickly as possible to one of
the majors. Oil wells could make an ordinary Texan rich, but typically only
after being sold to Gulf, Magnolia, Texaco, or the 'umble, as Humble was
known. Yet the symbiosis between independents and majors went deeper.
The majors amassed their own acreage and drilled their own wells, but they
couldn't be everywhere, not in a state as vast as Texas. They sat on millions
of acres they hadn't the time or money to drill, and often the easiest way
to test a plot of land was to let a wildcatter drill it. The independent bore
a portion of the cost—and the risk—the two split any profits, and if serious production was found, the major swooped in and bought it all. In many
cases a major would actually advance an independent the cash—known as
"dry hole money"—to drill a promising tract. Wildcatters thus became the
woolly frontiersmen of oil, discovering new pockets of petroleum the majors
could then develop into rich fields.

"The major companies...were eager to assist independents," the Houston wildcatter George Strake remembered. "They were happy to see every
wildcat that was drilled. They'd give you tips that would send you on a lease-
buying spree. When you needed something you could borrow it from one of
the majors, and if it looked like you had a good thing, they'd come in and
buy you out before you finished the well." Everyone knew the game. The
challenge was finding the oil.

Of all the thousands of men who swarmed into the muddy tent-camps at
Ranger, Burkburnett, and other boomtowns in those boisterous years after

World War I, four would find the most. Two did it the old-fashioned way, drilling holes deep in the earth. One did it with his mind. The fourth did it with a fountain pen. If Texas Oil had a Mount Rushmore, their faces would adorn it. A good ol' boy. A scold. A genius. A bigamist. Known in their hey-day as the Big Four, they became the founders of the greatest Texas family fortunes, headstrong adventurers who rose from nowhere to take turns being acclaimed America's wealthiest man.

II.

On an autumn afternoon in 1920, two men ambled across a weed-strewn pas-ture seven miles south of downtown Houston.* The younger man, a thirty-nine-year-old would-be oilman named Hugh Roy Cullen, led the older man, his lead investor, Judge R. E. Brooks, across their acreage, pointing out the features he felt made it a smart spot to drill. Cullen was a serious man, just under six feet, with wide-set blue eyes, a thatch of black hair he combed to one side, and enormous hands. He was looking for a dip in the land, a sign, he felt, of oil beneath.

"See, Judge, that's dippin' there," Cullen said, pointing to a patch of dirt.

"I can't see it," the judge said. "I think it's rising."

That the judge couldn't find oil in a barrel didn't matter to Cullen; all that mattered was that Brooks and a dozen other leading Houston citizens had committed money to this drill site. Cullen politely suggested Judge Brooks pick the spot to break ground, and he did. When Cullen realized they had nothing to mark it, he walked to a mound of what Texans tastefully call "cow chips," picked up a few dry chunks, and piled it on the ground—which is how the man who would become Houston's greatest wildcatter came to drill his first actual oil well beneath a pile of handpicked cow manure.

History has not been kind to Roy Cullen, a fifth-grade dropout who in his heyday was probably America's richest man. If he is remembered at all out-side Houston, it is usually as an early champion of Texas ultraconservatism,

---

*The site is located about one mile south of where the Houston Astrodome stands today.

a compulsive letter-writer who jousted with politicians from Franklin Roosevelt to Dwight Eisenhower. Stern, humorless, and a bit of a scold, Cullen at his zenith was a Faulkneresque figure in a white summer suit, a man who detested Communists, "pinkos," and especially Roosevelt, who preferred "niggers and spics" and "New York Jews" to know their place, and whose favored politician was the red-baiting Joe McCarthy. The image is not unfair, but there was more to Cullen than his political views, which were hardly unusual for mid-century southern millionaires.

A man whose life spanned the years from Jesse James to Elvis Presley, Cullen grew up poor in San Antonio. By his own account—the only one that survives—he endured a difficult childhood, marked by family turmoil, financial reversals, and frequent fistfights with other boys. His mother, Louise, whom Cullen idolized, was a slender South Carolina woman who moved to Texas after Union troops burned her family's plantation during the Civil War. She married a cousin named James Beck, settled in San Antonio, and gave birth to five children. After Beck's death in the 1870s Louise wed an itinerant cattle buyer named Cicero Cullen and moved to a farm outside Denton, north of Dallas. Roy, the first of the Cullens' two sons, was born there in 1881. When he was two the family returned to San Antonio.

Cicero Cullen's father, Ezekiel Cullen, had fought during the revolution against Mexico in 1836 and had played a part in the formation of the state's first public schools. Unfortunately, his lineage was Cicero's only apparent asset. He abandoned the family when Roy was four and reappeared only once, two years later, when he arrived in San Antonio unannounced, persuaded Louise to let him take the two boys for a photograph, then promptly spirited them off to Dallas. Louise hired a lawyer and gave chase, but Cullen fled, taking Roy and his brother, Dick, in a covered wagon all the way to Phoenix. Not for several weeks did he return, sheepishly, after an episode in which Roy fell from the wagon and was run over. Though the boy was unhurt, Cicero Cullen apparently realized he was unsuited for fatherhood. He returned the children to Louise. Roy didn't see his father again for years.

The kidnapping episode imbued the Cullen household with a bunker mentality, drawing Roy even closer to his mother. So frightened was the family of a repeat incident that Louise kept Roy out of school till he was eight, making him at least two years older than other children when he finally began attending classes. From stories he told in later years, Cullen appears to have been a stubborn, prideful child, qualities that would follow him into adulthood. Ashamed of the family's poverty, he clung to his mother's stories of his grandfather Ezekial Cullen's prominence and her memories of the antebellum South. The latter instilled in Cullen a distrust of most things eastern and northern, a mind-set that also stayed with him throughout his life.

A lonely boy, Cullen hung a blanket over his bed and at night retreated inside with a lantern, maps, and dozens of books: Sir Walter Scott, Thomas Carlyle, Shakespeare, Dickens, Blackstone, and plenty of history. He daydreamed of traveling the world, of owning his own business. His favorite dream was of the massive white plantation home he would build someday, with porticoes and trellises and gardens, just like the beloved family plantation the hated Union men had burned. Someday, he promised his mother, they would live in it together.

His older half brothers left the house in their teens, and by the fifth grade, when Cullen was twelve, the money ran out. Defying his mother, he dropped out of school to work ten hours a day in a candy factory. Yet he itched to see the world. At sixteen, hearing his father was ill, Cullen left home for Dallas, where he lived with a half sister and attempted a rapprochement with his father. It didn't take. Searching for a path in life, Cullen joined the military to serve in the Spanish-American War but was rejected when his father informed an officer his son was underage. When Cullen's half sister and her husband moved to Schulenberg, a town of German immigrants east of San Antonio, Cullen followed, taking a job in a cotton-buying firm, Ralli Brothers. At eighteen he knew enough about cotton to become a buyer, the roving company representatives who negotiated with farmers for their crop. A Houston company hired him and dispatched him to the town of Mangum in western Oklahoma, where his rehabilitating father had resettled.

In 1900 Oklahoma Territory was still the Wild West, and Cullen did his business on horseback, almost dying at one point during a blizzard. In his spare time he expanded his father's forlorn farm, building up a herd of cattle until rustlers stole most of the animals one dark night; Cullen strapped on a Colt revolver but, to his relief, found his livestock unguarded in a remote pasture. In December 1903 he returned to Texas long enough to marry a quiet Schulenberg girl named Lillie Cranz he had been courting for five years. They settled back in Oklahoma. Cullen bought land and built a shack, where their first child, Roy Jr., was born followed by two girls, Lillie and Agnes.

For seven years Cullen made his way as an independent cotton broker, but he lost most of his savings in the financial panic of 1907, and cotton markets were slow to recover. By 1911, when he turned thirty, Cullen was itching for a change. He made an impromptu study of southern cities, and was impressed when he read in the newspaper that city fathers in Houston, hoping to lure shipping after a hurricane devastated nearby Galveston in 1900, were planning to dredge a ship channel to the Gulf of Mexico. As it happened, Lillie's parents owned a parcel of land on its path. Liquidating his holdings in Oklahoma, Cullen took a leap of faith and moved his family to Houston. He rented a bungalow on Hadley Street, leased an office downtown, and set his sights on learning about real estate.

Houston in 1911 was a sleepy bayou city of seventy-eight thousand men, women, and children who spent their days swatting mosquitoes, mopping their brows, and sipping iced tea. It was oppressively humid, so hot diplomats at the British consulate received hardship pay. Few streets were paved—the main avenue downtown, Travis, was a bed of seashells—and when it rained the roads were often impassable, stinking green water sloshing out of the ditches into the roadways. The air was so damp that bedsheets stayed moist nine months a year. There were oil fields scattered north and east of the city but few real oilmen, with the notable exception of the Hughes family, whose odd little boy, Howard, could be seen tooling his tricycle around his south-side neighborhood. The wealthiest men in Houston, magnates like John Henry Kirby and Big Jim West, made their fortunes in East Texas lumber and dabbled in oil, cotton, and cattle.

Cullen managed to sell his wife's family parcel to an oilman named R. E. Brooks, a friend of his new neighbors, but beyond that he bought little and sold less. "The real-estate business," he noted years later, "was not exactly booming in Houston in those days." He stuck it out for four frustrating years before giving up and returning to the only business he knew, trading cotton, buying a seat on the Houston Cotton Exchange and placing advertisements in Texas papers announcing his willingness to buy bales sent to Houston. He had bank credit behind him and what remained of his savings, but he also had children to support, and business, at least initially, was torpid. He was brooding on his dismal prospects one day in 1915 when, as he trudged into his office, a man named Jim Cheek stuck his head into the hall. Cheek was a real estate developer who was busy building houses around town.

"Roy, I've got a proposition for you," Cheek said. "Can you come into my office for a minute?"

Like many businessmen across Texas, Cheek was thinking about getting into oil. He didn't know the first thing about it, but he knew real estate, and he figured if he bought enough mineral rights someone might drill the land. He asked if Cullen would come work for him, traveling the state to acquire lease rights. It was easy work; most farmers would turn over their rights for pennies on even a remote possibility someone might find oil beneath it. "I don't know anything about oil, Jim," Cullen said. "I've never read an oil lease in my life." Cheek's offer, though, was attractive: all expenses paid, a solid salary, plus one quarter of anything they brought in. Cullen talked it over with Lillie, and felt he couldn't say no. Which is how, at the advanced age of thirty-four, Roy Cullen became an accidental oilman.

He first headed to the Houston Public Library, where in time he read every book he could find on the geology of oil. What he found was a mishmash of fairy tales and guesswork; the first genuine geophysical equipment, using technology developed during World War I to bounce sound waves off underground structures, wouldn't be in wide use until the 1920s. Some thought oil flowed in underground rivers or pooled in subterranean caverns. Oil fields were thronged with characters who claimed to have special oil-finding powers, preachers who swore they had X-ray eyes, and drifters who used everything from divining rods to psychic powers to direct drillers. In the 1910s major oil companies began to hire staff geologists, but for years their use was met with skepticism. As one oilman complained, "Rocks! Rocks! Sam, all they talk about is rocks. Do they think we're running a stone quarry?"

What oil had been found in America was discovered, as at Spindletop, near visible seaps and existing wells. A few companies had begun "surface mapping," reasoning they might deduce what lay beneath the land by what lay atop it. Studying the land was known colloquially as "creekology," from the analysis, such as it was, of creek beds and hillsides. Later, when he began drilling wells on his own, Cullen became a renowned creekologist, sinking wells in low areas and other sites he thought promising. He embraced geology's new tools, especially the seismograph, but much of his work was pure instinct.

Cullen's maiden trip for Jim Cheek was to Coryell County, in Central Texas, where Cheek had arranged a meeting of farmers at the school. "Gentlemen," Cullen said, nervously addressing the crowd, "I'm not an oilman, I'm a cotton man. But I'm going into the oil business, and if you'll give me leases on your land, I'll do everything possible to get the oil rights developed." By morning Cullen signed forty-three leases, each for one dollar "and other valuable considerations." If Cullen and Cheek could find someone to drill a lease, and if oil was found, they would share percentages of any profits with the landowner. Those were big if's; few Texas farmers ever saw a nickel from oil.

For the next five years Cullen roamed West Texas, leasing oil rights everywhere he went, roving as far west as the Pecos River and south to the Rio Grande. It was good money, though it kept him away from home for long periods. In time Cullen saved enough to buy a two-story house at the corner of Alabama and Austin. Soon his mother moved in, and Lillie gave birth to two more children, both girls. Of all the hundreds of leases Cullen signed during those years, he and Cheek managed to attract investors who drilled exactly three oil wells at a total cost of $250,000. Every penny was squandered; all three wells came up dry. What Cullen got, though, was an education and, through Jim Cheek and other new friends, a wealth of contacts in downtown Houston.

By 1920 Lillie was making noises about how much he was traveling. Cullen was almost forty by then, and his five children barely knew him. Couldn't he find work in Houston like other fathers? Which is how, armed only with his library books and a reputation for hard work and honesty, Cullen decided to drill a well on his own. For backing he enlisted Judge Brooks, who helped him assemble an investor group of a dozen leading citizens, including John Henry Kirby and Captain James Baker, grandfather of the James Baker who served as secretary of state during the first Bush administration. The group gave Cullen forty thousand dollars, he chipped in twenty thousand dollars from his savings, and after selecting a forty-acre tract south of Houston, began laying plans to drill for oil beneath his pile of cow manure.

In truth, the drill site Cullen chose lay not on some farmer's pasture but in the decrepit old Pierce Junction oil field. Gulf and other majors had been poking holes in the salt dome beneath Pierce Junction for a decade—fifty-two wells in all—and had very little to show for it. There was a good natural-gas well on the dome's southeast flank. Cullen had been studying it for months when one morning he took out his handkerchief and placed it over a gas-release valve. It showed no color. Still, on a hunch, Cullen returned every day for two weeks, and one morning he saw a faint hint of yellow on his handkerchief. Each day the color grew deeper, until Cullen saw a rich amber. It was just a hunch, but he was willing to bet the amber was a sign of oil.

He secured an appointment with Gulf's Houston geologist, L. P. Garrett, and spread a map of Pierce Junction on his desk. He pointed to an area known as the Howe lease. "Let me have a lease on this land," he said, "and I'll drill you a well."

"On the Howe land?" Garrett said. "It's too far off the dome, Roy. You won't do any good there."

"Give me the lease anyway," Cullen said. "If the well is no good, you won't lose anything."

There were delays getting started. The driller Cullen selected, Judge Brooks's son Emory—he was part of the deal—was busy on another job. Meanwhile, an oil scout caught wind of Cullen's plans, and before Cullen could break ground another independent set up a derrick on an adjacent lease. Just as Cullen's drill bit finally chewed into the prairie, the neighboring crew struck oil. Cullen's confidence soared. On the day he expected to reach oil-bearing sand—the depth where the competing crew had found oil—Cullen called out his entire family to watch. It was a Sunday afternoon, and everyone wore their church clothes, Lillie and the girls in dresses, Roy Jr. dressed in a white suit with an immaculate blue shirt. Judge Brooks came out, along with several other investors and their families.

Years later, Cullen described the scene as if it had been staged for Hollywood. Minutes after he gave the signal to drill, the well erupted in a geyser of oil, blowing the "Christmas tree" of steel valves into the sky and spattering the spectators, who laughed and whooped as Cullen and his drilling crew struggled to control the bucking well. Roy Jr. ran to and fro in the black rain, ruining his Sunday clothes. "We're rich! We're rich!" one of the wives kept shouting. "We'll never have to work a day again!"

Once under control, the well flowed twenty-five hundred barrels a day, a strong producer. Pierce Junction not only assured Cullen's future as an oil-man, it cemented his ties to Houston's downtown elite and thus to the inves-tors who might fund his future efforts. Captain Baker was the first to cash out—at a five-to-one return—while the others endured three more wells, all dry, and settled for 300 percent returns.

Cullen had found his oil by drilling beside, or on the flank, of a salt dome, a strategy larger companies had been pursuing with some success since 1914 as production atop the older Gulf Coast domes began to peter out. Cullen thought it an idea worth pursuing. By the time the Pierce Junction well came in, he already had another dome in his sights. Damon's Mound, a lonely finger-shaped hummock a mile in length, barely eight hundred feet wide, rose thirty yards above the coastal scrub fifty miles south of Houston, near the Brazos River and the town of Rosenberg. Others had drilled it but man-aged only a handful of meager producing wells. Cullen thought his flanking strategy might find oil others had missed.

To Cullen's irritation, the first well came up dry, as did the second, and the third; his investors, he was acutely aware, were quickly losing their earlier profits. For months Cullen rose before dawn at the house on Alabama Street, kissed Lillie good-bye, and drove down to tramp Damon's Mound in his greenish-white Witch Elk boots and khaki jumpsuit, eyes studying the scrub for anything that might suggest oil beneath; many nights Cullen didn't come home at all, grabbing sleep on the ground beside the drilling rig. More than one new hand, spying his slumbering figure, mistook him for a bum. When he did make it home, streaked with mud and sweat, his skin pimpled with insect bites, Lillie would be waiting on the veranda, and they would walk up the stairs in silence. "Help me get my boots off, Lillie," he would say, slumping on the bed. "Let me get a shave and a bath. Tomorrow's another day."[2]

But the days stretched into months, then into years, and every well came up dry. Cullen spent thirty-six months at Damon's Mound, the most frustrating period of his life—"three years in hell," he called it. Finally, in 1924, after drilling a dozen holes around the dome, he gave up. In the meantime, his Pierce Junction well had played out. Cullen had several thousand dollars in the bank, but no cash coming in. Roy Jr. had gone off to college in upstate New York, the two older girls were in high school, and now they had two more little girls to pay for. At night, lying in bed, Lillie wondered how long this could go on. "We got to keep going a little longer, honey," Cullen would say. "I want the children taken care of. Tomorrow'll be another day."

He decided to take another crack at Pierce Junction. In his first well back, Cullen drilled down to the Miocene sand at four thousand feet and found nothing. Rather than try again nearby, he decided to drill a thousand feet deeper, to the Frio sand. He hit a small producer, sixty-five barrels a day, and it got him thinking. Every prospect in the area had been drilled to the Miocene. Why not drill deeper? There were technical challenges involved, and deeper drilling cost more, so Cullen put the matter to his backers in the Second National Bank's boardroom. "The trouble with this business," he told them, "is that everybody expects to find oil on the surface. If it was up near the top, it wouldn't be any trick to find it.... You got to drill deep for oil."

It was an iffy proposition; all of Cullen's investors, Big Jim West and John Kirby and the others, had lost their profits at Damon's Mound. Still, they backed his new strategy. Heading straight for the Frio this time, Cullen hit the second gusher of his career, a pool almost as big as the first. It came in strong on a rainy night, blowing off the Christmas tree, the spewing oil coating the glasses of his driller so thoroughly he couldn't see to cap the well for hours. The investors got their money back and more, and if Cullen's financial future wasn't yet certain, his reputation as an oilman was. "That man had guts," one investor, J. E. Duff, said years later. "If he thought there was oil under a tract, he'd spend his last dollar drilling it, regardless of what anyone thought. And he had an uncanny nose for oil. But when he missed, it didn't faze him. Nothing ever discouraged him."[3]

The second Pierce Junction well created a new mantra for Cullen. Hit the flanks of the old abandoned salt domes, and drill deeper. If he didn't find oil, drill deeper still. Many of his field hands, a number of whom would work with Cullen for the next thirty years, could imitate his laconic instructions in their sleep: "Boys, let's go a little deeper." His longtime operations manager, Lynn Meador, once said, "When they start to lower Mr. Cullen into a grave, I'll bet he'll sit up and say, 'Boys, dig her a couple of feet deeper.'"[4]

Among those who liked the sound of Cullen's strategy was Jim West, the cantankerous lumberman who had invested with Cullen from the beginning. West dabbled in oil himself, but his wells had all come up dry. One Saturday in 1927 he called Cullen to his downtown office and made him an offer. As Cullen remembered the conversation twenty-five years later, West said: "I've got three million dollars lying around. And I've got the West Production Company, which don't amount to much. I'll put the three million in my oil company, and give you one-fourth interest—if you'll go in with me. You'll be president—and have complete charge of the company." Cullen said he'd think it over.

The following Tuesday, having heard nothing from Cullen, West telephoned him again. "I made a proposition to you last Saturday, and since then I haven't heard a damn thing from you," West said. "I offered to give you almost a million dollars, and you haven't taken the trouble to reply. What's your answer, Roy?"

"Not interested," Cullen said.

For a moment West was speechless. "Not interested!" he barked. "My God—what do you want!"

"Tell you what I'll do," Cullen said. "I'll go in the oil business with you—fifty-fifty. For every dollar you put in, I'll put in a dollar. But only on condition that I have full charge—no interference." West went quiet for a moment. "Roy," he said after a moment, "I didn't know you had that kind of money."

He didn't. "I'll put up five thousand dollars," Cullen said, "and you put up five thousand. We'll each have half-interest."

West didn't understand. "I offered you three million dollars, Cullen—and you'd get a quarter of that (outright). You'll turn that down and put up five thousand of your own money?"

"That's right," Cullen said. That way, he went on, "I won't be working for you."

Later they flipped a coin to determine whose name went first. Cullen won; the new partnership was named Cullen & West. Cullen purchased two big Union Tool drilling rigs—the first he owned outright—and moved them to the east side of the old Blue Ridge Field in Fort Bend County, south of Houston. The Blue Ridge dome had produced some good wells beginning in 1913, but had since dwindled. West objected. "I've drilled wells on that side of Blue Ridge," he said. "There ain't any oil there. You'll be throwing the money away."

Cullen looked balefully at his new partner. "I'm supposed to be in charge—isn't that it, Jim?" he said.

"Sure you are," West said. "But—"

"Then that's where we drill," Cullen said.

Drilling on a site known as the Bassett Blakely tract, Cullen again headed straight for the Frio—and brought in a good well, nearly sixty thousand barrels a day. In quick succession he brought in four more by the end of 1927, all solid producers. Increasingly confident he could find oil in untouched sands deep below the abandoned fields around Houston, Cullen then announced his intention to enter the derelict Humble Field. Sixteen miles northeast of downtown Houston, Humble was one of the state's oldest oil fields, a salt dome discovered in 1903, two years after Spindletop. Its production, like that of many older fields, had dwindled. By the time Cullen began studying its geophysical data, Humble was an oil ghost town, littered with the skeletons of abandoned derricks and pockmarked with unused waste pits.

Jim West thought Cullen had lost his mind. "They've drilled so many wells out there that there isn't room to put down another hole," he protested, "even if there was any oil left, and there ain't." Laying out a map, Cullen pointed to a spot he wanted to drill near the field's southeast corner. West asked when he planned to inspect the site. Cullen said he didn't intend to. "You're gonna lease it without even looking at it?" West asked. "Are you crazy, Roy?"

"We have geophysical reports on the entire area," Cullen said. "Looking at it won't add anything to that. It won't mean anything."

"Well, it will to me!" West barked. "I'm going out there!"

"Go ahead, Jim," Cullen said. "I'd go with you, but I'll be too busy getting ready to drill it."

West returned from his inspection trip more determined than ever to stop his partner from drilling. "You'll have to tear down old derricks to find a place to drill a hole," he said. "You'll have to fill in the slush pits. That field is like a piece of Swiss cheese!"

Cullen ignored him. The two men were temperamental opposites, and he suspected their partnership wouldn't last long. Out at Humble, his men cleared away the old derricks and readied a drill site. As they drilled down toward the Miocene, Cullen began studying the old wells nearby. Most had filled with salt water. Poring over well logs and his library books, Cullen understood that three thousand feet below the field lay a band of blue mud, what the drillers called gumbo. Below that was five feet of salt water, followed by more gumbo. It flooded the wells, and previous drillers had given up trying to get through it.

This was the first challenge a deep driller on the Gulf Coast faced, but Cullen thought he had found a way through it. Once the drill bit sank into the lowest layer of gumbo, he instructed his tool-pusher to set their casings, an outer pipe that shielded the hole from water, all the way down through the salt water to the gumbo. It worked; both the salt water and the gumbo were kept out of the well. Free to go deeper, Cullen hit a pool of pure "pipeline" oil—oil so free of impurities it could be pumped directly into a big company's pipeline.

But Cullen was determined to go even deeper, into virgin dirt beneath the Humble Field. That, however, brought him face-to-face with the most serious obstacle a deep driller could encounter, the dreaded "heaving shale." Jackson shale, as geologists termed it, lay about thirty-five hundred feet below most Gulf Coast fields, a tier of crumbled rock that pressed in around a drill bit, freezing it. No driller had defeated it, and with other shallow fields to attack, few had tried very hard.

Cullen ordered his driller, Dalton Brown, to "thin up," that is, pump water into the hole instead of drilling mud, the better to clear the broken shale from the drill bit. After what Cullen characterized as "a startled glance," Brown tried it, and the drill began to spin more easily. Still, a thick "bridge" of shale lay at the bottom of the hole, and they needed to get through it. Cullen had his plan ready. It was late at night when he instructed Brown to use the drill bit as a pile driver, lifting it and dropping it onto the shale bridge in an effort to break through. He kept water pumping into the hole to keep the drill bit spinning freely. It took several hours but it worked, and by midnight, when Cullen lay down by the derrick to grab some sleep, they had penetrated the slushy, oil-bearing sand underneath, the sought-after Yegua sand. Four hours later Cullen heard Brown shout: "Wake up, Mr. Cullen. I think she's coming in!"[5]

In the dim predawn light Cullen rose to see oil and gas spewing so violently from their hole that the four-inch flow line, which connected the well to a storage tank, had come loose and was wildly slashing the air. Cullen dashed to the storage tank, leaped atop it, and held the thrashing line until his crew could tie it down. It was the kind of daredevilry other operators might have avoided, but Cullen did it time and again, and his crews loved him for it. Years later, Lynn Meador remembered a dangerous blowout a Cullen crew encountered in Fort Bend County. One spark meant an explosion that would kill them all, but Cullen, who could easily have directed operations from the safety of his car, stayed in the thick of it on the derrick floor. When one young man panicked and ran, Cullen, covered in black oil, grabbed him by the belt and pulled him back. It took all day to maneuver a half-ton cement collar onto the well to cap it, and Cullen stayed through all of it.

"That was one of the many performances of his that have made those who know him admire him," Meador said. "He was never a man to tell his employee, 'You do this.' When there was a dangerous task to be done, it was 'Follow me.'"

Cullen brought in a series of big producers from the Yegua sand at Humble, opening the way not only for the field's renewal, but the renewal of other Gulf Coast fields as well. It was his victory over the "heaving shale," however, little known to anyone outside the oil business, that brought him the kind of recognition only a fifth-grade dropout could appreciate. Seven years later, in 1935, the University of Pittsburgh's engineering department, after launching a quiet investigation into the factors that led to deeper drilling in Texas, awarded Cullen a doctor of science degree for his achievement.

After Humble, Cullen headed back to Blue Ridge and hit another series of big producers. At night he and Lillie sat out on the veranda in their rockers, sipping iced tea and talking about the children and their lives. Cullen was almost fifty by then, graying at the temples, and the money in his bank accounts indicated he was a millionaire, not that anyone noticed. Unlike so many other independents, he had so far resisted selling his wells, despite Jim West's constant agitation to do so. In the fall of 1929, sitting there on his back porch in Houston, Roy Cullen was a happy man. However, 1930 would be another year.

THREE

Sid and Clint

I.

W hile Roy Cullen remained close to his home and family, working in the worn, discarded fields around Houston, wildcatters across the state were finding oil in new and uncharted areas. As at Ranger and Burkburnett in the 1910s, their discoveries triggered scrambles to amass acreage that overnight turned drowsy country villages into rollicking boomtowns: at Mexia, east of Waco, in 1921; at Luling, east of Austin, in 1922; then a series of gushers that pushed the oil frontier deep onto the empty plains of West Texas. The boomtowns became moveable feasts for young, energetic Texans who scurried from gusher to gusher, furiously buying and trading leases, drilling a well or three, then moving on to the next town when the frenzy died down.

For the first time a handful of wildcatters began to get seriously rich. Those first Texas oil millionaires, however, often found keeping their fortunes was tougher than making them. The classic case was a Massachusetts rubber heir named Edgar B. Davis, the mammoth 350-pound dreamer who found oil at Luling. Typical of the oddball adventurers drawn to Texas during the 1920s, Davis enticed Luling's city fathers to help him drill for oil following a séance with the noted mystic Edgar Cayce. Combining his savings with theirs, Davis drilled a dry hole, then another, then four more. Finally, on a steamy August afternoon in 1922, his last dollar spent, his office furniture sold, his telephone disconnected, Davis drove into the countryside to see how his seventh and last attempt was faring. He pulled up and stared. A geyser of oil was shooting into the Texas sky.

The Luling field stretched for twelve miles, and after drilling dozens of wells along its length, Davis sold out to Magnolia for twelve million dollars in 1926, roughly three hundred million dollars in today's dollars. He then embarked on a spending spree that has gone down as one of the strangest in the history of Texas Oil. He first threw a barbecue outside Luling, said to have been the largest in Texas history, to which he invited every citizen in three adjoining counties. Then he began handing out bonuses to his crew, a two-hundred-thousand-dollar check to each of five men. He built clubhouses for Luling—one for whites, one for blacks—then gave his hometown back in Massachusetts a check for one million dollars.

His fortune was already streaming through his fingers when, on a trip to San Antonio, Davis ran into an old friend, a onetime newspaperman named J. Frank Davis. Frank was down on his luck, so Edgar, on a whim, suggested he write a play. They came up with a topic, reincarnation, and Edgar pledged to finance the whole thing. In no time Frank Davis banged out a three-act drama he called *The Ladder*. It followed a group of characters from an English castle in 1344 through three reincarnations, the last in New York circa 1926. It was by all accounts a spectacularly awful play, which did nothing to dampen either Davis's enthusiasm for it. Edgar enlisted a Broadway producer and staged tryout performances that summer in Detroit and Cleveland.

Despite reviews that were at best lukewarm, Davis insisted on conquering Broadway. *The Ladder* opened at the Mansfield Theatre on October 22, 1926, with a stalwart cast led by Antoinette Perry and Hugh Buckler. It was a colossal flop, but Davis paid to keep the play open for two long years at a cost, it was said, of $1.5 million. It was the beginning of the end. As his money ran low, the state of Massachusetts sued Davis for back taxes. He couldn't pay, and the Depression wiped him out. He went bankrupt in 1935. A writer for *The New Yorker* found him living in Luling in 1948, broke, a man in his late seventies passing his last days playing bridge. He died in 1951, forgotten.

In the 1920s Texas was littered with men like Edgar Davis. Most remain forgotten. Some made millions. A few would make history.

II.

Neck-deep in this adrenalized rush were two lifelong friends, keen-eyed country boys from the town of Athens, sixty miles southeast of Dallas. During their boyhoods Athens was a dirt-road-and-buggy village deep in the East Texas pines, three thousand or so farmers struggling to pry cotton from the infuriating sandy soil. The area had been settled only in the 1840s, and by 1900 its first family was the Murchisons, whose patriarch, Thomas Frank Murchison, migrated from Mississippi to East Texas and finally in 1855 to Athens, where he clerked at the general store.* T.F., as he was known, soon started his own store and, after years of loaning money to men who couldn't pay their bill, founded Athens's first bank in 1890. The two-story brick building went up right on the square, adjacent to the Murchison store.

T. F. Murchison had six children, and when he died in 1902 he left everything to his three sons. Over time the second boy, John Weldon Murchison, took control of the bank. Wed in 1893, John and his wife, Clara, raised eight children in a Victorian home that took up an entire block on Tyler Street, known as the "street of plantation homes." A whiff of the antebellum clung to all the Murchisons. Though he hadn't fought in the Civil War, T.F. liked to be called "Colonel Murchison," and he famously had little use for Yankees, easterners, or railroad men, attitudes that permeated his clan. In a town where dusty overalls were the rule, the Murchison children and grandchildren could be seen walking to the private Bruce Academy in polished shoes and tailored woolen clothes. The Murchisons "were sort of snobby," one family acquaintance recalled. "They thought they were better than the rest of the people in Athens because they were Murchisons."[1]

Except, that is, for John and Clara's third child, small, homely Clint, saddled with the body of a snowman—big head, beanbag nose, no neck to speak of—and a face like a dish of melted ice cream. But what Clinton Williams Murchison lacked in physical appeal he made up for with a mind that whirred like a Swiss timepiece. Headstrong and independent, disdainful of

---

*The Murchison family pronounces the name "Murkison."

his father's stuffy ways, young Clint was Tom Sawyer with an abacus, the kind of seven-year-old who skinned squirrels and sold the little pelts for nickels. He loved the outdoors, spending lazy afternoons fishing with a Negro man outside town, ignoring the disapproving clucks of his neighbors. While his brothers took jobs at the bank, teenaged Clint was drawn to the excitement of the Athens lifestock pens, where roving traders wheeled and dealed for the best prices on cattle and horses. He found the give-and-take thrilling, and as a teenager he made extra money trading livestock.

He was joined by an older boy named Sid Richardson, whose father, a bar owner who also owned a peach orchard outside town, was one of the bank's customers. Clint and Sid established a lifelong friendship during impromptu cattle-buying jaunts into Louisiana, where they purchased cows they sold for meager profits. Prewar Athens, in fact, was home to any number of teenagers who would one day emerge as Texas millionaires, many of them Murchison's running buddies. Several worked alongside him at the Richardson orchard, betting their earnings against one another in running games of poker and gin rummy that, in many cases, would still be going on fifty years later.

In 1915, when he turned twenty, Clint joined his brother Frank at Trinity College, a Presbyterian school in Waxahachie, south of Dallas, whose graduates typically joined the ministry. Chafing at the classroom structure his brother embraced, Clint took to organizing craps games; when school officials found out, Clint found himself on the first train back to Athens. Downcast, he reluctantly took a job at the bank. A natural with numbers, he could add, subtract, and multiply large sums in his head while other tellers did it on paper, but he found life in a teller cage just that, a cage. He complained he could make more money in a week trading cattle than he made at the bank in a month. Finally, to his father's consternation, he quit. Within days America entered World War I and Clint, impatient and eager to see the world, enlisted.

Assigned to a motor transport division in the Quartermaster Corps, Murchison longed to go overseas. It was not to be. He was shuffled between army camps in Texas, Arkansas, and finally Michigan, where, on the war's completion in November 1918, he was handed his mustering-out papers. He was twenty-three by then, eager to tackle the world and certain of his plan.

He was heading to Fort Worth to work with a young oilman who had bombarded him with letters of the money to be made in North Texas, his old peach-picking pal Sid Richardson.

III.

For a man who would one day be proclaimed America's richest citizen, who at his death controlled more petroleum reserves than three major oil companies, Sid Williams Richardson left few footprints on history. He attracted no biographer. In life he earned exactly one magazine profile of note, and while he gave newspaper interviews over the years, they consisted largely of aphorisms and apocryphal stories. Oil-industry histories ignore him; a mammoth, 1,647-page history of American oil exploration, 1975's *Trek of the Oil Finders*, mentions Richardson all of three times. A lifelong bachelor who lived before the age of prying reporters, Richardson disdained letter-writing, preferring the telephone or making assistants author important communications. One protégé, the evangelist Billy Graham, once said, "Sid Richardson told me years ago, 'Don't put anything in writing. If you use the telephone, they can never use it against you.'"

Since his death, Richardson's heirs have adorned several Texas universities with Sid Richardson buildings: there is a Sid Richardson Hall at the University of Texas, a Sid Richardson College at Rice University in Houston, a Sid Richardson Physical Science Building at Baylor University in Waco, and a Sid Richardson Science Center at Texas Christian University in Fort Worth. Yet his family went out of its way to obscure the facts of Richardson's career. A portrait of Richardson hangs in the Permian Basin Hall of Fame and Museum in Midland, but Richardson's is the only biographical file at the facility that is restricted—reviewable only with the family's approval.

Much of what's known about Richardson's early years comes from stories Richardson himself told to friends, family, and the occasional reporter—any listener knew to take them with a grain of salt. He came from humble beginnings, that much is sure. Born in 1891—his mother named him Sid Williams after an itinerant evangelist—Richardson was one of seven children born to

Nannie and John Isadore Richardson; three of his siblings died before the age of seven. A 1903 directory lists the Richardson residence as the family saloon a half block west of the square. Family lore suggests the business was so profitable it made neighboring stores envious. The reality was probably not so rosy. In later years Richardson joked that his family was so poor he sometimes slept on the pool table. Friends joked that Richardson, a heavy drinker in his youth, had probably passed out.

Family stories suggest that Richardson, unlike his friend Clint Murchison, was not exactly a go-getter. When he was sixteen he took a dollar-a-day after-school job at a cotton compress, but was fired for laziness. According to Athens lore, Richardson had a reputation for failing to pay his debts; one story has it that a drugstore manager told his soda jerks they would be fired if they sold Richardson one more Coca-Cola on credit. In later years, some of Richardson's favorite stories were of the ways his father tried to straighten him out.

At the age of eight, Richardson said, his father gave him a downtown lot to learn about business. When John Richardson offered to take back the lot in return for a bull, Sid took the bull—only to realize he now had a large male cow with nowhere to put it and no cows with which to breed. "My Daddy taught me a hard lesson with that first trade," Richardson once said. "But he started me tradin' for life." When he was eleven, another of Richardson's stories goes, his father suggested it was time for him to own a horse.

"Great," Sid said. "When are you gonna give it to me?"

"I'm not going to give it to you," his father said. "You're going to buy it from me."

Sid said he worked all summer crating peaches to raise the money, but once he purchased the horse he discovered it was blind.

"Daddy, you cheated me!" he exclaimed.

"I did not," his father said. "People will try to get at you any way they can, and you might as well learn now."

John Richardson traded cattle in his spare time, and by his teens his son was trying it as well. When he wasn't crating peaches, the younger Richardson began taking Clint Murchison along to buy cattle in Louisiana. Virtually every story told of Richardson and Murchison's early years emphasizes what great boyhood friends the two had been. No doubt that became true. But Richardson's career would be marked by an ability to befriend those who could help him most, and one suspects that sixteen-year-old Sid Richardson's primary interest in eleven-year-old Clint Murchison was his father's money. The elder Murchison, in fact, later lent Richardson several thousand dollars to buy cattle. Taking Clint under his arm wasn't just a good deed. It was smart business.

In later years, Richardson's favorite story of the cattle-buying expeditions with Murchison revolved around a trip the two made to Ruston, Louisiana. As Richardson told it, he decided to buy a natty suit and masquerade as a clueless city-slicker, a charade he insisted somehow allowed him to buy his cattle for cheaper prices. Whatever his tactics, Richardson and Murchison proved able cattle buyers. They found "trading" to be a thrilling pastime. During his senior year of high school in 1909, Richardson claimed he made thirty-five hundred dollars in profits.

At some point the Richardsons briefly relocated to—or perhaps vacationed in—the West Texas town of Mineral Wells, where Sid's sister Annie began dating a sharp young doctor named E. P. Bass, who was to have a profound influence on Richardson's life. Bass had a medical degree from Tulane, and after the family returned to Athens, he married Annie Richardson, in 1909. Bass proved his merits after Richardson was badly injured when a buggy he was driving overturned, crushing one of his legs below the knee. As with so many Richardson stories, details of precisely what happened are sketchy. In a note to the Henderson County Historical Society in Athens, a nephew said the accident broke Richardson's right leg when he was nineteen; in a 1954 interview, Richardson said he was fifteen and the injured leg was his left.

Whatever the case, doctors wanted to amputate the leg. But E. P. Bass managed to save it during an operation in which he removed two inches of bone and built a "trough" to connect the remnants. In time Richardson managed to walk unaided, but for the rest of his life he limped. "I practiced me a walk that wouldn't make me limp," he once said. "Took me a year. Now I take long steps with the left laig, short steps with the other. That swingin' walk of mine is my own invention."

In September 1910 Richardson enrolled at Baylor University in Waco, but after two semesters of classes he did not return. An Athens man was registrar at Simmons College in Abilene, so Richardson enrolled there, in the fall of 1911. A tradition on campus holds that he was a bright but lazy student, far more interested in whiskey than classwork. The dean of students, J. D. Sandefer, is said to have called Richardson into his office on several occasions after he returned to his dormitory long after curfew. According to this tradition, Sandefer repeatedly lectured Richardson that he was squandering his abilities. "You have the brains and the personality to do whatever you want to do, and be what you want to be," he is quoted saying. "If you would just lay aside this foolish waste of time, and set your heart on being a man."

There is ample evidence that Richardson was a heavy drinker, which may explain a tendency to engage in fistfights. Once, explaining why he disliked the game of golf, he quipped that the single time he played eighteen holes he drank an entire bottle of bourbon. Carousing had no further effect on his studies, however, because in January 1912, after only four months in Abilene, Richardson's father died. There was no more money for Richardson's education. His brother-in-law, "Doc" Bass, was dabbling in the oil business, and it was probably on Bass's suggestion that Richardson decided to find work in the oil fields.

He began as a laborer, hauling pipe by day and apprenticing on derrick floors at night. Richardson never said where he first worked, but it was likely the new Electra field west of Fort Worth in 1911. Some of his favorite stories emerged from this period. In one, he was working alone one night, shoveling coal into a derrick furnace, when he was suddenly surrounded by coyotes. He spent the hours until dawn atop the red-hot furnace, hopping from one foot to another, until rescued by the arrival of day-shift workers. In time his education attracted notice, and he was hired as an office boy for the Oil Well Supply Company in Wichita Falls. Richardson once said this job came to an end after he engaged in a fistfight with a bookkeeper. The fight, however, impressed one of his bosses, who decided to send him back out into the field, this time as an oil scout in Louisiana. Scouts are the oil industry's happy spies, spending their days driving from well to well, checking production trends, gauging competitors' strategies, and picking up rumors. It's a job where charm and likeability matter more than subterfuge, and Richardson, a natural raconteur, was good at it.

A career in oil, however, was never Richardson's dream. What he wanted to do was trade cattle. After two years in the oil fields he returned to Athens in 1914, borrowing money from Clint Murchison's father to purchase a herd. The venture didn't last long. As Richardson told a Fort Worth newspaper in 1954, "my herd died of tick fever, and I lost my taw. What's more, I owed Mr. Murchison's bank six thousand dollars. I went back to Wichita Falls to get me some oil money."

Perhaps Richardson's favorite story was returning to Athens one year—to the day—later. Scouting was good money, and Richardson entered the town square at the wheel of a new Cadillac. "I swung back around that dusty square twice so's all the bench warmers would see me good, and then I marched into the bank and paid Mr. Murchison his money in cash," he recalled. "Then I drove out of town again. 'Fore the dust had settled, all those old boys got off their benches and started for the oil fields. They said, 'If that dunce can make so much money, we'll go, too.'" One of those impressed was young Clint Murchison.

IV.

Richardson was waiting the day Murchison, still wearing his army uniform, stepped off the train in Fort Worth in the spring of 1919. Murchison intended to head next to Athens, but Richardson insisted they go right to work. The first thing he did was march Murchison to the Washer Brothers men's store and buy him a pair of nice suits. "You gotta get outta that uniform right now," Richardson said. "You wear that and when you go around to talk to people they'll want to talk about the war. We aren't talking anything but oil." Murchison didn't make it home to visit his mother for another six weeks.[2]

Despite their common backgrounds, they were a mismatched pair. Murchison was energetic, impatient, and, like many country boys before him, intellectually insecure. His favorite book was the dictionary, which he employed to adorn his vocabulary with ever-larger words; during drives he loved nothing more than challenging a fellow traveler to query him on word definitions. Richardson, meanwhile, hated nothing so much as pretension. A nifty hat, a pocket square, a dropped name—anything could prompt a cutting remark from Richardson, usually delivered with a wry smile. When Murchison used a big word, Richardson would wrinkle his brow and say, "What's that word again, Murk?"

Murchison was shy and would remain so all his life; if he didn't absolutely have to talk to someone, he avoided it. Though capable of warmth around family and friends, strangers found him standoffish and occasionally rude. In sharp contrast, Richardson presented himself as the essence of the Texas good ol' boy, joshing, laughing, and cursing in a thick backwoods accent. In later years, if a subordinate or family member made a mistake, Richardson would scowl and call him a dunce or a knucklehead; then, just as his target appeared crestfallen, he would grab him around the shoulders for a hug. "Sid," says one longtime friend, "could just make you feel great."

In the summer of 1919 the hottest oil play in the country was centered around the raucous boomtown of Burkburnett, on the Red River border with Oklahoma. Richardson and Murchison, taking rooms at the YMCA, dived headlong into the thick of it, using their savings—and, it appears, a good chunk of money from Murchison's father—to join the hectic trade in oil leases. It was a thrilling ride for two young country boys on the make, with muddy streets and prostitutes, wads of leases exchanged between grimy oilmen on every corner, and gunshots echoing in the night. Lease trading was all about oil field intelligence; the value of a lease fluctuated largely on rumor—that the land held oil beneath it, that a major oil company was set to drill an adjoining lease, that a nearby test well had come up dry. When completing a trade, Murchison and Richardson usually made sure to retain a minority interest in the sold lease, allowing them to cash in on other men's wells months and sometimes years after cutting the original deal.

While Murchison could calculate royalty payments in his head, it was Richardson who did much of their snooping. Throughout his career, Richardson augmented his down-home charm with tricks that old friends call crafty but a neutral observer might consider sneaky. According to an oilman who knew him well in later years—hereinafter referred to as the Old Family Friend—Richardson once said he made his most daring bet at Burkburnett as he was studying a highly anticipated test well Gulf Oil was drilling on the Texas side of the Red River. It was what oilmen call a "tight hole," that is, everything about it was top secret. If Gulf found oil, though, nearby leases would skyrocket in value. When Richardson heard that a team of Gulf executives from far-off Pittsburgh was to visit the well any day, he hustled into town and pulled Murchison out of a poker game. They piled into a car and drove to the drill site, told the night crew they were the Gulf men, and quizzed them on the well, which, as it turned out, the drillers were expecting to be a gusher. By the next morning Richardson and Murchison had bought up every available lease nearby—by one account, $50,000 worth. When the well came in not long after, they managed to quadruple their money.

They did well in those early months; in later years, Richardson claimed, probably inaccurately, that he made his first million at Burkburnett. Whatever he made, he didn't keep it long. The two pals had been trading leases for barely nine months when disaster struck, at a time when almost all their money was tied up in drilling blocks along the Red River. In early 1920 the overheated commodities markets collapsed, forcing the price of oil down from $3.50 a barrel to a dollar. Richardson and Murchison awoke one morning to find all their capital invested in land no one would be drilling anytime soon. Worse, they had borrowed money—probably from Murchison's father—to assemble the block, and both men now faced their first serious debts.

Unable to afford even room and board, the two sheepishly moved into Doc Bass's house in Wichita Falls, which soon became a clubhouse for their oil field friends. Murchison, meanwhile, used the idle time to court the girl he hoped to marry, Anne White, the charming, petite daughter of one of Tyler's wealthiest families. He had proposed to Anne as a teenager, but despite a plea from his father, Anne's father had judged her too young to marry. Now, during a visit to Wichita Falls, she accepted his proposal, and this time her father consented. The wedding, representing the union of two of East Texas's most prominent families, was the social event of the year in Tyler. Richardson limped down the aisle as an usher. Either Murchison's fortunes had improved overnight, or his father had given him more money, because the newlyweds left the reception in a yellow Rolls-Royce, Clint's wedding gift to Anne. For Christmas he gave Anne's father a mink coat.

By the time Murchison returned to Wichita Falls oil prices had recovered, and Clint went to work buying new leases. It was then he began to display his true genius. For the first time he actually began drilling his own oil wells. Chronically short of cash—like most wildcatters—he would trade a share in one lease for a rig to drill another; once he got the rig, he would trade shares in its production for another rig, and so on. He called it "financing by fina-glin'"; other oilmen watched him in awe. Murchison's instinctive mastery of banking and lending practices translated easily into an understanding of oil field drilling and geology. Unlike older oilmen like Roy Cullen who still believed in creekology, Murchison put his faith in science. One of the first men he ever hired was a talented geologist named Ernest Closuit, whom he lured from Gulf. Within months the two began to find oil in commercial quantities—several of his strikes lay on the vast Waggoner Ranch—and Clint soon moved Anne into a rented home of their own. They needed it. Between 1921 and 1925, Anne gave birth to three children, all boys, John, then Clinton Jr., then Burke.

By then Murchison was no longer working with Richardson. Exactly why has never been explained, although family members speculate that as a bachelor Richardson was willing to take more risks. The fact was, Murchison no longer needed Richardson; he knew the oil game now, and, unlike Richardson, he had the family money to play it. County leasing records suggest it took years for Richardson to unload the last of his land along the Red River, at which point he was all but broke. A single yellowed clipping from a Dallas newspaper indicates he returned to East Texas to try to drill a well of his own in 1922. Land records there show he did it by going into partnership with a dozen of his relatives, who turned over their mineral rights for a song. Richardson got one of Doc Bass's crews to drill the hole. It came up dry.

Murchison, meanwhile, remained in North Texas and thrived. He partnered with a local wildcatter named Ernest Fain, and through the early 1920s they hit strike after strike. The partnership grew prosperous enough to open offices in a Wichita Falls building, and eventually generated enough cash that they were able to add a side business that drilled wells for other oilmen, called "contract drilling." By 1925, when he turned thirty, Murchison was already a wealthy man, taking in about thirty thousand dollars a month. But the North Texas boom was waning, and he began to cast about for something new. When Ernest Fain balked at drilling outside the area, Murchison dissolved the partnership.

He took his proceeds, an estimated five million dollars, and moved Anne and the boys to cosmopolitan San Antonio. He joked to friends that he was retiring, but in truth he just wanted a settled life, one where he could work finite hours in a clean office, making it home for dinner while Ernest Closuit and a group of new employees worked in the oil fields. There were new fields popping up around San Antonio, and Murchison invested in them, all the while casting envious eyes at the massive cattle ranches that stretched south to the Mexican border; like Richardson, what Murchison really wanted was to be a gentleman rancher.

The easy life he envisioned in San Antonio, however, was not to be. That winter Murchison took Anne and her sister to New York for a vacation, embarking from New Orleans on a ship. On her return Anne noticed faint brown spots on her skin. Doctors diagnosed yellow jaundice, probably caused by contaminated shipboard water. Her condition quickly deteriorated; she entered the hospital and died in May 1926. Murchison was stricken. He left the children in the care of relatives and disappeared, driving around the state, alone, for weeks at a time, a whiskey bottle usually at his side. What remained of his business began to decay. "When Anne died," Murchison told his secretary many years later, "people said I stayed drunk for a year."

V.

From the moment the first American settlers crossed into Texas in the early 1800s, no one wanted much to do with the western half of the state. Out beyond Fort Worth, for six hundred miles all the way to El Paso, stretched little but arid, lifeless plains, much of it flat as a frying pan and just as hot. Once the Indians were run off, West Texas proved good for little but cattle ranching, and a drought during the 1910s forced many small ranchers back east. By the 1920s there was no reason to go to West Texas and every reason to leave; most counties had few if any paved roads, a single town, and maybe a few hundred people. To most Texans the entire region was an afterthought, Hell with cows.

In the twenty years after Spindletop, oilmen began venturing out onto the plains, buying leases here and there; every rancher between Fort Worth and Pecos was certain there was oil beneath his land if only someone would drill a hole. For the most part, geologists scoffed. A few wells were drilled; none found anything but the faintest showings of oil. Then, much as happened at Spindletop, a local attorney named Rupert G. Ricker began buying leases around his hometown of Big Lake, a flyspeck located in the high mesa country two hundred miles west of San Antonio. When Ricker ran out of money, the leases passed to one of his old army chums, who with a partner hoped to sell the land to a major company to drill. Finding no takers, and facing the expiration of their leases, they were forced to actually drill a well. As fate would have it, it was a gusher, the fabled Santa Rita No. 1, and it triggered a massive land rush across West Texas.

All the majors plunged in, leasing millions of acres from the small towns of Midland and Odessa all the way south to the Rio Grande and west to El Paso. In 1926 a rancher named Ira Yates, having pestered oil scouts for years to drill a hole beneath his land in Pecos County—Roy Cullen had turned down the opportunity—finally succeeded in having a well drilled; it, too, was a gusher, opening the legendary Yates Field, one of the largest ever found in Texas. The same year a fast-talking Fort Worth promoter named Roy Westbrook, obliged to drill a well he hadn't planned to satisfy his suspicious investors, struck oil even farther west, in remote Winkler County, which wraps around the southeast corner of New Mexico. The Hendricks Field, as it was called, lured scores of oilmen into the farthest corners of West Texas.

The most desolate spot in which Texans would ever find serious quantities of oil, Winkler County, was to figure prominently in the careers of both Clint Murchison and Sid Richardson. There was no actual town there. The only settlement, Kermit—named after Theodore Roosevelt's son, who had visited on a hunting trip—was a smattering of houses. There were no paved roads, no post office, no hotel, no telephones. There were barely any people. The 1920 census put the population at eighty-one; by 1926 there were exactly six registered voters. There were no rivers and no lakes, just mile after mile of yellowy grass, a belt of sand dunes, and a hot wind that blew its grains into every nook and cranny. The opening of the Hendricks Field, however, triggered the birth of a consummate Texas boomtown, dubbed Wink, which sprouted in a cattle pasture and within months was home to ten thousand oil workers, speculators, prostitutes, gamblers, and merchants to feed them.

The new gushers in West Texas roused Murchison from his struggle with alcohol and depression. By chance he already owned several leases in Winkler County. His geologist, Ernest Closuit, was already analyzing data from the new field when Murchison dispatched him west to drill a test. Murchison, meanwhile, hit the phones. Not for him the muddy boots and windblown tents of a remote drill site: Clint Murchison found more oil on the telephone than most of his peers would ever draw from the ground. He first began buying leases. By the end of 1926 he had put together eighty acres on the edges of the Hendricks Field.

One night while he and Closuit were meeting in San Antonio, their test well came in strong. Within weeks they had a dozen more just like it. The problem was, there was no place to put the oil. For the moment, Murchison did what oilmen had always done: he built two giant, five-hundred-thousand-barrel storage tanks. The nearest railhead was at Pyote, south of Wink, but there was no way to get the oil there. Though he knew nothing about pipelines, Murchison decided to try to build one. A lumberyard worker at Pyote said he could locate secondhand pipe and lay the pipeline if Murchison paid. Murchison arranged a line of credit and work got under way, but before the pipe reached Pyote he received crushing news: Humble Oil was building a pipeline of its own. Murchison didn't have enough production to feed the pipeline himself. If other producers sold to the Humble line—as they would—he would face a massive loss.

Then, walking down Wink's muddy main thoroughfare one evening, Murchison had a thought: Why not offer gas heating and light to the locals? He already had the pipe; it took a matter of weeks to lay it down one side of the street. Residents were invited to tap into it anywhere they could, five dollars a month for a home, ten dollars for a business. Natural gas had been used to heat homes and factories in England for a century but had never caught on in the United States; most Texas oilmen simply allowed the gas they found to escape into the atmosphere. Murchison was amazed how simple the business was; once a pipeline was built, all he did was sit back and collect monthly checks. When other West Texas towns expressed interest in having gas lines of their own, Murchison incorporated the Wink Gas Company and built lines to Pecos, Barstow, and Pyote. A friend in Oklahoma said his town could use one, too, so Murchison sent work crews north and by mid-1928 had a pipeline furnishing gas to the towns of Kingfisher and Hennessey. He sent salesmen fanning out across South Texas and soon had contracts to supply gas to Navasota, Sealy, Bellville, Eagle Lake, and Columbus.

In late 1928, after taking expansion loans from a pair of Dallas banks, Murchison moved into an apartment in a fashionable Dallas building called Maple Terrace; his roommate, a fastidious, nearsighted boyhood chum named Wofford Cain, ran the Oklahoma side of the pipeline business. The two would remain in business together for decades. In January 1929, in an effort to consolidate the chaotic piles of paperwork in his apartment and San Antonio office, Murchison leased space in the fifteen-floor American Exchange Building in downtown Dallas and, together with his brother Frank and Ernest Closuit, officially merged their far-flung gas operations into a single company they decided to call the Southern Union Gas Company. Murchison had big plans for Southern Union, the kind that occurred to few if any of his peers in Texas Oil. He wanted to make it a national company, with stocks and bonds sold in northern stock markets. Murchison would arrange and build the pipelines, Closuit would drill for the gas, and Frank Murchison was sent to Chicago to begin raising money.

Clint and Closuit easily held up their end. During a New Mexico vacation that spring with Wofford Cain, Murchison found their next customers when he realized neither Albuquerque nor Santa Fe used natural gas. Instead of hunting and fishing, the two pals ended up spending weeks negotiating the acquisition of a small oil company that had found gas in the mountains near Farmington. Once the gas supply was secured, Murchison

had little trouble obtaining a franchise to supply Santa Fe. Albuquerque
was another story. A half dozen competitors sprang up to bid against him.
At a city council meeting the mayor asked whether any bidder could supply a twenty-five-thousand-dollar cash bond to insure its financial viability.
Everyone raised their hand. When the mayor asked for fifty thousand, Murchison and another man raised their hands. When the bidding went to one
hundred thousand dollars, only Murchison raised his hand. He scribbled out
a check and left with the franchise.

As they walked outside, Cain shot him a glance. "We don't have that
kind of money in the bank," he said.

"We'll worry about that when we get back," Murchison said.

Murchison operated this way the rest of his life; as the son of a banker, he
knew he could always find a gullible loan officer somewhere. In this case he
took a train directly to Dallas and met with one of his father's oldest friends,
Nathan Adams, president of First National of Dallas. Adams was a crucial
building block in the budding Murchison empire and would remain so for
years. One meeting was all Murchison needed to get the one hundred thousand dollars. "If you are honest and you are trying, your creditors will play
ball," he told Cain afterward.

Once the franchises were secured, the challenge became building a pipeline 150 miles across the Continental Divide to link Santa Fe and Albuquerque to the gas wells in the mountains. It was an engineering effort
that would have daunted lesser men. Murchison surveyed the route from
an airplane, dropping flag-tipped bags of flour to mark the route he wanted.
Roads needed to be laid across canyons and mountainsides, then huge sections of pipe trucked in and buried, often in rain- and snowstorms. The
pipe alone cost three million dollars, all of which Murchison got on credit.
He had hoped his brother Frank could raise money to repay the loans in
Chicago, but their brokerage firm, Peabody & Co., ran into management
problems, and Frank was forced to step in and actually run Peabody himself. Still, Murchison was confident that once the pipeline was complete, gas
sales would allow him to repay the banks and his main trade creditor, the
Oilfield Supply Company.

Construction had just begun in the fall of 1929 when the stock market crashed. In a matter of weeks America sank into a national depression. Murchison watched in dismay as his cash flow sputtered, coughed, then finally stopped altogether. He couldn't pay his workers, endangering the entire pipeline project. One week he made payroll only by borrowing forty thousand dollars from one of his father's friends. When the pipeline reached a point seven miles outside Albuquerque, they ran out of money once more. Only when Wofford Cain appealed to the mayor to return a portion of their cash bond was the pipeline finally completed.

All through the worst months of the Depression during 1930 and 1931 Murchison signed up new customers for Southern Union, and barely two years after its founding he could boast service to forty-three towns in six states—this despite the weekly struggle every Friday to meet payroll. What saved him was the fact that he knew more about banking than any other oilman in Texas. He coaxed every last dollar he could out of the Dallas banks, then pushed back repayment, all but daring the bankers to foreclose. By 1932 his debt had grown to more than four million dollars, far more than his net worth. "Aren't you concerned about owing all this money you can't pay?" Ernest Closuit asked him.

"No," Murchison said with a smile. "If you're gonna owe money, owe more than you can pay, then the people can't afford to foreclose."

VI.

The town of Monahans, thirty miles south of the new Hendricks Field in far West Texas, was a cluster of sandblown shanties clinging to the railroad line out to El Paso. Though the seat of Ward County, it had barely two hundred people, many of them trying to eke out a living on the ranches outside town. By 1926 West Texas was in the tenth year of an agonizing drought, and a number of ranchers had given up, leading off their remaining cattle and letting the desert reclaim their land.

One of the most desperate was a man named George Washington O'Brien. In 1928, according to his grandsons, O'Brien was on the verge of bankruptcy. His only hope, one he shared with every rancher west of the

Pecos, was oil. He had sold leases to Gulf in the rush of the mid-twenties, but Gulf had shown no inclination to drill. O'Brien was certain there was oil under his dirt, a conviction that grew when Ward County's first well, a small one, came in that November. Unable to mobilize Gulf, O'Brien decided to drill his own well. He gathered his sons, borrowed a derrick from a water-driller, and managed to get down several hundred feet before giving up, the apparent victim of a broken drill bit. Desperate, his bank threatening to foreclose, O'Brien drove into Monahans and found a doughy, down-on-his-luck character who said he was an oilman. He introduced himself as Sid Richardson.

In the fall of 1928 Richardson was thirty-seven years old, flat broke, and deeply in debt. Between the day he drilled a dry hole in Henderson County in 1922 and the day he agreed to drill the well for George O'Brien, Richardson all but vanishes from the pages of history; more is known about sixth-century Byzantine kings than Richardson's life in the mid-1920s. In all likelihood he had continued working for a time with his brother-in-law, Doc Bass, but at some point he struck out on his own. One account has him trading leases in the Mexia boom. Records at various county courts show he leased land in several spots in West Texas, which took little money. He presumably attempted to drill some of that land, given the astounding two hundred thousand dollars or so he had somehow managed to borrow from the First National Bank of Fort Worth.*

Whatever holes he drilled, however, came up dry. He had no cash flow, and no way to repay the bank. But he still had credit at the equipment-rental compounds in Winkler County, and with a borrowed rig he struck oil on George O'Brien's ranch in early 1929. It was a small well, but it got his name known around the county, and a number of neighboring ranchers happily signed over leases he promised to drill. Other independents had begun wells on the Estes ranch in northern Ward County, and Richard-

---

*Exactly how Richardson coaxed that much money out of a bank is unclear, but Richardson cited the two-hundred-thousand-dollar figure more than once in later life. Maybe he ran into a gullible loan officer. A more likely explanation is that he received help arranging the loan—a letter of recommendation, maybe even a loan guarantee—from the banker he knew best, John Murchison. If so, it wouldn't be the last time he sought the Murchisons' help in coming years.

son, paying his drill hands in groceries and promises, plunged into the play with vigor. In the coming months he struck a half-dozen good wells in what became known as the Estes Field. He began paying off his debt and looking for more land. For the first time in his thirty-seven years he was a successful independent oilman.*

By the end of 1929, sale of oil from Ward County was putting $25,000 a month in Richardson's pocket. He was wealthy enough to rent one of Fort Worth's finest hotel suites, the $750-a-month penthouse at the Blackstone Hotel, with a balcony looking south over the city. Most Fridays he drove his maroon Chevrolet in from West Texas, washed the dust from his clothes, then headed to Dallas, where Clint Murchison had purchased an old polo club and was transforming it into a family compound. There Richardson joined a revolving group of young oilmen who drank and cursed and played poker through the long weekends. Monday morning he would head west again. It was a good life, soon to end.

---

*Richardson drilled several of these first wells in partnership with an oilman named Eugene Kelsey.

FOUR

The Bigamist and the Boom

*You can't hold it against a man for possessing
and being possessed by all the component and
conflicting parts of being a genius. It's just that
sometimes it is difficult for mortals to live with.*
MARGARET HUNT HILL, *H.L. AND LYDA*, 1994

I.

In the 1920s the eastern quarter of Texas, a triangle of thick pine forests loosely bounded by Dallas, Houston, and Shreveport, Louisiana, was about as backward a region as America knew. Dotted with tar-paper villages and the odd sawmill, the area had few paved roads and fewer telephones, little indoor plumbing and not many people. The white farmers, many of whom tried to grow cotton, corn, and yams, were suffering through a decadelong drought that had banks knocking on their doors. Teetering on the brink of depression, East Texas, its inhabitants mostly poor, suspicious, Bible-thumping fundamentalists, was Sherwood Forest with a drawl. Its unlikely Robin Hood, the man who promised to find oil beneath the pine needles, arrived in 1926.

His name was Columbus Marion Joiner, and he was sixty-six years old when he drove into East Texas that autumn. A thin, kindly man, bent at the waist, he said he was a famous oilman. He wandered from farm to farm talking about oil prospects, but seemed to spend much of his time tending to several elderly widows in Rusk County, over toward the Louisiana line. The ladies gave him tea and afternoon conversation, and their mineral rights. What the old gent didn't mention was that he was down to his last forty-five dollars and had found the widows after reading their husbands' obituaries in the Dallas newspapers.

Depending whom you talked to, C. M. Joiner was either a con man, a down-on-his-luck wildcatter, or some combination of both. Born on the eve of the Civil War in Alabama, he had earned a law degree and served a term in the Tennessee legislature before settling near his sister in Oklahoma Territory; her husband, a Choctaw Indian, introduced him to tribal leaders who, impressed with his legal acumen, hired him to negotiate sales of their mineral rights. Joiner thrived for a time, using his earnings to amass twelve thousand acres of prime farmland. But just as it had Roy Cullen, the Panic of 1907 wiped him out. Penniless, he drifted into the Oklahoma oil fields, where he bought and sold leases on credit, scraping by, living in boarding-houses and begging for food.

It was then he met Doc Lloyd. Lloyd, whose real name was Joseph Idel-bert Durham, was an aging, rumpled three-hundred-pounder who had wandered west from Cincinnati, leveraging a modicum of training in phar-macology into a career conducting "Dr. Alonzo Durham's Great Medicine Show," in which he hawked worthless medicines. When that venture ended, Lloyd searched for gold in Mexico, Idaho, and the Yukon, eventually reading enough about geology to try his luck in the Oklahoma oil fields. By the time he began teaming with Joiner in the late 1910s, Lloyd was billing himself as a "nationally known geologist," which of course he wasn't. Together Joiner and Lloyd raised money for several drilling ventures, much of it from farmers and widows. Both were good with the ladies; Lloyd had six ex-wives himself, thus his penchant for pseudonyms. Joiner was the salesman. Lloyd was his "expert," the one who compiled data into thick geological reports, much of it nonsense. They never found serious amounts of oil, but—and this was the point—they did manage to raise enough money to make a living. Barely.

In the early 1920s, after a succession of dry holes in Oklahoma, Joiner began venturing into East Texas, picking up his first leases in Rusk County. Oil scouts had been sniffing around the region for twenty years. A well or two had been drilled, but no serious oil had been found. Joiner, though, had a hunch. Hoping to assemble a five-thousand-acre lease, he moved to Dal-las in 1925, taking a one-room office in the Praetorian Building and slowly wooing a series of East Texas widows to sign over their mineral rights. By 1927 he had enough land to think about drilling, at which point he brought in Lloyd. The onetime snake-oil salesman cranked out another thick report—"Geological, Topographical and Petroliferous Survey, Portion of Rusk County, Texas"—that portrayed the county's geology as a maze of anticlines, salt domes, and geological faults harboring the greatest oil field in the world. Not a word of it was true, but in Joiner's smooth hands it impressed scores of Rusk County widows and a few of their friends. Joiner and Lloyd came away with mineral rights to almost four thousand acres, along with the thousand dollars or so they needed to start a well.

By August 1927 Joiner had his equipment in place, a few sections of rusty pipe, a rickety 112-foot wooden derrick, and two boilers, one for use in cotton gins. He lured a driller down from Dallas, filled out his crew with local farmboys at three dollars a day, and broke ground in a clearing belonging to one of his favorite widows, a cheery fifty-four-year-old named Daisy Bradford, known as Miss Daisy. After six months Joiner's team had drilled down to 1,098 feet but there the pipe stuck. Joiner used the last of his funds to ignite a dynamite charge in the well, but even that failed to dislodge the pipe. He gave up. It took another six months to raise the money for a second well, which Joiner spudded 100 feet from the first. This one reached 2,500 feet before the pipe broke. Joiner told Miss Daisy not to worry. Doc Lloyd predicted oil at 3,500 feet. The next well, he promised, would be the gusher.

At that point, Joiner got lucky. A Humble Oil team struck oil in a new field barely sixty miles southwest. The strike came from a previously unknown stand, the Woodbine. Suddenly all anyone in East Texas wanted to talk about was the Woodbine. Joiner wasted no time capitalizing on the news, selling twenty-five-dollar shares in a third Daisy Bradford well to dozens of local families, raising enough to lure a first-class driller over from Shreveport, a man named Ed Laster. In May 1929, nearly two years into the project, Laster began drilling a new well in Miss Daisy's clearing. In just two days he reached twelve hundred feet; then the boilers gave out. By the time they were fixed, Joiner had once again run out of money. Laster quit, then returned. Joiner began driving prospective investors down from Dallas to show off the well. Laster did his part, drawing up samples of dirt he swore held signals to oil.

The stock-market crash in October 1929 made it all but impossible for Joiner to raise more money; in the end, it took a solid year to gather enough to resume drilling. But in the summer of 1930 Laster went back at it, eventually, by late July, reaching a depth of 3,456 feet—just 44 feet above the level where Doc Lloyd predicted they would hit oil. At that point Laster retrieved a sample from the bottom of the well—and was stunned to find it saturated with oil. He began to get excited. A week later, Laster drew up a second, larger sample, seven inches thick, and found it, too, slathered with oil. Curious oil scouts had begun appearing at the site that spring, and Laster was terrified one would discover what he had found; if news got out, the majors would snap up every remaining acre in the county. In fact, a Sinclair Oil scout found Laster's sample in a barrel that same night, but dismissed it as a common wildcatter's ruse called "salting the well," aimed at enticing a larger oil company to buy the well.

All through that steamy August Joiner and Laster worked to prepare the well—the Daisy Bradford No. 3, it was called—for a formal test, which would use a piece of equipment called a drill stem to suck up any measurable amounts of oil. Dozens of scouts began lingering around the drill site, standing off to one side to debate whether Joiner had found traces of oil or was simply faking it.

Finally, on September 5, 1930, everything was ready. Then, just as preparations neared completion that afternoon, a car drove up. A man got out, a hefty six-footer wearing a shirt and tie and a straw boater.

His name was H. L. Hunt.

II.

He was a strange man, a loner who lived deep inside his own peculiar mind, a self-educated thinker who was convinced—absolutely convinced—that he was possessed of talents that bordered on the superhuman. He may have been right: in the annals of American commerce there has never been anyone quite like Haroldson Lafayette Hunt. At a time when itinerant wildcatters like Sid Richardson couldn't find time for a wife let alone a family, Hunt would build three, two in secret. If they made a movie of his life, no one would believe it could be true.

The man who came to embody all the myths of the Texas oilman, whose madcap family dramas would one day captivate a nation, was neither raised in Texas nor introduced to oil until well after his thirtieth birthday. He was born in 1889, the youngest of eight children reared by an aging Confederate veteran who had moved north after the war, to a farm in downstate Illinois, seventy miles east of St. Louis. His family called him June, short for Junior, and he was barely walking when his parents realized his intelligence bordered on that of a prodigy. In later years his siblings swore he could read the newspapers aloud at the age of three. His capacity for mathematics became a local legend; people marveled how the child could multiply large sums in his head. Early on, June's math skill manifested itself in a fascination with card games, a passion he would indulge to his dying day.

His mother, Ella, doted on her youngest, lavishing him with praise and, as Hunt recounted with an odd pride in later years, breast-feeding him until he was seven years old. All the flattery and attention imbued Hunt with a keen sense of entitlement, a feeling that he possessed a unique intellect, exponentially more insightful than anyone he met, and this, too, became a lifelong trait. By the time Hunt was born his father, a stern but savvy man known as "Hash" Hunt, had built his initial eighty-acre farm into a five-hundred-acre spread, among the largest in Fayette County. June helped on the farm, but by his sixteenth birthday he was showing signs of restlessness. He and his father weren't getting along. When June said he had no interest in college, his father and older brothers pressed him to become a bank clerk, a natural fit for a boy so good with numbers. But June had no more interest in a teller's cage than young Clint Murchison a decade later. He yearned to see the world.

And so, one day in 1905, sixteen-year-old H. L. Hunt packed a deck of cards in his bedroll and ran away from home. From St. Louis he took a train into western Kansas, planning to try his luck as a laborer. He ended up a dishwasher in a railroad restaurant in the town of Horace. That lasted a month. From Kansas he headed on to Colorado, where he cut sugar beets. South of Salt Lake City, he signed on as a sheepherder. Riding the rails into

Southern California, he took a job driving mule carts brimming with road gravel. Outside the town of Santa Ana, he took his first ranch job, again driving mule teams. For two years Hunt ambled from job to job as the spirit moved him, planting cattle feed outside Amarillo, lumberjacking in northern Arizona, and narrowly missing the San Francisco earthquake of 1906—he had left the city just days before to try out for a semipro baseball team in Reno.

For a time his brother Leonard joined him. They ranged across the Pacific Northwest through 1908 and 1909, working the harvests in Washington, Montana, and then the Dakotas. When Leonard returned to Illinois to take a teaching job, Hunt headed north into Canada looking for work. He had just arrived at a small town outside Moose Jaw, Saskatchewan, when he found the telegram from his parents. Leonard was dead; of what it wasn't said. It was February 1910. Hunt returned for the funeral, then returned west for a year, until March 11, 1911, when he received the call that his father had died. Once again he went home.

In the days after his father's funeral, Hunt finally confronted his future. He was twenty-two at that point, with a five-thousand-dollar inheritance, enough in 1911 to get him started in just about any business he wanted: farming, ranching, tavern owner, anything. He couldn't sit still for college, he knew that. All Hunt knew was that farm life bored him. He had already gone west. Then, remembering his father's stories of the lush farmland around a southern Arkansas town he had visited during the Civil War, he decided to head south. Like Roy Cullen, Hunt was proud of his southern heritage. The leisurely life of a plantation planter appealed to him, and in late 1911 he arrived at the town his father had remembered, Lake Village, in the heart of Arkansas's best cotton country, just inland from the Mississippi River in the state's southeast corner.

Cotton prices were booming, and so was Lake Village, a bustling town on the shores of Lake Chicot. The population had grown to fifteen hundred in the last few years. Taking a room at the hotel, Hunt used most of his inheritance to buy a 960-acre farm called Boeuf Bayou five miles south of town, then headed to Little Rock to buy horses and mules. That first year his cotton crop was wiped out by the first Mississippi River flood in thirty-five years, but Hunt recovered nicely. Negros worked his land, allowing Hunt to spend much of his time playing cards. In time he made as much gambling as he did from the cotton fields.

Now in his mid-twenties, Hunt had grown to be a serious, solitary young man, quiet, focused, and disciplined. He had a clerk's face, a soft nose, and wide-set eyes. He dressed neatly. He didn't drink. Around Lake Village he was considered a touch odd, a deep thinker, a man who read the newspapers and could sometimes be seen writing poetry or song lyrics. Still, people liked him. He developed a reputation for honesty. When he borrowed money for a harvest, he repaid it on time. He dated a girl here and there, but for the most part he kept to himself, drawing about him a sense of mystery in the little southern town. When he was flush, Hunt took a train to the big-money poker games in Memphis and New Orleans, where he adopted the moniker Arizona Slim, a nickname he kept the rest of his life. Then, in the fall of 1914, Hunt's mother died, and within weeks, perhaps unsurprisingly, he proposed marriage to a Lake Village girl who was very much like the departed Ella Hunt.

Then twenty-five, Lyda Bunker was a plain, plumpish schoolteacher from a prominent family, a quiet, stable woman who was about to be married to another man. Hunt had been seeing Lyda's sister. Love came quickly, though, and both broke off their relationships to be married, in a simple ceremony at the Bunker home. The newlyweds moved into one of the Bunkers' rental houses and began a family. Their first child, a daughter they named Margaret, arrived in November 1915. Two years later came a boy, Haroldson Lafayette Hunt Jr. They called him Hassie.

Hunt's fortunes continued to rise and fall on cotton prices and his poker winnings, and Lyda, now with two mouths to feed, began urging him to find something more stable. "June," she pleaded more than once, "why don't you get a regular job?" But Hunt was addicted to the adrenaline he found dealing cards, parcels of land, and, when the war ended in 1918, cotton futures. That's when his luck ran out. Believing the cotton boom was poised to end, he placed a massive bet that prices would go down; when they didn't, he lost everything, including what little money he had put away. According to family lore, Hunt only saved his farm at a high-stakes poker game in New Orleans, during which he managed to turn his last hundred dollars into one hundred thousand dollars. He kept his land, but much of it was now hamstrung with bank liens, and when the recession he expected finally hit in 1920, the price of both cotton and land went into free fall. For the first time Hunt began to question his style of living. He turned thirty-two that year. He wasn't a kid anymore.

His epiphany, as Hunt remembered it years later, came in January 1921, as he was negotiating to buy land from a family named Noell. During a long afternoon of talks at the Noell home, Hunt found himself listening as one of the other men discussed the manic scene at El Dorado, seventy miles west of Lake Village in south-central Arkansas. Oil had been found there the previous spring, and thousands of people were flooding into town in hopes of finding more. It sounded exciting—far more exciting than another year of praying for cotton and land prices to rebound. Hunt stepped outside onto the porch and gazed at a setting sun. "What is it that you are trying to do?" he asked himself. "Are you going to bury yourself here for the remainder of your life?" Why not rent out the land and try something new?

Hunt headed into Lake Village determined to raise enough cash to give El Dorado a try. Both town banks, however, strapped for cash themselves, turned him down. All Hunt could raise was fifty dollars from a trio of gambling buddies. It was enough. He took his pals and boarded the train to El Dorado. When they arrived, they disembarked into a roiling boomtown thronged with gamblers and prostitutes and hustlers of every stripe. From the steps of the train depot, Hunt peered down the muddy main thoroughfare, South Washington Street, which was packed with people. The town's population had exploded from maybe a thousand people to something over fifty thousand in less than a year. There were no rooms to rent. People were sleeping in tents all over town; the last available berths, the barber's chairs, rented for two dollars a night. Space was at such a premium that the city council had begun renting space on the sidewalks, tranforming the eastern side of South Washington into a row of outdoor grills dubbed Hamburger Alley. This, Hunt saw, was a place a smart man could make easy money. "All I need," he muttered, "is a deck of cards and some poker chips."

The next day Hunt hit the tables. The gambling halls of El Dorado teemed with professional card players, but there were so many new marks flooding into town, Hunt made a killing. Again and again he raked in big pots, three times taking the biggest game in town. He quickly amassed enough cash to buy his way into one of the town's few hotel rooms, and within days he had enough to rent a shack at the foot of South Washington where he opened his own dingy cardroom. Within weeks he had saved enough to take over the first floor of a nicer building up the street, a large single room he packed with card tables, chairs, and floor areas to throw dice. Hunt's ad hoc casino was a tidy, safe place in a violent, dirty town; he earned a good reputation and caused no trouble. Soon, though, the city council began a cleanup campaign, closing down the brothels and Hamburger Row.

This being Southern Arkansas in the early 1920s, El Dorado's antivice campaign was augmented by the forces of the Ku Klux Klan. As Hunt told the story years later, a group of twenty or thirty white-robed Klansmen arrived outside his establishment one night that summer. "Shut this place down," the leader yelled. "Shut it down or else...."

Hunt stayed open, but once again he began thinking about a change in careers. El Dorado was an oil town now; it was natural that he would consider oil as a line of work. He decided to start small, throwing in with a partner and several friends to lease a half-acre plot outside of town; he secured a drilling rig by paying the overdue freight on an old rig abandoned beside the rail depot. In the first instance of what came to be known as the "Hunt luck," that first well, the Hunt-Pickering No. 1, struck oil, a decent amount, before petering out several weeks later. Rather than buy a thirty-five-hundred-dollar pump jack to restart it, they sold the well to another independent, who promptly went out of business before paying. It was a dispiriting experience, but as Hunt noted years later, "it served the purpose of getting me started in the oil business."

Closing the gambling hall, Hunt set his sights on establishing a viable oil company. He raised money from friends in lots of two hundred dollars, then had one of the casino's floor men, a character known as Old Man Bailey, sweet-talk a farmer named Rowland into assigning Hunt a lease on his forty-acre farm in return for a twenty-thousand-dollar IOU—a sum far larger than Hunt had access to. Rowland grumped a bit, to the point that Bailey was obliged to take a room in his farmhouse to placate him. Hunt, meanwhile, began drilling in Rowland's fields, and in January 1922 his second well came in strong, five thousand barrels a day. He soon started two more, and both proved good producers as well.

In early 1922, thanks to the efforts of local authorities and the Klan, El Dorado was safe enough for Hunt to send for Lyda and the children. They moved into a rented house on Peach Street, where a year later Lyda gave birth to a third child, a daughter they named Caroline. Hunt's luck in oil ebbed and flowed. He hit a good well or two, then watched in dismay as production fell to a trickle. He began to borrow from the El Dorado banks, always repaying in full and on time. Soon he had drilled enough wells that he found himself with employees, who became known around town as "Hunt men." When Hunt men had troubles of their own, their boss listened, nodded his head, and loaned them money. They proved loyal and indefatigable workers, in time helping Hunt hit a string of strong producers in the fields outside El Dorado.

In three short years Hunt had transformed himself from a gentleman planter into a professional gambler and then, finally, at the age of thirty-five, into a successful oilman. By the end of 1924 he controlled nearly four hundred thousand barrels of proven oil reserves, valued at roughly six hundred thousand dollars, about seven million dollars in today's dollars. That year he moved Lyda and the children into a new three-bedroom brick home in El Dorado. Usually busy in the oil fields or, at night, at a poker game, Hunt was rarely at home. Lyda didn't complain. A pleasant, religious woman, she struck up friendships with many townspeople, raising the children and pulling them to the Methodist church without their father on Sundays. About the only thing Lyda nagged Hunt about was money. He was always willing to bet their nest egg on a new well, while Lyda urged him to save. Their new house was as much a receptable for their savings as a place to live.

In February 1925 Lyda gave birth to the couple's fourth child, a girl they named Lyda Bunker Hunt. But a month later a radiator in the new home malfunctioned, leaking fumes that smothered the infant in her crib. Lyda was devastated. Hunt did what he could to assuage her grief, whisking her off to New York for their first real vacation together, breaking ground on a big new house outside town and once again getting her pregnant. But in the weeks after his daughter's death, it was Hunt who underwent the most profound change. Maybe it was the grief. Maybe at the age of thirty-five, it was an early midlife crisis. Maybe it had been his plan all along. Whatever it was, Hunt decided he had no interest in living out his days a country oilman in the Arkansas boondocks, monitoring oil flows and depositing checks.

Four months after his daughter's death he suddenly sold all his holdings for a six-hundred-thousand-dollar promissary note, then had a bank discount it for cash. He was leaving the oil game, he told Lyda. With the money in hand, Hunt announced he was taking an extended trip—alone—to Florida. He said he was thinking of investing in the state's postwar real estate boom, and maybe he believed it. But it wasn't real estate he was looking for, as an operetta he began composing on the train to Tampa suggests. He called it, "Whenever Dreams Come True, I'll Be with You." "Up to the time I met you," he wrote, "life was as drab as can be. . . . Something was missing for me. You are my love now forever . . ."

Leaving his pregnant wife and three young children behind in Arkansas, H. L. Hunt headed to Florida in search of romance.

III.

Her name, once he found her, was Frania Tye. She was twenty-one years old, and when she sat down on a park bench outside the Tampa Terrace Hotel that hot September afternoon in 1925, she had been in Florida for only a few weeks. Raised in Buffalo, Frania was the daughter of Polish immigrants; her father, a carpenter, had changed the family name from Tyburski. She had reached adulthood with no career plans or obvious skills, taking jobs at department stores and hair salons, but her dark beauty, enhanced by an ample bosom and slender hips, probably persuaded her she wouldn't need any. She was small, an inch over five feet, and spoke with the hint of an accent. As a teenager she had fallen in love with a Polish boy and followed him when he took a job in Cleveland. It didn't last, though, and by the time she returned to Buffalo the only thing Frania knew for certain was that she didn't want to stay in Buffalo. Eager for a new start, she persuaded her father to help her resettle in Tampa, where he moved her into a room at the DeSoto Hotel. The owners agreed to look after her.

She had just taken a job in a real estate office that day when she began talking to the older man sitting on the park bench. He said his name was Bailey, and just her luck, he was interested in real estate. Frania gave him her phone number. Later, she took a call from a man who identified himself only as "Hunt." Mr. Hunt said he was interested in seeing some property. When he drove by to pick her up, Frania realized it was the man from the park bench. Not wanting to spook one of her first clients, she didn't say anything. Together they drove out and looked at the property. Hunt said he'd think about it. When they returned to Tampa, he asked her to dinner. She politely declined. He called again a few days later. Once again they drove out to see the property, and this time when they returned Hunt asked her to his hotel room. She slapped him, saying she wasn't that kind of girl.

Frania Tye didn't see Mr. Hunt again for several weeks. When he next called, in October, he pressed her for dinner, and she relented, embarking on what she later described as a whirlwind courtship. Every night they dined out, sipping Cokes and ginger ale, as Hunt regaled her with stories of his youth wandering the West, felling trees and boxing and playing semi-pro baseball. He said he was a Louisiana oilman and that his full name was Major Franklyn Hunt, and when she asked if he had been in the military, he said no, that everyone in the South was a colonel or a major. In no time Frania, rootless and lonely, began to weaken beneath Major Hunt's romantic onslaught, so much so that when Hunt, during a long drive to St. Petersburg on the night of November 10, mentioned marriage, she was helpless to deny him.

The next morning Hunt took her to a pawn shop and bought her a simple gold band. Afterward they drove to a white stucco bungalow in Tampa's bustling Cuban quarter, Ybor City, where Old Man Bailey was waiting with a justice of the peace, a Latin gentleman. When Frania asked about a marriage license, Hunt said Florida didn't require one. Dazed, she signed some kind of ledger, as did Hunt, and before she knew it, the Latin gentleman was reading aloud from a Bible.

"Do you take Frania Tye as your lawful wife?" he asked Hunt.

"Yes," Hunt said.

"Do you take Franklyn Hunt as your lawful husband?" he asked Frania.
"Yes," she said.

The justice pronounced them man and wife, and if that status was legally
questionable without a marriage license, it didn't matter to the newlyweds,
not yet anyway. The idea that her husband might already have a family never
entered Frania's mind; what was in Hunt's, other than a rapid consummation
of their union, was anyone's guess. For the rest of his life he never spoke or
wrote a meaningful word about Frania Tye. But from the moment they walked
out of that Ybor City bungalow, Hunt treated her as his wife. After honey-
mooning in Orlando, Hunt said he needed to return to Louisiana for business.
He promised to send for her and, after another brief return to Tampa he did, in
February 1926, telling her to meet him in New Orleans for Mardi Gras.

On the steamship across the Gulf of Mexico Frania became seasick,
and by the time she reached Louisiana she realized she was pregnant. Hunt
appeared overjoyed. After a few days enjoying the French Quarter, he spirited her to Shreveport, where they lived briefly in a hotel before Hunt rented
an apartment on Hearndon Street. Frania settled in to prepare for the baby,
who was due in October. Hunt disappeared on frequent business trips, but
for now she was happy. He didn't make the birth, a boy who bore his father's
initials, Howard Lee Hunt.

For the next three years Frania remained content in her new, if slightly
bizarre, life. Major Hunt traveled constantly, never quite making it back to
Shreveport for Thanksgiving or Christmas. On his return he was always
sweet and apologetic, always making sure she had plenty of cash on hand, but
after the couple moved into a small brick home on Gladstone Street, a few of
the neighbors appeared to grow suspicious, wondering aloud just where the
peripatetic Major Hunt was actually going. For her part, Frania would later
insist she never suspected a thing. Her husband traveled for his oil field job,
simple as that. She was too busy with their growing family to give it much
thought. In October 1928, two years after little Howard's arrival, she gave
birth to a second child, a girl Hunt named Haroldina. Then, in the spring of
1930, Frania became pregnant for a third time.

By that point Hunt was doing a good deal of his traveling in Texas, and
when Frania was deep into her pregnancy that summer, he explained it

would be necessary for the family to move to Dallas. There Frania unpacked her things in a two-story brick home at 4230 Versailles Avenue, in the city's nicest residential area, Highland Park. Not long after, she gave birth to their third child, a girl they named Helen Hilda Hunt. Once again Hunt couldn't make it to the hospital. He was too busy, off on important business in the remote East Texas pines, where he had stumbled on a down-on-his-luck wildcatter named Columbus Joiner, a man with whom he was about to make history.

IV.

After secretly settling Frania Tye in Shreveport, Hunt had returned to El Dorado to find his first family slowly recovering from his infant daughter's sudden death. Their new house was finished, a three-story, eight-bedroom English Revival mansion that was easily the largest in the area. They named it The Pines. Hunt bought a mammoth Packard limousine whose chauffeur drove Margaret, Hassie, and little Caroline to school each morning. The family continued to grow. In February 1926, just as Frania was moving into her new apartment in Shreveport a hundred miles to the southwest, Lyda gave birth to a second son, a twelve-pound butterball they named Nelson Bunker Hunt. Three years later came a third, Herbert, named for the president Hunt favored, Herbert Hoover. Still later came a fourth, Lamar.

Hunt's life as a covert bigamist did nothing to mar the idyllic days at The Pines. If anything, he spent more time with his children than before. Hassie operated a lemonade stand in the driveway. Margaret learned to drive. Meals were a time for Hunt to educate his growing brood, lecturing the children on everything from politics and music to the complications of serving on a bank's board of directors. Every night after dinner Lyda played their new grand piano, and everyone would gather around and sing, Hunt theatrically draping his arms around Lyda, crooning, "I can't give you anything but love . . . baby." It was as close as the Hunts ever came to displays of intimacy. No one in the family kissed—ever. Once, when Hassie went to kiss his mother's cheek, Hunt shooed him away. "Stop that," he said. "Don't be kissing people." No one was quite sure what Hunt had against kissing, but his authority in the family remained unquestioned, as it always would.

On his return to El Dorado, Hunt wasted no time getting back into oil. Though not unwilling to bet money on a rank wildcat, he preferred to improve his odds by investing in areas where oil had already been found. Mostly he drilled his own wells, but sometimes he bought ones other men had started; under both strategies, Hunt, like most oilmen, put a premium on intelligence-gathering, seeking to learn everything he could about existing and aborning wells. For the first time he incorporated his own company, Hunt Oil, and gathered around him a half-dozen seasoned oil scouts, sending them nosing around southern Arkansas and especially several new fields opening in northern Louisiana. Hunt's men, led by the ever-present Old Man Bailey, proved top notch. Time and again they identified the best spots to drill. Hunt struck oil in the Tullos-Urania Field outside Monroe, Louisiana, then at several places near Shreveport. By 1929 he had opened offices in El Dorado and Shreveport and was operating more than one hundred wells, though with oil selling at barely $1.25 a barrel he was forced to plow most of his profits back into the search for still more acreage. He had emerged as one of the region's largest independent operators, a wealthy man by Arkansas standards, but still far from serious riches.

For the first time his eyes turned to Texas. The West Texas boom that lured Clint Murchison and Sid Richardson out onto the plains also attracted Hunt. He and Old Man Bailey drove west in 1927, putting together a drilling block near Ballinger in Runnels County, but the few wells they attempted came up dry. The West Texas fields were large compared to those in Louisiana, however, and Texas continued to intrigue Hunt. When oil was found in 1929 at Van, east of Dallas, he tried to lease land but found the majors had already snatched up the best acreage; the one well he managed to drill came up dry. Then came the stock market crash, which overnight gave way to an economic depression the likes of which America had never before seen. Hunt survived with little trouble, but money remained tight, and by that following summer of 1930 he still hadn't found a drop of oil in Texas.

Then, on September 5, Hunt took a call from an El Dorado oil field equipment man named M. M. Miller. Miller was marketing a new drill stem that

Hunt used; Hunt, in fact, was one of his best customers, and Miller wasn't above freshening the relationship with the odd tip.

"There's a wildcatter working down in East Texas, and he may have something going," Miller said. "He might call on us to run a drill-stem test on his well, and I thought you might be interested."

He was. This, in fact, was exactly the kind of inside information Hunt valued most. Still, he was short on cash that week, so before driving to Texas he put in a call to the owner of an El Dorado clothing store, a squat, bald character named Pete Lake. Lake had lent Hunt money in tight spots before, and was always interested in new action. They left that same day, taking Lake's car into Louisiana, where they stopped for maps in Shreveport, then headed into East Texas to find the drill site in remote Rusk County. That afternoon, six miles past the hamlet of Henderson, they turned down a rutted dirt road that ran deep into the pines. A mile down the road they arrived at a clearing where a crowd of about twenty people was watching as Columbus Joiner and his crew prepared for the test. One of Joiner's men, M. M. Miller's brother Clarence, saw the car drive up and scrambled down from the derrick.

After a minute Joiner ambled over.

"H. L. Hunt," Miller said, "meet C. M. Joiner."

There was an instant rapport between the two wildcatters, and not just because they were dressed in almost identical outfits; slacks, ties, matching white shirts and straw boaters. Joiner needed money; Hunt wanted to invest. He had thought he might offer to pay for Joiner's casing in return for a share of the lease. But Joiner squelched the idea the moment Hunt brought it up. "All taken care of," he said. "Got a string of used casing on the way any time now."

What Hunt didn't realize was that Joiner stood at the middle of a spreading net of intrigue. To raise money for his third well on the Daisy Bradford land, he had sold three rounds of investment certificates to dozens of local people. In return for a hundred dollars, each of the certificates entitled the buyer to 4 of the 320 acres Joiner had set aside for investors; this meant only the first eighty certificates were valid. Never one to let ethics stand in the way of fund-raising, however, Joiner had wildly oversold the certificates, selling rights to the same land to multiple buyers; one lease went to eleven different people. It was only a matter of time before someone found out. Worse, Joiner's own driller, Ed Laster, had betrayed him to a Kansas oil company, selling off his drilling data and rock samples. The Kansas company had already begun leasing up nearby acreage.

As Hunt sat back to await the drill-stem test that afternoon, he knew none of this. He watched as Laster, standing up on the derrick floor, lowered the twelve-foot drill stem to the bottom of the hole, where it began boring into the earth. When they reached the planned depth it pierced a pocket of natural gas, which whistled out of the hole in a rush. Laster quickly pulled it up from the hole. A moment later, the rig began to shudder. The ground rumbled so violently, one of the derrick's supports broke with an audible crack. Suddenly a shower of mud erupted from the hole, shooting as high as the top of the derrick, before calming into a small fountain, flooding the derrick floor. When the fountain ebbed to a stop, Laster dipped his finger into it and put it to his mouth. "Whaddya think?" someone shouted.

Laster tasted oil. "It ought to make a pretty good well," he announced, "if we can bring it in."

The crowd surrounded Joiner, farmers in overalls thrusting their hands forward in congratulations. Joiner closed his eyes and leaned against a pine tree. "Not yet," he said. "It's not an oil well yet." The hole would need to be stabilized, a process that could take several weeks, before Joiner could determine how much oil he really had.

Still, there was no containing the rumors, or the excitement that began to spread through the downtrodden villages of Rusk County. By the next morning, in fact, people had begun streaming into Henderson. A line of shacks sprouted down at the main road, the locals selling hamburgers and cots; the little encampment was dubbed "Joinerville." Each day that September more scouts and lease traders poured into the area. Taking a room in Henderson, Hunt called in his men from El Dorado and waded into the thick of the trading, swapping IOUs for four hundred acres of leases south and east of Joiner's well; Joiner had almost everything north and west. As the fervor spread, Joiner was celebrated as a local hero. The newspapers proclaimed him "Daddy of the Rusk County Oilfield"; after that everyone called him "Dad." The town of Overton feted Joiner with an all-day parade. When the inevitable finally occurred—one of his investors, comparing his certificates to others, realized he had been cheated and sued to place Joiner's acreage in receivership—the *Tyler Courier-Times* rushed to his defense, denouncing the "slick lawyers" who dared to attack their savior. "It's high time the independent operators had their inning," an editor wrote. "Now, if this be bolshevism, then we're bolshevists."

It took almost a month for Ed Laster to ready the Daisy Bradford No. 3 for completion, laying the casing and cementing it into place. When they ran out of wood, they took to stoking the boiler with old tires. After the last drill test, Laster had blocked the hole with a cement plug, and by Friday morning, October 3, word had spread that he was preparing to drill out the plug and see what lay beneath. By nine o'clock that morning nearly eight thousand people had tromped through the woods to the drill site. Kids sold soda and sandwiches, while bootleggers hawked bottles of white lightning. All morning the clearing thrummed with anticipation. All that was missing was Joiner himself, said to be recovering from a bout of flu in Dallas.

By midmorning Laster had drilled through the cement plug and begun to run the bailer, which cleared mud and water from the bottom of the hole in hope of freeing up any oil. The bailer ran all afternoon and when night fell lanterns were lit so Laster and his crew could work into the evening. The crowds reassembled when they resumed work in the morning, but by noon there was still no sign of oil. Joiner appeared then, climbing through the wire fence around the derrick with a young oilman named D. H. Byrd, known as "Dry Hole" Byrd for a string of fifty-six straight dry holes he had once drilled; Byrd, like so many of the eager young men who flocked to Rusk County that fall, would later become one of the wealthiest oilmen in Dallas. But Joiner's appearance failed to bring any signs of oil. By nightfall the crowds began to disperse.

The next day, Sunday, October 5, Laster and his crew, now glassy-eyed with fatigue, continued swabbing the well. The farmers and townspeople began arriving after church services and remained through the day, but by the time darkness fell, there was still no oil. Again lanterns were lit, and again Laster and his crew worked into the darkness. Finally, around nine o'clock, Laster detected a faint gurgling sound deep in the well. For a fleeting second he smelled gas. "Put out the fires!" he hollered. "Put out your cigarettes! Quick!"

The ground began to rumble. A roar came from the hole. As Laster and his men dived for cover, a jet of black oil suddenly exploded from the derrick floor, arcing into the night sky, falling like rain on everyone. The crowd went wild, dusty farmers hugging their wives, grown men rolling in the oil-soaked mud, one worker firing a pistol into the air again and again. After a minute or two, Laster spun a set of valves, diverting the gush of crude into the waiting storage tanks.

Suddenly everything went quiet. Dad Joiner asked D. H. Byrd to measure the flow on the gauges. "Whisper it to me," he said.

Byrd checked the gauges, then leaned in close to Joiner. "She's flowing at sixty-eight hundred barrels a day," he whispered.

For a split second Joiner lost his composure. "SIXTY-EIGHT HUNDRED BARRELS!" he yelled. "UNBELIEVABLE!"

By any standard, it was a massive well. The people of Rusk County went to bed that night convinced they were sleeping atop an ocean of oil. The professionals, however, weren't so sure. Within days the flow from the Daisy Bradford No. 3 eased to 250 barrels an hour; worse, it was "flowing in heads," that is, in uneven spurts, 100 barrels one hour, 500 the next. Within days many scouts were dismissing it as a freak.

Not H. L. Hunt. Great fortunes are built on great convictions, and from the moment he watched Joiner's drill test Hunt was certain this was a giant field. On October 20, two weeks after the initial strike, he began drilling his first lease, south of the Daisy Bradford. Over the years several dry holes had been drilled east of the Joiner leases. It got Hunt to thinking. The more he studied the land, the more he became convinced that the field stretched north and west of Joiner's well, on the four thousand acres Joiner had already leased. An idea began to form. Maybe, Hunt mused, the play wasn't to drill near the Joiner leases. It was to *buy* the Joiner leases.

His opportunity, he sensed, lay in the old wildcatter's legal troubles. Already his investors were beginning to sue. One lawsuit, seeking to force Joiner into involuntary receivership, had been filed in Dallas; if Joiner lost, all his leases might be thrown into the hands of a court-appointed receiver. Joiner tried to hide from process servers at the Hotel Adolphus, but a lawyer slipped a hundred-dollar bill to a bellboy, flushing Joiner out, then forcing him to attend a hearing on October 31. It was a confusing session, with Joiner's attorney asking for a voluntary receivership, and by the time the judge gaveled it to a close a receiver had in fact been named, though Joiner retained control of his leases, at least for the time being. As the crowd filed out, no one was quite certain what had happened. It was then that Hunt, standing outside the courtroom, approached Joiner in the hallway.

"Mr. Joiner," Hunt said, "I'm offering to buy you out lock, stock, and barrel."

"Boy," Joiner said before walking off, "you would be buying a pig in a poke."

Hunt returned to East Texas more determined than ever to buy Joiner's land. If he could only raise the money, he felt Joiner could be persuaded. Unfortunately, at the moment Hunt had a grand total of $109 in free cash. For several days he scurried between Henderson and Dallas, talking to every oilman he could find, including scouts representing several of the majors, but none were interested in a partnership with him, much less advancing him any cash. Finally he arm-twisted his Arkansas pal Pete Lake into a deal; Lake agreed to supply $30,000 in return for a 20 percent stake in Joiner's leases. It was enough to get discussions started. He and Lake then returned to Dallas, determined not to leave until they had Joiner's land.

Before leaving, Hunt called in three of his men from El Dorado. If the Daisy Bradford No. 3 did herald the opening of a vast new field, and if that field stretched north and west, then a discovery well being drilled by the Deep Rock Oil Company, northwest of the Joiner well, would tell the tale. Hunt had to know what Deep Rock was finding, and he had to know it before anyone else. Despite Hunt's later denials, court documents would show that he cut a secret deal with the Deep Rock driller to supply his men with inside information in return for twenty thousand dollars in cash. Hunt positioned his three scouts near the Deep Rock drill site, where they stood ready to relay information to him as he dealt with Joiner. The price he would offer for Dad Joiner's leases would depend largely on their reports.

When Hunt reached Dallas, he and Pete Lake checked into the Baker

Hotel, just across Commerce Street from the city's other fine hotel, the Adolphus, where Joiner was staying. The morning papers brought good news; two new wells drilled southeast of the Daisy Bradford No. 3 had come in dry, providing fuel to those who felt the Joiner well was a freak. Lease prices in Rusk County went into free fall. This was the best possible situation, Hunt sensed. Joiner, he wagered, might now be eager to sell. First, though, they had to smoke him out. Hunt tracked down an acquaintance named H. L. Williford, an oil field character who had located Joiner for him once before. Williford scurried across to the Adolphus, where he was told Joiner was again sick with the flu. Williford found him hiding in one of the bedrooms, and all but dragged him back to the Baker, where Hunt was waiting on the mezzanine.

It was Monday evening, November 24, 1930, seven weeks after that first gusher. The three men sat at a table. Hunt quickly came to the point. He was prepared to offer Joiner $25,000 in cash for his leases, plus $975,000 from the proceeds of whatever oil was found. It was more money than Joiner could expect to make in a lifetime. But if Joiner was impressed, he didn't show it. They talked through the details, but in time the old wildcatter appeared to have second thoughts. He got up and returned to the Adolphus, promising to meet Hunt in the morning.

When the sun rose Tuesday, Hunt was ready. One of his scouts, Robert V. Johnson, was standing by at the Deep Rock well, which was fast approaching the coveted Woodbine sand; whatever the Deep Rock crew found, Hunt was confident he would know first. He was waiting in his suite when Joiner arrived. Pete Lake was there, as was H. L. Williford. "Boy," Joiner said, addressing Hunt, "you're gonna have to pay me more for these leases than what we've been talking about."

"Well, Mr. Joiner," Hunt replied, "what do you think they're worth?"

Joiner wanted $50,000 up front, plus more on the back end. They began to talk, about life, about their families, about East Texas, about the Daisy Bradford No. 3. Lunch came. They ate in the suite. At breaks Hunt would call for whispered updates on the Deep Rock well. By nightfall they were still negotiating. Dinner came. At midnight they were still talking. By dawn they had the outlines of an agreement, and Hunt telephoned his attorney, J. B. McEntire, and two stenographers. When they arrived, he and Joiner began dictating the terms. In return for all four thousand acres of Joiner's leases, including the Daisy Bradford No. 3 and acreage surrounding the Deep Rock well, Hunt agreed to pay $30,000 upfront—all of it out of Pete Lake's pocket. The back end, to be paid out of production, would be $1.305 million. The paperwork was exhaustive, and McEntire and his stenographers spent all morning hammering out the agreements. They were still writing late that afternoon when, around 4:30, Hunt received the call from East Texas.

It was his man in Henderson, Charles Hardin, relaying news from the Deep Rock site. "Mr. Hunt," Hardin said, "I think they're right on top of the Woodbine sand now." Hunt returned to the paperwork. Four hours later, at 8:30, as they haggled over final details, Hardin called once more. Even over the telephone line, his excitement was palpable. "Mr. Hunt," he said, "they've cored sixteen feet of material from the Deep Rock well, and ten and a half feet of it is saturated with oil."

Hunt later insisted he had shared this news with Joiner; almost certainly he didn't. Hunt now knew that the Daisy Bradford was indeed the discovery well of a large new field, perhaps a monumental one. Had Joiner known, he might have doubled or even tripled his asking price. But he didn't, and the price remained unchanged. At midnight the paperwork was ready. Hunt and Joiner signed, then shared a celebratory plate of cheese and crackers. As Hunt recalled the moment, Joiner appeared thrilled. "Boy," he said, "I hope you make fifty million dollars." Even when news of the Deep Rock well shot through East Texas later that week, sending lease prices into the stratosphere, Joiner seemed happy. His troubles were over. He was rich. And if Hunt found even a fraction of the oil they expected, he would be richer still.

News of the historic deal broke on the front pages of the Dallas papers that Sunday. The men of Texas Oil were left speechless. It was, by wide acclaim, the most astounding business deal the state had ever seen; as the enormity of the East Texas field became apparent in coming months, it would be hailed as the deal of the century. An obscure interloper, a closet bigamist, a man just nine years removed from life as a professional gambler, and from *Arkansas* of all places, had seized the heart of the greatest oil field in history, a field that in the next fifty years would produce four *billion* barrels of oil. H. L. Hunt had snatched East Texas from beneath the noses of the slumbering majors, and most incredible of all, he hadn't used a cent of his own money.

V.

For the moment, though, while Hunt was certain East Texas was a massive field, the rest of Texas Oil wasn't convinced. In mid-December all eyes turned toward the northern edge of Rusk County, just outside the town of Kilgore, where a penniless wildcatter named Ed Bateman was sinking a well called the Lou Della Crim No. 1. On December 28 the well, a full fifteen miles north of the Daisy Bradford No. 3, erupted in a gusher of oil straight from the Woodbine. The majors, now ready to believe the field extended from the first well to the second, rushed in to buy leases. The next site to draw attention loomed a full twenty miles farther north, where a team of wildcatters was drilling a well they called the Lathrop No. 1 in Gregg County, near Longview. On January 26, 1931, a crowd of eighteen thousand people hooted and hollered as the well exploded into the third East Texas gusher in as many months.

The Lathrop well confirmed what no one had dared imagine: it was all one field. One vast oil field, shaped like a forearm, Lathrop high on the northern fist in Gregg County, Daisy Bradford at the southern tip in Rusk, a single pool of oil forty-five miles long and five to twelve miles wide. At the depth of the worst depression in the nation's history, this realization set off a boom unlike any America had ever seen—as the field's biographers, James Clark and Michel T. Halbouty, put it, "the California gold rush, the Klondike, the Oklahoma land rush and the wildest of past oil booms rolled into one."

Overnight the people came in waves, hundreds of them, then thousands, by train, automobile, horseback, and on foot. The sleepy hamlets of Rusk and Gregg Counties—Kilgore, Henderson, Gladewater, and Overton—were overrun. When the hotels filled, the townspeople rented out rooms; when all the rooms were let, the newcomers threw up tents; when they ran out of tents, men slept in the open fields. The first to arrive were oilmen and scouts, every single one arriving with cash to buy mineral rights and trade leases. Farmers who two years earlier couldn't lease their rights for $1.50 an acre were suddenly demanding and receiving $1,800 to $3,000 an acre. In the oilmen's wake came hungry and unemployed men who took thousands of new jobs, as drillers, toolies, and errand boys. Behind them came the entertainment infrastructure: prostitutes and gamblers, bar owners, fry cooks, musicians, thieves—anyone and everyone who could make a buck.

Within hours the derricks began to go up—everywhere. The epicenter of the boom was Kilgore, in Gregg County, which within days was transformed into an oil field. Scores of buildings, even a bank, were torn down to make way for derricks. Forty-four separate wells went up on a single city block; by the middle of 1931, it was said a man could leap from derrick to derrick and never touch the ground for six miles. East Texas dwarfed any boom since Spindletop; in 1931 alone, an amazing 3,067 new wells would be drilled, an average of 8 a day, and every one seemed to find oil. New fortunes sprouted on a weekly basis. The first new millionaires included Ed Bateman, who sold his acreage to Humble for $1.5 million in cash—fifty times what Hunt paid Joiner—plus $600,000 in future oil payments.

At least initially, Hunt feared he might miss out. While he now controlled the Daisy Bradford No. 3 and four thousand acres at the heart of the new field, $109 wouldn't drill many wells. On his way back to El Dorado to break the news to Lyda and the children, he stopped in Shreveport to ask a banker there for a $50,000 loan, the cost of two wells. Louisiana bankers, however, like their brethren across the country, had yet to appreciate the intricacies of the oil business; if they couldn't physically see their collateral, their vaults stayed shut.

"You are broke," the banker said, as Hunt recalled the conversation, "and your statement shows you are broke."

"I've got the Joiner leases," Hunt replied. "It's a proven field. There's oil in the ground, and that's a bankable asset."

Not in Louisiana it wasn't. Dallas, however, was another story. The city's two largest banks, First National and Republic National, were among the first in the United States to see the wisdom of lending against proven reserves; it was their vision that would transform Dallas into the mecca of

Texas oil banking and fuel the city's future growth. All Hunt needed was a single meeting with First National's president, Nathan Adams—the friend of Clint Murchison's father—to walk away with the fifty thousand dollars he sought. It was the beginning of a relationship between First National and the Hunt family that would last for decades.

Within days Hunt began sinking new wells on Daisy Bradford's land. To sell the oil, he quickly laid a three-mile pipeline to the nearest railhead and was soon moving not only his own oil but his competitors' as well, all of it sold to the Sinclair Oil Company at sixty-two cents a barrel. By the time Hunt completed his first wells, hundreds of seasoned drilling crews were already at work across East Texas. Unlike the Dad Joiners and Ed Batemans of the world, who took months scraping together enough cash to complete a well, the majors and their professional crews could reach the Woodbine in fifteen days, sometimes nine or ten. The furious pace of drilling was driven by what was called the "rule of capture," a legal term roughly translated as finders keepers. Oil beneath the earth doesn't conform to the niceties of lease boundaries, and a well drilled beside a competitor's lease might well suck up oil from both. No matter. What you got, you kept. Those who tarried might end up with nothing. Thus, every well was a race.

The problem, as oilmen had found at the height of several booms during the 1920s, was that too much drilling risked reducing the subterranean pressure oil needs to rise to the surface; if a field was "overdrilled," no amount of effort would coax oil out of a hole. An entire oil field could, in effect, be destroyed. As far back as the Winkler County boom of 1926 the majors had argued that drilling should be regulated so that a field wouldn't be permanently damaged. That was fine for large companies, but wildcatters, most of whom drilled on a shoestring and needed immediate returns to stay afloat, strongly objected; they saw drilling limits, or "proration" as it was called, as a plot by the big companies to squeeze them out of business. Prorationing had been tried in Winkler County and elsewhere with some success; Clint Murchison had been among the independents who protested loudest. The problem was that enforcement fell to the Texas Railroad Commission, a toothless state agency widely regarded as a joke. The Railroad Commission had field agents, but many were corrupt. By and large, proration only worked where oilmen policed themselves.

The debate whether to institute proration in East Texas began almost as soon as the drilling. Governor Ross Sterling, a onetime chairman of Humble Oil, had issued proration orders even before Dad Joiner brought in the Daisy Bradford No. 3. But the Railroad Commission was all but powerless to enforce them, and they were widely ignored. Hunt, a stranger to most Texas oilmen, quickly emerged as one of proration's strongest proponents; though an independent, he was sitting on millions of barrels of oil, which he would need time to drill. His first public appearance came at a meeting of oilmen in Longview on February 5. Tensions between pro- and anti-proration oilmen were already running high. "We don't want rules that favor major companies over independents," Hunt told the crowd. "We want long-term conservation measures that will benefit all operators in the field."

"Sell out!" someone yelled.

"I haven't sold out to anyone," Hunt replied. "I favored prorationing in Arkansas. I favored it in Louisiana. And I favor it in Texas, too."

Arguments over proration stretched on into the spring, even as oil prices began to plummet. Before the Daisy Bradford No. 3, prices stood at $1.10. By May, the rushing tide of East Texas oil had driven them down to fifteen cents a barrel, as low as two cents on the spot market. At these levels, it was almost impossible for anyone to make a profit. The Railroad Commission issued its first proration order for the field on May 1, fixing the field's production at 160,000 barrels a day at a time when it was already producing a half million; the order was widely ignored. The Texas Legislature got involved, arguing proration bills all through the summer. Finally, on August 5, a day after the governor of Oklahoma declared martial law and sent in troops to shut down and begin regulating his state's oil fields, Hunt and thirty-six other large East Texas operators sent a telegram to Governor Sterling urging him to follow suit. Sterling gave in. On August 16, declaring East Texas oilmen to be in open "rebellion" against the state, he declared martial law and sent in the National Guard to shut down the oil field.

The next morning, more than twelve hundred Guardsmen, many of them on horseback, marched into Rusk and Gregg Counties as the frantic drilling whirred to a stop. Scores of small oilmen, unable to pay their bills, abandoned their wells or sold out to larger outfits. Three weeks later, on September 5,

Governor Sterling announced the field was being reopened, but with sharp limits: no well could produce more than 225 barrels a day. With prices so low, it was impossible for many wildcatters to make money. Only the wealthiest drilling groups, including the majors and large acreage holders like Hunt, could afford to keep drilling, and they did. When they found oil, they simply left it in the ground, confident that they would be able to sell it at some future point.

Slowly but surely, prices began to rebound, hitting ninety-eight cents a barrel in early 1933 before once again dropping like a stone, eventually falling to a low of four cents a barrel that May. The problem was obvious: many wildcatters were staying in business by openly flouting the proration limits. This illegal oil, much of it piped into tanker trucks that made nightly smuggling runs into Louisiana and Oklahoma, quickly became known as "hot oil," and the fall of 1931 marked the beginning of a four-year struggle between feisty independents and state and federal regulators known as the "hot oil wars." All through the Depression the Railroad Commission and its allies sued, seized, arrested, and prosecuted hot oil operators, who fought back with bribes, secret pipelines, and a blizzard of lawsuits.

To the big operators in East Texas, including Hunt, hot oil not only depressed prices, it threatened to damage the field itself. Stopping the flow of hot oil, however, proved beyond the Railroad Commission's limited capabilities. At the end of their rope, Humble and other majors appealed directly to the new president, Franklin Roosevelt, to appoint a federal oil czar to squelch the hot oilers. Roosevelt appointed a top aide, Harold Ickes, and on July 14, 1933, the president signed an executive order not only upholding proration but sending hundreds of federal agents into East Texas to enforce it. It took three more years of hot oil prosecutions and legal challenges, some going all the way to the Supreme Court, but by 1936 the federal government had brought order to the East Texas field and stability to oil prices.

VI.

All through the hot oil wars Hunt drilled wells like a madman. Shirtsleeves rolled above the elbow, khaki pants splattered with mud, a cigar habitually jammed in one corner of his mouth, he worked from dawn till late in the evening seven days a week, driving from well to well to well, often with his teenage son Hassie at his side. Every cent he took in—from oil sales, from the new First National loans, occasionally from selling part of a lease—he plowed back into the search for more oil. By the end of 1932, despite proration and chaotic conditions and drenching rains, he managed to drill an astonishing 145 wells on his new East Texas leases. More than seven hundred thousand dollars in profits went to Dad Joiner, but even with oil prices at less than a dollar, Hunt himself took home his first million dollars and probably much more.

Closing his business in El Dorado and Shreveport, Hunt opened new offices in the People's National Bank Building in downtown Tyler, a genteel southern town twelve miles west of the oil field's western boundary. One of his brothers, Sherman, quit his Montana ranch to become Hunt's right-hand man. Hunt had brought Lyda and the children from Arkansas in June 1931, moving them into a three-bedroom red-brick rental on Woldert Street. Though jammed with seven people—Hassie was now sent to Culver Military Academy in Indiana—the household ran with precision. Meals were served at eight, noon, and six, on the dot. The children were enrolled in schools. Other than a vague kidnapping threat or two—after one scare Texas Rangers guarded the family—life was good.

The foundation of Hunt's little empire, however, remained far from secure. The problem was legal title to the Joiner leases. The old wildcatter had sold and resold shares of his mineral rights so many times that, as Hunt's attorneys analyzed the situation, he actually held unobstructed titles to barely two of the four thousand acres he sold Hunt. One by one Joiner's investors sued, until by mid-1932 Hunt found himself facing some three hundred separate legal challenges. Rather than fight each one in court, Hunt told his lawyers to offer the plaintiffs cash settlements. It worked. Dozens of the plaintiffs, mostly poor farmers, took the $250 or so Hunt offered and withdrew their suits. The whole mess would still take nearly a decade to clean up, but Hunt's strategy saved him a fortune in legal fees and potential judgments.

The real danger, however, was Joiner himself. The trouble began in the autumn of 1932. On a rainy night outside Tyler, Hunt and his brother

Sherman came upon an overturned car. The driver was trapped inside. Hunt joined a crowd of people trying to push the car into an upright position, but in the process he badly wrenched his back. The pain was so bad, he could barely move. A doctor gave him a steel back brace and ordered him to remain in bed for the next six months. It was just days later, on the night of November 19, while lying in the brace in Tyler, that ominous rumblings reached Hunt's beside: Dad Joiner was unhappy.

Hunt cursed. He had done everything in his power to placate the old man. Every month he made sure Joiner received his scheduled oil payments, usually between thirty thousand and fifty thousand dollars. But Joiner was now living high, squiring a new girlfriend around Dallas, and pouring much of his earnings into questionable new ventures. Every month, it seemed, he would track down Hunt and ask for an advance or a loan. Hunt gave it to him, usually with a little lecture on the importance of saving, but even though Joiner continued testifying on Hunt's behalf in the lease-challenge lawsuits, it was clear Joiner resented it. *He* had discovered East Texas, not Hunt, yet he was the one obliged to beg for money.

A lawsuit's statute of limitations, November 26, the two-year anniversary of Hunt's historic deal, loomed just a week away. Hunt dragged himself from bed and drove to Dallas, where he summoned Joiner to his suite at the Baker Hotel. There remained genuine affection between the two men. They had so much in common—their love for oil, the years in obscurity, the sudden riches—and they passed several long days at the Baker reminiscing, Hunt lying in bed in his brace, Joiner on the opposite bed, smiling and bringing him food.

At nightfall on November 25, with the statute of limitations scheduled to expire the next day, Joiner rose to leave. Hunt rolled out of bed and limped to the door. For the first time he addressed the gorilla in the room. "Mr. Joiner," he said, "I think efforts are being made to get you to sue me. I hope you don't fall for that."

Joiner looked at Hunt for a long second. Tears appeared in his eyes. "My boy," he said, throwing an arm over Hunt's shoulder, "I would never do a thing like that. I love you too much."

When Joiner left, Hunt crawled back into bed to think. *Tears.* He kept thinking about the old man's tears. There was real sentiment there. But the more Hunt thought, the more he became certain what Dad Joiner's tears actually meant. They were tears of sadness.

He was going to sue.

And—far worse—Hunt was defenseless. It was the leases. His banker, Nathan Adams, had been so confident of the East Texas field's potential, he had never demanded the Joiner leases as collateral. No one had. If Joiner sued, his attorney could place a lien on everything. From his bed Hunt reached for the phone, called his attorney, J. B. McEntire, and told him to bring every stenographer he could find to Dallas that night. Next he telephoned his principal trade creditor, Continental Supply Company, and told them to send down a credit executive. Everyone reached the Baker by midnight. Under Hunt's supervision, they began writing out dozens of mortgages, giving Continental and Hunt's banks first liens on every acre of his land. By dawn they were almost finished. When the final papers were signed, Hunt had them messengered to Henderson with orders to file them at the courthouse the moment it opened. Everything went off without a hitch—and not a moment too soon. An hour later Joiner's attorney filed suit.

Hunt's all-nighter prevented Joiner from seizing or freezing his wells, but the lawsuit still threatened everything. Joiner claimed fraud. He charged that Hunt had lied to him about the Deep Rock well and had bribed Deep Rock's superintendent; the former was probably true, and the latter certainly was. Valuing the leases at fifteen million dollars, Joiner argued that he should have gotten three to five times what he had received; if a judge agreed, Hunt was staring at three million to five million dollars in damages. While Joiner put his best face forward for the press—"I have nothing against Mr. Hunt personally," he told reporters—the legal maneuverings quickly grew heated.

Hunt was in a dangerous position. Joiner was still a folk hero to many in East Texas, and the Deep Rock "bribe" would look very bad to a local jury. As the January 1933 trial date approached—lawsuits moved quickly in those days—rumors swirled fast and furious, that Hunt was promising Joiner big money to settle, that he had unearthed dirt from his past, that threatening phone calls and letters had been sent to various Hunt men involved in the original deal. What happened next has never been explained. On January 16, 1933, just as a crowd of reporters and oil scouts took their seats in a Henderson courtroom to watch opening arguments, Dad Joiner stood and began to read a statement. After a "thorough investigation" of the case, Joiner announced, he had "determined to my satisfaction that the allegations of fraud in my petition are not true." He was dropping the lawsuit.

Around the courtroom, jaws dropped in disbelief. What had just happened? No one knew. Joiner hustled out and brushed away all questions, as he would for years afterward. Somehow, just about everyone in East Texas agreed, Hunt had gotten to Joiner. What he did, or what he promised, remains a mystery to this day. In the months to come, the two men silently returned to their old routines, Joiner accepting his monthly payments, then pestering Hunt for an advance. They remained that way for the next fourteen years, until, in 1947, Joiner, after years of fruitless drilling ventures throughout Texas, finally died, virtually penniless, at the age of eighty-seven.

"Much has been written about the Joiner deal," Hunt wrote years later, "and some with overly active imaginations have implied machinations all the way around, but the fact is it was a sound deal for both Joiner and me, and Joiner received through the cash, notes and production payments more funds than he would have received by trying to operate the properties in the face of his legal difficulties."[1]

More than two long years after his arrival that fateful day at the Daisy Bradford No. 3, Hunt was finally secure. Life, in fact, was just about perfect. Joiner's challenges were over. He had nearly two hundred wells pumping oil in East Texas now, with more on the way. His two families remained cared for, content and, most important, oblivious of each other. When he found time, Hunt swung by the house on Versailles Avenue in Dallas to visit Frania and their three young children. His "first" family, meanwhile, was putting down roots in Tyler. Chafing at life in a cramped rental house, Lyda had wasted little time selecting a permanent home, a two-story, white-columned Greek revival mansion known as the Mayfield Estate. She spent a year remodeling before finally moving into the house in October 1932. The Hunts were now, officially, Texans. And though no one realized it at the time, Hunt himself was fast on his way to becoming not only the wealthiest Texan of all, but the richest man in the world.

FIVE

The Worst of Times, the Best of Times

I.

East Texas made fortunes for scores of Texas oilmen, many of whom were from or later settled in Dallas. Few plunged into the backwoods chaos with the fervor of thirty-five-year-old Clint Murchison who, having grown up in East Texas, knew the piney woods well. Unfortunately, like Hunt and so many of his peers, Murchison had no free cash to invest; every last cent was going to Southern Union's banks and his main creditor, Oilfield Supply. Lack of funds, however, never stopped Murchison. All that oil needed pipelines to reach market and, just as he had in West Texas, he burned to build one. A single meeting with Oilfield Supply executives persuaded them he was right, and they put up the money, which in three short months Murchison used to build an eight-inch pipeline from the field's southern edge thirty miles to Tyler. By the summer of 1931 it was complete. Murchison even prevailed upon a partner to erect a refinery there as well, and entered the gasoline business.

That was just the start. To that point, while Murchison had drilled his own wells here and there, much of his money sprang from Southern Union and lease trading. But there was too much oil in East Texas not to grab some himself. Ernest Closuit and his brother Frank demurred; they had too many responsibilities already. So Murchison entered into a new partnership with an old Wichita Falls pal, Dudley Golding, who had a pair of rigs and money from his wife's family. With new cash coming in from the pipeline, Murchison sent men to buy every lease they could find, paying as little as possible up front, then dispatched Golding and others to begin drilling. In the next three years they hit dozens of producers around Henderson and Kilgore.

The collapse in prices hit Murchison hard; his pipeline contracts obliged him to buy the oil delivered to Tyler at rates far higher than ten cents a barrel, which was all he could sell it for. The banks understood and allowed him to delay a payment or two; he managed to make payroll by coaxing cash advances out of the pipeline's biggest buyer. But proration loomed as the real killer; the only way for Murchison to offset his mounting losses was to pump more oil, but the federal government was now saying that was against the law. Murchison was apoplectic. This was un-American, he told anyone who would listen. It was all a scheme devised by the majors to squeeze the independents out of East Texas. As one of his peers put it, "It's my oil, and if I want to drink it, it's none of your damned business."

It was this line of thinking that led Murchison to become an outlaw, a defiant hot oiler. From 1932 until 1934, in fact, he may have been the biggest hot oiler in all of East Texas, and he didn't especially care who knew. He even renamed his partnership with Dudley Golding American Liberty, because, he said, it represented freedom against regulatory tyranny; the new company would soon become Murchison's largest. No stories of how Murchison managed the shady side of American Liberty's business have survived, but running hot oil was a cat-and-mouse game. Federal inspectors, Railroad Commission agents, and Texas Rangers were everywhere; lookouts had to be posted. Most hot oil was pumped and refined late at night and loaded onto trucks that crept down the rutted dirt roads in caravans for Dallas or Shreveport; when federal agents were in the area, decoy caravans were sometimes used. In a pinch, many inspectors could be bought off with bribes.

This was a dicey game. Every night he ran hot oil Murchison risked court action or, worse, his own arrest. The Railroad Commission knew what was going on and repeatedly sent in raiding teams, which resulted in a string of state lawsuits against Murchison-owned companies. Murchison fought back with a maze of paperwork, repeatedly switching the ownership of questionable oil from one of his companies to another. By the time the Railroad Commission realized what had happened, it would take months to rebuild its case. When the commission did manage to haul him into court, Murchison sent in one of his boyhood pals, a canny lawyer named Toddie Lee Wynne, to fight on his behalf. Wynne succeeded in getting several irksome actions dismissed.

In the early 1930s, how you felt about Clint Murchison depended on how you felt about hot oil, to many wildcatters he was a hero, to others a selfish criminal. In time his notoriety grew to the point that, when East Texas oilmen gathered in Washington to meet with Harold Ickes, his peers turned Murchison away. "I didn't [want to] have anything to do with Clint, because I knew damned well his hands weren't clean," the oilman J. R. Parten once recalled. "A fine person—I liked him very much—but Clint was a hot oiler in every respect. Clint didn't want any rules. He had made a fortune by disobeying the rules. He showed up in Washington and asked what he could do to help us. [We] said, 'The thing you could do most to help us is to get on that airplane and get out of town as quick as you can'. And he did." Murchison resented it; he felt his efforts to fight proration should be applauded. "If I had been guilty of all the things they say I have done," he once remarked, "I'd be under the jail, not in it."

All through the hot oil wars Murchison's mind thrummed with ideas for new pipelines and refineries and oil fields, all limited only by his chronic shortage of cash. His genius, though, was not so much finding oil as finding money. In 1933 he heard of a young loan officer at the Bank of Manhattan named Rushton Ardrey, a native Texan who had opened the bank's first energy department. In those days the big New York banks steered clear of anything but the shortest-term loans to wildcatters, deeming them too risky. But Ardrey had made a five-year-loan to an Oklahoma independent, and when Murchison heard about it, he flew to New York. When Ardrey asked how much he wanted, Murchison replied, "All I can get." He left with a one-million-dollar loan, the promise of more to come, and a crucial new ally; he called Ardrey "the Big A." In time Murchison would introduce him to dozens of other oilmen, including Sid Richardson, and in 1937 Murchison would hire Ardrey to work for American Liberty, where he worked full-time raising money.

Rushton Ardrey's loans enabled Murchison to plunge back into East Texas at a fever pitch, drilling dozens of wells, throwing up small refineries, building an asphalt plant when low-quality crude was found, even buying up giant storage tanks for surplus oil outside New Orleans. His most ambitious project was a pipeline built from East Texas all the way to the town of Conroe, north of Houston, which allowed independents to move oil directly to the tankers in Houston's ship channel. By the time the hot oil controversy finally died down in 1936, Murchison had at least tripled his fortune, more than two million dollars alone coming from the sale of 125 wells. He finally stopped carping about proration, claiming he saw the wisdom in preserving Texas oil fields, and most oilmen assumed his days as a hot oiler were over.

In late 1936 came news they weren't. Federal prosecutors in Houston indicted his Conroe pipeline subsidiary for exceeding legal limits on transport of oil. It was a serious situation, with rumors that Murchison himself might be indicted, and dealing with it would take the young Texan all the way to the White House.

II.

The day Dad Joiner discovered the East Texas field, there were wealthy oilmen in the state, but there were no true oil *fortunes*. The only Texan who could be said to control an oil fortune—actually an oil-service fortune—was Howard Hughes, but Hughes had long since left for Hollywood. It was East Texas and other fields discovered during a single five-year window—1930 to 1935—that created the state's great family fortunes. The magnitude of wealth initiated in those sixty months would not become apparent for years and remains underappreciated today. In fact, the spigot of cash Texas Oil opened in the early 1930s ranks among the greatest periods of wealth generation in American history, in size perhaps the largest creation of individual wealth between the Gilded Age and the Internet boom of the 1990s.

The irony was that it happened during the worst economic depression in American history. But the explosion of Texas oil wealth didn't happen *despite* the Depression. It happened *because* of the Depression. There is a notion in Texas, repeated in several oil-industry histories, that wildcatters grabbed the lion's share of the East Texas field because the majors ignored its potential and, once it was found, pooh-poohed its size. In fact, the majors had a good sense of what Joiner had found. The problem wasn't lack of vision. It was lack of money. The Depression hit oil companies hard, and it was their wholesale retreat from exploration that allowed men like Hunt to amass fantastic oil reserves. More than any other single factor, it was the majors' retreat that created the fortunes that came to define Texas.

The day East Texas oil first spewed into the skies, all the major oil companies were in the process of drastically scaling back their exploration budgets. Shell, saddled with heavy debt after a late-1920s expansion, had all but stopped acquiring Texas leases and rarely drilled the ones it had; in 1931 alone, it let lapse leases for which it had paid eight million dollars. Gulf was in even worse shape. Beset with financial losses, unable to secure bank loans, it simply lacked the money to buy new acreage. Many of Gulf's leases lapsed as well. The Texas Company, meanwhile, had refocused its operations, deemphasizing exploration in favor of refining and marketing. Only the sharp executives at Houston-based Humble, while slashing their exploration budget, remained actively in the market for new reserves. Its men, led by the geologist Wallace Pratt, instituted a new policy that emphasized the acquisition of proven oil fields, that is, fields found by others.

It was Humble's new policy, in turn, that laid the foundations of the three greatest oil fortunes in Houston. The first recipients of Humble's largesse were Roy Cullen and his obstreperous partner, Big Jim West. By 1931 West had been pestering Cullen to sell their newfound properties for almost two years. At least sell one field, he pleaded—Rabbs Ridge in Fort Bend County. To West, it was "just another field." Cullen wouldn't hear of selling. Without telling Cullen, West sounded out Gulf Oil's Pittsburgh headquarters, but found no interest. In March 1932 he telephoned Wallace Pratt at Humble in Houston, who had been following their progess. The previous November one of Pratt's geologists, estimating that the field might contain 127 million barrels, had recommended buying it "on anything like reasonable terms." Pratt proposed a deal in which Cullen and West would receive $3 million in cash and an additional $17 million to be paid from future production; in today's dollars, the offer was roughly $250 million.

Cullen scoffed. "The stuff in that field is worth a hundred million dollars," he told West. "Maybe two hundred million."

West gave no ground. They had suffered two blowouts, West argued; the field was unstable. The argument stretched on for days, becoming bitter. Finally Cullen gave in, if only to find a way to dissolve the partnership. "After this deal is closed, we're through as partners," Cullen said. "I'm not going to find any more oil fields for you to give away." The Humble deal was struck in March 1932, and Cullen was right about the field's potential; by the late 1940s Humble would pump more than one hundred million barrels of oil from the field. A Humble corporate history regards the purchase as a great bargain.*

The Humble deal made Cullen and West two of Houston's richest men. Cullen wouldn't have to work another day in his life, but he was only fifty-one and retained the itch to find more oil. Striking out on his own, he hired a onetime neighbor, an attorney named Harry Holmes, to be his No. 2 man. They called their new company Quintana Petroleum, after a Gulf Coast ghost town Cullen spied on a fishing trip. Cullen brought in his daughter Agnes's new husband, Isaac Arnold, to be chief engineer, and his son, Roy Gustave, who was now married, as well.

By the mid-1930s the last great untested lands in Texas lay beneath the state's largest ranches, and during the Depression Humble competed with a series of wildcatters to lease them. Humble secured the biggest prize of all in 1933, the rights to drill beneath the million-acre King Ranch, a vast coastal prairie of more than thirteen hundred square miles stretching from Corpus Christi nearly all the way to the Mexican border; it was the largest oil lease in American history. Roy Cullen, meanwhile, set his sights on the King Ranch's neighbor to the north, the fabled Tom O'Connor Ranch, a five-hundred-thousand-acre spread outside the town of Victoria. The O'Connors had been running cattle since before the Texas Revolution, and their feisty paterfamilias, Tom O'Connor himself, had made clear to any number of oil-

---

*Humble renamed Rabbs Ridge the "Thompsons" field, the name it is known by today.

men he wanted nothing to do with the kind of chaos and controversy he saw playing out in East Texas.

Cullen wangled a meeting with O'Connor by calling him directly. The crusty old rancher said he'd get back to him, then didn't. After several weeks, Cullen sent in a mutual friend, Chad Nelms, who persuaded O'Connor to make the deal. O'Connor warily instructed his lawyer to draw up a set of contracts. "You tell this feller, Cullen," O'Connor warned, "that if they change one damn word, or dot an 'I' or cross a 't,' they can tear up the paper." Cullen signed the contract exactly as written. In it, he agreed to pay one dollar an acre to survey the ranch, plus the right to select two seventy-five-hundred-acre plots to drill. To reduce his risk, Cullen sold a half-interest to Humble for fifty thousand dollars.

The first hole they drilled, the "O'Connor A-1," hit salt water at 4,450 feet. Cullen dispatched his son Roy to rework the hole in preparation for—his mantra again—drilling deeper. A few weeks later, O'Connor was standing by when the rig reached oil sands. Cullen dipped his finger into a sample core, smiled, and turned to the old rancher. "There's a mile-deep pool of oil down there," he said. "This will be one of the biggest oil fields ever discovered."

It was an overstatement, but not by much. The O'Connor Field would produce more than seven hundred million barrels over the years, making it the tenth-largest field discovered in Texas. Though other companies were eventually allowed onto the ranch to drill, Cullen and Humble sank hundreds of wells in coming decades, much of the oil sold to the federal government and refined as jet fuel. Thanks to the O'Connor Field and his deal with Humble, by 1936, although practically no one knew it, Roy Cullen was the richest man in Houston.

III.

Roy Cullen had been lucky enough to make money during the 1920s that allowed him to drill during the Depression. For those who hadn't, and there were hundreds of penniless oilmen in Texas during the early 1930s, the only way to drill an oil well was by "poor boying," that is, scraping, borrowing, begging, and even stealing equipment, then paying drill crews with promises, IOUs and groceries. In West Texas roustabouts who worked for poor boys like Sid Richardson called the work "bean jobs," because they were literally paid with beans.

The third of Houston's great oil fortunes fell to one of the unlikeliest of poor boys, a thirty-six-year-old named George W. Strake. Orphaned at the age of eight and raised by his sisters in St. Louis, Strake was a wiry little man who wore glasses, a quiet, deeply religious Catholic who, seeking romance and excitement after college, had taken a job in the Mexican oil fields in the early 1920s. Striking out on his own, trading leases and arranging the odd well himself, he had managed to amass nearly $250,000 by 1924. Hoping to start an integrated company that not only found oil but refined and sold it as gasoline, he took his new wife and resettled in Havana, where he set his sights on becoming Cuba's biggest oilman. In the meantime, to keep money coming in, he started a Hupmobile car dealership. But Cuban sugar prices collapsed shortly after his arrival, plunging the country into depression. Strake found no oil, sold no cars, and within two years was running out of money.

In 1927 he washed up at his in-laws' home in Houston, all but broke. He knew no one in Texas, nor the first thing about the state's geology. But he had a car, and for two long years, while he scraped by buying and selling a few oil leases around Houston, he drove the backroads of East Texas and Louisiana looking for signs of an oil-bearing clay known in the Mexican fields as Lagarto-Reynosa. He would stop his car and walk through the woods, swinging a geologist's pick, looking for Lagarto rocks. Finally, one afternoon in 1929, he was wandering down a creek bed outside Conroe, a village thirty miles north of Houston, when he noticed a crack in the bank where a flood had washed away the soil. He chipped off several bits of exposed rock and rolled them between his fingers. They were Lagarto. He was sure of it.

No one had found oil anywhere near Conroe, although traces had been found in a well drilled years earlier on the far side of town, which had led several majors to send in geophysical crews. Back in Houston, Strake tracked down their maps and studied them. He grew convinced there was a major field at Conroe. He took the last of his savings and leased every available plot of land southeast of town, nearly eighty-five hundred acres. He had no money left to drill himself, but if he could interest one of the majors, they might do it for him, splitting the proceeds. Humble, however, turned him down. So did Gulf. And six other large companies. "You're out of your mind," one scout told him. "There's no oil over there. You're on the wrong side of Conroe."

Just one well, Strake begged. Just one well, and you'll see. But after the stock market crashed that fall, no one wanted to hear it. All through 1930 Strake prowled the bars and hotels of downtown Houston, buttonholing every scout and geologist he could find, arguing for Conroe; once the scramble began in East Texas, people thought Strake was daft to be looking anyplace else. He offered to "checkerboard," that is, give up alternate tracts of his land, to any driller who would poke a single hole in his ground. He found no takers. By the spring of 1931, with his leases scheduled to expire on August 31, Strake realized he would have to drill the land himself or lose it. There were only three problems. He had no rig, no money to get one, and, despite ten years in the oil business, only the vaguest idea how to drill a well.

Still, he would try; he had no other choice. His wife was a bank clerk, and Strake tried to raise money from her co-workers in exchange for land; no one was interested. There were breadlines in Houston that year; banks were closing; no one had an extra nickel to spend on food, much less on an oil well. A loner, far more comfortable in a church pew than a bar stool, Strake had joined an Elks Club to make friends; none of the Elks wanted in either. Strake went ahead anyway, finally renting a rig big enough to reach the target depth of six thousand feet; the owner handed it over in return for a piece of the action. Unfortunately, the rig was at a field one hundred miles away. Strake was obliged to truck it in; the hauler, too, took a piece of the action.

Once he had the rig, though, Strake had to find people to run it. Unfortunately, he didn't know anyone who had worked in a Texas oil field, much less drilled an actual well. The only person he knew in Conroe was a timber-cutter who worked on one of his leases. The man gave him the name of a tool-pusher in Humble. The tool-pusher agreed to work the job, and bring along some friends to round out the crew. Now he needed a driller. Unfortunately, in August 1931 just about every driller in Texas was busy in East Texas. The tool-pusher gave Strake a name, Harvey Lee, an old driller now down on his luck. He was supposed to be up in East Texas somewhere.

Strake set out for the piney woods. At each town he stopped and asked for Lee. Nothing. In town after town, no Harvey Lee. He had been at it a solid week when, heading home after another sweltering day, he coasted up to a road crew. A man said he knew Lee. In fact, he lived in a hilltop shack just above the road. Strake trekked up the hill, found Lee and his young wife nursing a newborn, and left with a driller.

Everyone convened in a clearing deep in the pines three miles southeast of Conroe on August 13, just eighteen days before the leases' expiration. It took two days to get the rig into place; Strake hired his lumberjack pal to begin felling trees to feed the boilers. They would need water to cool them, but after two weeks of work a water-well driller had been unable to find water beneath the land. On the final day, August 31, with no water in sight, Harvey Lee told Strake they would need to break ground by hand. With Lee showing them what to do, Strake and seven of his men took three chain tongs and, as the sun began to set, physically pulled the drill around and around until it bit into the dirt. They had made the deadline.

When water was finally found, the drilling began in earnest. Townspeople came out from Conroe to watch and shake their heads. To a man, everyone thought Strake had lost his mind. Every day was a battle. The boilers threw off showers of flaming bark, and the men spent half their time scrambling off the derrick to shovel dirt on fires that erupted in the dry pine needles, sometimes three and four of them a day. A forest ranger appeared one morning and ordered them to shut down until Strake installed wire screens atop the boilers to catch the burning bark. It worked fine until the screens filled with debris, at which point the boilers shut down and the drill bit sagged to a stop.

"Mr. Strake, I can fix those boilers for you," one of his men offered.

"You can?"

"Yes, sir, if you'll turn your back a minute."

The man picked up a shotgun, aimed at the screens, and, with three rapid shots, shot them all off. "Nice shooting," Strake said.

When they reached 4,125 feet on October 13, they took a core sample. It showed a slight sheen of oil. Strake divided the core into three sections, threw each in a bag, and dropped them at the Houston offices of Humble, Gulf, and the Texas Company. If the core was valid, all three realized, Strake had the makings of a good well. But a number of oil scouts who knew the Conroe area felt it was a trick, that Strake must have "salted" the sample with outside oil to get them interested. As the drilling continued, more scouts began dropping by the site, eager to see whether "Strake's folly" was real.

In the ensuing two months Strake fought every kind of mechanical problem, from broken drill bits to temperamental boilers. Finally, on December 5, 1931, the well struck oil—sort of. It was what oilmen called a "gasser," that is, a small show of oil accompanied by an immense rush of hissing natural gas. Most of the scouts immediately dismissed it as a freak; the sand Strake had struck, the Cockrell, wasn't known as an oil producer. Hundreds of lease traders and oil field workers came down from East Texas, but almost all returned within weeks. Intent on proving the doubters wrong, Strake began preparations for a second well, two thousand feet from the first.

As he did, he received a call from Humble's Wallace Pratt; alone among the majors, Humble was willing to bet Strake was sitting on a bonanza. "Well," Pratt said, "Humble was too big and smart to see what you saw. You brought us this thing and we turned you down. I guess we're getting ready to pay for that mistake." He asked how much Strake wanted for his acreage. Strake, still penniless, offered meager terms for half his leases: $500,000 up front, plus $3.5 million out of oil proceeds. Humble took the deal and began a discovery well.

On June 5, 1932, the Strake No. 2 well struck oil at five thousand feet, a geyser of black crude that arced high over the surrounding pines. Strake had been right about everything. Within ninety days another seventy-five operators had begun drilling in the area, and by the end of the year the Conroe field would be recognized as the state's second largest, behind only East Texas, and the third-largest oil field in the United States. Even after the

Humble sale, Strake held on to nearly a third of its acreage, holdings worth tens of millions of dollars in the coming decades. Overnight he became the third-wealthiest oilman in Houston.[1]

IV.

While East Texas made fortunes for Hunt, Murchison, and other oilmen, it ruined Sid Richardson. With prices for the new, high-quality crude driven as low as ten cents a barrel, the majors saw little reason to buy the remote, sulfur-laden West Texas oil Richardson was selling. Cash from his best wells in Ward County shriveled to almost nothing. In January 1930 his income was twenty-five thousand dollars a month; by December it had fallen to sixteen hundred dollars a month, and the bank took every cent.

In a matter of months Richardson went from the penthouse to the outhouse—literally. He moved out of his top-floor suite at the Blackstone Hotel into a forty-dollar-a-month maid's room. When he couldn't afford that, he moved into a twenty-five-dollar-a-month room at the Texas Hotel; when he couldn't afford even that, he was evicted and sued for back rent, at which point one of his closest friends, Amon Carter, publisher of the Fort Worth newspaper, gave him a room at the Fort Worth Club for free. When he was evicted from his office, Richardson set up business at a downtown drugstore. If he was out, the soda jerk, a man named Jack Collier, answered the pay phone, "Sid Richardson's office."

His only hope was to find more oil. Murchison begged him to come to East Texas, but he wouldn't; all his leases and contacts remained in remote Ward and Winkler Counties on the New Mexico border. He was determined to keep drilling, but his credit at West Texas oil-supply outfits was running low. When the inn in Monahans threw him out, he fled to the new hotel in Kermit. When the Kermit hotel threw him out, Richardson resorted to bunking at a ranch outside town. In desperation he resorted to poor boying, paying his men with groceries and the promise of an eventual paycheck. During the Depression poor-boys like Richardson paid drillers and toolies ten to twelve dollars a day in oil, if it was found. If it wasn't, they received whatever a driller could come up with. Richardson, who kept a chronically overdue account at the general store in Kermit, became reknowned for paying his men in bread, eggs, or milk—whatever they needed to eat.

In 1931 Richardson managed to drill another well on the O'Brien Ranch in Ward County, but the gusher he needed turned out to be a fountain; whatever profits he saw went to his banks. Worse, the well took the last of his money. By early autumn Richardson was flat broke. He was convinced there was more oil in Ward County, as he had argued in vain as he made the rounds of the banks in Fort Worth and Dallas. Temporary relief came in late October when, with Clint Murchison's help, Richardson managed to coax $110,000 out of Nathan Adams at First National of Dallas. The money allowed Richardson to spud a series of new wells, on the Estes and Scarborough ranches in northern Ward County. Several were meager producers, but by mid-1932 Richardson's cash flow had slowed to a trickle. His creditors, including the oil-supply companies that leased him his rigs and piping, began to sue. There was no way he could begin to repay his debts.

Subpoena servers began hounding him everywhere. Whenever Richardson stepped out onto a Fort Worth sidewalk, he watched for one. In later years he told of one creditor who tracked him down while he and Murchison were out hunting. In desperation he turned to Murchison for more relief, and Murchison began covering his loan payments. A year later Murchison would take over the First National loan entirely.* For collateral Richardson signed over his last leases on the O'Brien Ranch.

In later years Richardson often said he had been so destitute he was forced to abandon the oil business for several years. Like so many of his tales, this was an exaggeration. If he gave up wildcatting at all, it was likely only for several months, probably during the winter of 1932–33, when reports of his drilling disappear from the San Angelo newspaper. According to Bass family lore, Richardson was able to resume drilling only after his brother-in-law, Doc Bass, refused his entreaties for one last loan. It was his sister Annie who dipped into her purse and gave Richardson forty dollars. It was with

---

*Details of Richardson's various loans and lawsuits are contained in records filed in the Winkler County Courthouse in Kermit.

this money, Richardson claimed, that he bought train fare for one last trip west.

The story is apparently true. According to a rare interview his nephew Perry Bass gave the *Dallas Morning News* in 1984, the Bass family was vacationing in New Orleans in March 1933 when Richardson appeared, asking for a loan. When Doc Bass refused, Annie gave him forty dollars her husband had given her to bet on a horse; the "Old Family Friend" says it was actually four hundred dollars. Whatever the amount, Richardson used it for train fare to Ward County. As it happened, it was the last time he and Doc Bass ever spoke. One week later, having returned to Texas himself, Bass dropped dead of a heart attack at an Austin hotel.

While the story may be true, however, one suspects the $165,000 Clint Murchison persuaded First National of Dallas to lend Richardson a month later, in April 1933, was a tad more important. Richardson used the money to spud a series of four new wells on the Scarborough Ranch in Ward County. In the fall of 1933 all four struck oil, but not much, and not nearly enough to repay his debt. Once again he had found a babbling brook. What he needed was the Amazon. For officers at the First National banks of Dallas and Fort Worth, Richardson was a nightmare. His debt now approached $1 million. Yet they couldn't simply foreclose; his assets were minimal. They could either write off their loans and take a huge loss, or find a way to make Richardson a success.

The ideal solution would be to locate a wealthy partner, someone who could fund drilling that might produce enough oil to repay the banks. For most of 1933 loan officers from both banks queried their largest customers, but none proved interested in backing Richardson. Then, in late 1933, Eugene McElvaney, a vice president at First National of Dallas, came up with a new name: Charles E. Marsh, co-owner of several Texas newspapers, including the politically influential *Austin American*. Marsh, like his Austin neighbors Herman and George Brown of the Brown & Root contracting company, was using his spare cash to bankroll several Texas wildcatters. When McElvaney sent him a telegram at Washington's Mayflower Hotel, Marsh consented to a meeting. "Mr. Richardson, the party I had in mind, seems anxious to get together with you," McElvaney wired Marsh on November 27, "and I see no reason why you couldn't work out an advantageous deal."*

It is a measure of how totally Sid Richardson cloaked his business in secrecy that the name of Charles Marsh, the man whose backing made Richardson's fortune possible, remained unknown to Richardson's family until they learned it during research for this book. In all likelihood, in fact, Marsh's name would be lost to history were it not for his subsequent sponsorship of an obscure Texas congressman named Lyndon Johnson, a relationship chronicled by all of Johnson's biographers. A large man with an even larger ego, Marsh was by every account a self-important blowhard who expected his lessers to endure endless lectures on his favorite topic, politics. "Marsh was a high-rolling, J. Rufus Wallingford type," a Johnson aide named Welly Hopkins told an oral historian for the Johnson Presidential Library. "He could be very rude at times because of his overweaning ego. Charles Marsh, to himself, would think he could do anything.... He always wanted to be the great manipulator behind the scenes."

That Marsh and Richardson were polar opposites was obvious. Richardson didn't care; he just needed money. Marsh's initial commitment was modest, at least to him. He guaranteed a thirty-thousand-dollar line of credit at First National of Dallas. Like a starving dog, Richardson drew the money in gulps, ten thousand dollars in February 1934 and another ten thousand a month later. "He wants an additional ten now against same commitment feeling that he can deal more advantageously with his creditors," a bank officer then wired Marsh. "We are willing to make the advance... but would like to have your permission and approval beforehand."

When Richardson drew down that first thirty thousand dollars, he wanted more. In fact, if he was to pay off his debt and renew drilling, he might need close to one million dollars. Marsh resisted, finding himself in the same quandary as the banks: risk losing more money, or losing his investment to date. In a search for capital he began canvassing the New York

---

*What little is known of Richardson's dealings with Charles Marsh can be found, in part, in various corporate files Marsh left after his death in 1966, and which are now deposited at the Lyndon B. Johnson Presidential Library in Austin.

banks, several of whom were opening oil-lending departments. Years later the young bank officer Clint Murchison had befriended, Rushton Ardrey, recalled an incident that probably took place in 1934.

> Charles Marsh, who owned a newspaper in Austin, came to see me in New York, said he was doing a little oil business on the side, and that he needed $1,000,000. He had an interest in what Sid owned.
>
> "How much does Sid owe?" I asked.
>
> "Come on down to Texas and talk with Sid" [Marsh said]. "I don't think Sid knows what he is doing." I knew what he meant. Sid owed a total of about $1,000,000 to everybody in the area. "If he has the reserves—let me check the reserves—I will lend him $1,000,000." I came down to Fort Worth [and met with Sid. But] Sid was [drilling] wildcats. I wouldn't lend the money on wildcats, but if the well came in, I would make a loan.[2]

Unable to entice the New York banks, Marsh took a deep breath and decided to raise his bet. By August he had begun negotiating a complicated deal involving First National of Dallas. From what fragments of legal documents remain, it appears that Marsh agreed to guarantee Richardson's debt to the bank. In return, the bank agreed to loan Richardson an additional $210,000, followed by another $150,000 the following spring. Much of the money went to pay First National of Fort Worth, which, Rushton Ardrey recalls, was under pressure from banking regulators to reduce its exposure to Richardson's wells.

> I remember one time [Sid] was in so damned deep to the Fort Worth National Bank that they were just worried to death. It had reached the point where the Federal Reserve Bank was giving them hell about having Sid's paper. They said, "Well, we've tried to get him [to pay] off, but we haven't had any luck." Sid was one of the most skillful conversationalists I have ever known. The guy really had charm. The bank told the Federal Reserve guy, "You go talk to him and see what you can do." So this guy came over with fire in his eyes and went to work on Sid, and he

wound up by agreeing that it'd be all right for the bank to extend some
more credit to him![3]

While it kept his creditors at bay, the new partnership with Charles
Marsh coincided with the most brazen gamble Richardson had ever made,
what amounted to a Hail Mary pass for his career. It was in far northern
Winkler County, two miles from the New Mexico border. One or two wells
had been drilled in the area; they found natural gas but only a trickle of oil.
In the spring of 1935 Gulf Oil began trucking in rigs for a major drilling
effort it planned on leases just east of Kermit. While the majors generally
refrained from aggressive drilling during the Depression, Gulf was facing
the expiration of hundreds of ten-year leases it had signed during the West
Texas rush of the mid-1920s; it had to drill the Winkler leases or lose them.
Gulf's plans—and what it knew about the area's geology—were cloaked in
secrecy. The land itself was godawful, patchy sand and scrub crisscrossed by
cactus-strewn gullies, but Richardson knew it well. As Gulf moved forward,
he began buying up every available lease nearby. He even took a train to
Washington to wangle one from the Pure Oil Company.

Where Richardson found the confidence to place such a massive bet
remains a mystery to this day, even to his family. While tales of H. L. Hunt's
and Roy Cullen's first major discoveries became oil-patch legends, there
is not a single anecdote celebrating Richardson's—not one. In later years
there would be whispers he had bribed someone, maybe his close friend Jay
Adams, head of Gulf's West Texas operations; it's also possible Adams gave
Richardson a tip for free. Perry Bass, who was to serve as Richardson's No. 2
man for twenty years, once told the Old Friend that Richardson had devised
a more creative way of gathering intelligence. In the 1930s the main switching station for long-distance telephone calls made from West Texas was
in the dusty town of Mineral Wells. According to Bass's story, Richardson
became friendly with two of the telephone operators there. In Bass's words,
he performed "stud service" for these young ladies in return for information
they overheard. "He would get a call from one of the operators, you know,
saying, 'Oh, Sid, a Mr. Moore from Gulf Oil is calling his headquarters in
Pittsburgh, would you like to listen in?'" says the Old Friend.

However he learned details of Gulf's plans, by the summer of 1935 Richardson had used most of Charles Marsh's investment to buy land all around Gulf's drill sites. In July the first of the Gulf wells struck oil, one hundred barrels a day—not much. But each succeeding well found a bit more. That autumn Richardson began drilling nearby, on land he leased from a rancher, and in October his well came in strong, as did the next ten or twelve he drilled all around it. They were the first wells in what became known as the Keystone Field, not the largest oil field in Texas, but a good one, big enough to produce a half million barrels of oil in the first two years. Even after splitting profits with Marsh, Richardson was able to begin repaying his debts. What cash he had left over he plowed back into the earth; of the eighty or so wells he drilled in the Keystone over the next five years, he later swore every single one hit oil. He wasn't a millionaire—not yet—but he was well on his way. By early 1936 he was actually able to reopen a Fort Worth office and pay rent.

One day Richardson plunked himself down in his old "office," the downtown Fort Worth drugstore, and faced the soda jerk, Jack Collier, the man who had been taking his phone calls for years. Without Collier's help, Richardson felt, he might never have recovered. As the Old Friend recalls, "Sid said to Collier, 'If you had a lot of money, what would you do with it?' And Collier said, 'I'd like to own this drugstore.' And Sid said to him, 'You do.' He had bought it for him." The story is not apocryphal. Many years later, after Collier's death, Richardson took ownership of the drugstore, and another he bought Collier. He would own them until the day he died.

SIX

The Big Rich

*The Trents lived in a house on Pleasant Avenue that
was the finest street in Dallas that was the biggest
and fastest growing Town in Texas that was the
biggest State in the Union and had the blackest soil
and the whitest People and America was the greatest
country in the world and Daughter was Dad's
onlyest sweetest little girl.*
JOHN DOS PASSOS, *1919*

I.

The fortunes forged during the Depression created a new top layer of Texas
society, what came to be known in later years as the Big Rich. This was
wealth on a scale entirely new to the state, and during the 1930s Roy Cullen,
Clint Murchison, Sid Richardson, and H. L. Hunt, soon to be known as the
"Big Four" oilmen, laid the foundations of a flamboyant lifestyle that would
come to define the image of Texas Oil. There were mansions to build, presidents to meet, European vacations to take, islands to buy, and children to
raise; between them, the Big Four now had four families to support. Unfortunately, two belonged to Hunt.

Murchison, the first to earn his fortune, was the first to begin gathering the trappings of serious wealth. In 1927, two years after his wife's death,
he had taken an apartment in Dallas, the city closest to his hometown of
Athens. A year later he acquired a two-hundred-acre polo club on Preston

Road in a rural area fifteen miles north of downtown. The rolling fields and woods teemed with wildlife, and Murchison, in an effort to reproduce the country life of his boyhood, trucked in hundreds of new animals, cattle, pigs, and goats, as well as chickens and dairy cows for milk and butter. The clubhouse was a shack, but Murchison fixed it up, adding two wings with staff quarters and hiring a squadron of governesses and Negro servants. When finished, they called it "The Big House." Preston Road Farm, as Murchison dubbed the compound, was so vast that the Athens high school football team accepted his invitation to use it for summer training camp.

Murchison wanted his three boys to grow up as he had, trapping, riding, and fishing, and once he retrieved the boys they did just that. Already a primitive layout, the boys transformed Preston Road Farm into a backwoods paradise, their pet raccoons, skunks, and squirrels scrabbling over the furniture and unnerving the governesses. Murchison loved it; for the first time since Anne's death he began to relax. He was lonely though, and began reaching out to his old chums in Athens. They came in droves, along with Murchison's oil-industry pals, turning weekends at the farm into one long rollicking party of whiskey, poker, and dice games, the pots overflowing with not only cash but oil leases and royalty vouchers. The parties usually kicked off on Friday evenings when a dusty maroon Chrysler pulled up the long gravel drive, heralding Sid Richardson's arrival from West Texas. The gambling and drinking lasted late into the night, when the guests would wobble into the bunkhouse Murchison had erected just for his friends, a dozen beds arrayed in one large room, cowboy-style.

"Murk," as his buddies called him, blazed the trail for a generation of oilmen just learning how to be rich. He was among the first to begin traveling by private plane, a twin-engine four-seater that took off and landed on the old polo fields. He was among the first in Dallas to dig a swimming pool, an Olympic-size model that when finished in 1930 served as the focus of his entertaining, card tables splayed around its periphery as a haunch of beef or maybe a goat grilled on a spit by the cabana. Because his boys couldn't swim well, Murchison had the pool lined with extra ladders. He adored the boys and paid tuition for them to attend public schools in University Park; most days they arrived at school in a chauffeur-driven Pierce Arrow. Weeknights they all piled into one bed together with their father. Murchison swore he wouldn't remarry until they were grown.

In time Murchison began itching for more land, a place where he and his pals could hunt and fish. In 1933 he was trolling for tarpon in the Gulf when he noticed Matagorda Island, a thirty-eight-mile sandbar that lined the coast southwest of Houston. Its eastern half was a wildlife refuge—the island abounded in shore birds, white-tailed deer, and rabbits—but the western half was a sheep ranch, and Murchison hit on the idea of turning it into a personal retreat. He had American Liberty buy it in 1933. Construction on the compound began immediately. Because there was no causeway to the mainland, building materials were ferried to the island by barge. By 1936 Murchison's new spread was complete: a water tower, servants quarters, and a long, low clubhouse lined with bunks for thirty-five men. A veranda in front faced the beaches. It was a primitive layout in those early years, as if a dude ranch had suddenly been plunked down in the Hamptons.

Even before construction was finished, Murchison began leading hunting expeditions to Matagorda, the men sleeping on rented yachts as the clubhouse was built. While there was deer aplenty, the most prevalent animals were rattlesnakes, so many, in fact, that the Murchison boys weren't allowed to visit for months while their father and three of his old buddies, Dudley Golding, his lawyer Toddie Lee Wynne, and Sid Richardson, mounted a four-man eradication effort. Murchison designed a hunting car, a stripped-down Ford with bucket seats, for the task. The men laced on hunting boots, hoisted shotguns, and hopped in the car. Murchison drove. The ranch manager and a guide led on horseback. If the horses reared near a cactus patch, they knew rattlers were inside. The guide would dismount, fish out the snake with long prongs, and Murchison and the men would blast it with shotguns. Afterward everyone would repair to the yachts and, over tumblers of Wild Turkey, argue over who earned the most kills.

All in all, Murchison's was a splendid life, especially considering how most Americans endured the Depression. But it was not entirely free of sadness. On a chill morning at Preston Road Farm in April 1936, his youngest son, ten-year-old Burk, went out to check his animal traps, slipped, and fell into a creek. He went to school in wet clothes and by the next day had a high fever. By the time Murchison arrived home from a business trip, the fever had become pneumonia. Burk was rushed to a hospital, but his lungs, weak since birth, gave out. He died on April 11 and was buried beside his mother in Athens. Murchison was devastated. Friends said they never heard him, or any of the other Murchisons, mention Burk's name again the rest of his life.

The following year, as if to punctuate the family's loss, the boiler in the Big House exploded, burning much of the structure and all of Murchison's books. Murchison decided to replace it with a manor home he had begun to envision, and he moved his lead maid, Jewel Pfifer, and his two remaining boys, John and Clint Jr., now teenagers, into the Stoneleigh Hotel. When construction began in January 1938, Murchison oversaw every detail. It was to be the largest private home in Texas, thirty-four thousand square feet, a red-fieldstone colossus longer than a football field, its veranda overlooking White Rock Creek a half mile away. The master suite alone had eight full-sized beds for Murchison, his two sons, and whatever pals were staying over—"so we can stay up all night talkin' oil," he explained to his decorator. Murchison had seen the kitchen at the Waldorf-Astoria in New York, and wanted something bigger. He got it. A single first-floor hallway measured 256 feet long. A friend cracked, "Jewel's gonna need roller skates."[1]

The new home's centerpiece was the two-story game room. Murchison designed it himself. A mezzanine encircled the room, to stow his books, and for musicians at parties. Projection equipment was stored in a special closet, giving Murchision a 1930s-style home theater. Above the giant fireplace a motto was carved, "Sportsmanship above Pleasure." A local painter was hired to paint wildlife murals on each wall, the wildlife of North Texas on the north wall, East Texas whitetail deer on the east, West Texas mule deer, mountain sheep, and prairie dogs on the west wall, South Texas waterfowl on the south. Murchison's homey touches continued into the adjoining bar, where he banished wallpaper in favor of wall-to-wall tarpon scales.

The Murchisons and their Negro servants, now numbering nine, moved into the mansion in early 1939. Murchison was now forty-four. His eldest son, John, was in his second year at Yale, Clint Jr. at Lawrenceville Prep in New Jersey. Friends noticed their father was mellowing. Though the weekend parties continued, they were tamer, in large part because Murchison had quit serious gambling and cut back his drinking. The reason was a twenty-three-year-old blue-eyed blonde named Virginia Long. Clint called her Ginny; he thought she looked like a petite Lana Turner. They met when a girl named Effie Arrington, who was soon to wed Murchison's pal Wofford Cain, brought her to a pool party. Ginny was Murchison's dream girl, a spontaneous, energetic East Texas tomboy who could shoot ducks, reel in tarpon, and play gin rummy with the gusto of an oilman. Though clearly in love, they waited four years, until Murchison's sons were away at school, before marrying in a quiet ceremony at a friend's home. For their engagement, Clint gave Ginny a 16-carat diamond marquis. It cost $125,000.

For Murchison, it was pocket change. Cash from East Texas and the Southern Union pipelines was gushing into his accounts, and he leveraged his growing fortune fivefold by using every penny as collateral for new and ever-larger bank loans. He had been among the first oilmen to capitalize on the banks' new energy-lending practices in the early 1930s, and in the ensuing decade, years before any of his peers attempted it, he was the first to begin diversifying outside oil. Fearing inflation, he purchased his first non-oil asset, an Indianapolis insurance company, in 1939, then added another in 1942.

By the end of the war, he was working on his own. His American Liberty partner, Dudley Golding, died in a plane crash in 1938, an incident that frightened Murchison away from private planes for several years. Golding's death triggered an awkward situation with his widow, Georgia, who sought to remain active at American Liberty. When she agitated to appoint an independent board, Murchison turned to his attorney, Toddie Lee Wynne, to buy her out. Wynne took over Golding's half of the businesses. Their partnership, however, lasted barely six years. Its demise was a family secret for years afterward.

According to Murchison's authorized biographer, his longtime secretary Ernestine Orrick Van Buren, Wynne bowed out of the business after a heart attack. But according to his unauthorized biographer, Jane Wolfe, Murchison forced the dissolution after discovering that Wynne had secretly bought a group of Wyoming oil properties, a clear breach of the pair's gentleman's agreement to invest together. To split up their $150 million of assets, they decided to take turns selecting the businesses they wanted, like kids choosing up sides on a playground. They flipped a coin to see who went first. Wynne won, and chose American Liberty's oil-production division. The process left Murchison deeply embittered. As president of American Liberty and the one most attuned to its underlying values, Wynne chose wisely, and Murchison felt he did so by misrepresenting some of his own balance sheets. Worse, American Liberty owned Matagorda, which went to Wynne. It broke Murchison's heart. "The biggest mistake of my life," he told his sons years later, "was giving up Matagorda."

Their 1944 divorce left Toddie Lee Wynne a very wealthy man who in time would father his own colorful Texas clan. Matagorda remained under the Wynne family's control for decades.

II.

By 1936 Sid Richardson had been living in hotels for twenty years, and while his nomadic life was perfectly suited for the rooms he now kept at the Fort Worth Club—his widowed sister Annie lived in an adjoining suite—he yearned for something permanent. Once Murchison bought Matagorda, it wasn't long before Richardson began muttering about his own island. In later years, any number of stories circulated about how he came to buy St. Joseph's Island, a thin strip of barrier sand, twenty-one miles long and six miles wide, across the channel from Matagorda's southern tip. At low tide Murchison could drive across to St. Joe's, as it was called, and he pushed Richardson to buy it.

Richardson brought up the matter with Murchison's banker, Rushton Ardrey, one day at Murchison's home. As Ardrey remembered it:

Sid said, "Rush, you have been down to Clint's island. I am going to get me an island that is better than his. I want you to lend me the money."

"How are you going to secure it?" [Most of Richardon's oil wells were already being used as collateral on other loans.]

[Richardson said] "Well, I've got a lot of Brahman cattle I can put up for security."

At this point, Clint spoke up, "Rushton's not interested in cattle—he is interested in oil." I loaned Sid the money on some oil properties to buy St. Joe Island.[2]

The price tag was a paltry twenty-five thousand dollars. What Richardson got was an island lined with breathtaking white-sand beaches, flanked by dunes, its scrubby inland areas studded with cacti. For years its only occupants had been an absentee owner's cattle, and the only structure, other than corrals and fencing, was a broken-down bunkhouse. Richardson envisioned a compound like Murchison's. "He never had a home to live in, and he wanted one there," recalled his nephew, Perry Bass, who now entered his uncle's world. The only child of Richardson's sister Annie and the late E. P. Bass, Perry had grown up in Wichita Falls, where as a teenager he had become, of all things, a standout sailor. Intense and socially awkward, Perry raced boats called snipes on local lakes and in 1935 won the national championships, held in Dallas. In the spring of 1936, he recalled, "Uncle Sid came up to Yale where I was about to graduate, having studied engineering and geology, and said, 'Bass, you are an engineer, and I want you to build me a house.'"

It was the beginning of a lifelong partnership. With his own father dead, Perry Bass became Sid Richardson's surrogate son. Theirs was a complex, some might say mildly abusive, relationship. According to those who knew both men well, Richardson believed his sister had spoiled Perry; for a rough-and-tumble character like Richardson, sailing wasn't exactly a man's sport. During all their years together, Bass seldom received the bear hugs Richardson gave other men. And in all those years Richardson never called him anything but "Bass"; one wag joked that he thought Bass's first name was "Goddamnit." "Sid was very, very tough on Perry—always was," says the Old Friend. "He thought he was soft, self-important, and a tad pretentious."

Richardson's determination to "toughen up" his nephew began with the work on St. Joseph's. A year after his graduation from Yale, Bass arrived in Rockport, just across an open channel from the island. His "office" was the pay phone at the Magnolia gas station, where he relayed reports on his progress to his uncle. Richardson had budgeted thirty-five thousand dollars for construction, and in one of their first calls promptly took it back, saying he needed the money for a lease in West Texas. "Any son-of-a-bitch can build a house for thirty-five thousand dollars," Richardson told him. "It takes a genius to build it with nothing."[3] As the Old Friend puts it, "It was just one of the ways Sid tried to make a man out of him."

Bass "poor-boyed" construction much as Richardson had done in West Texas, buying everything on credit and promising his workmen an eventual paycheck. By autumn he had crews working on the island, casting concrete blocks from a mixture of cement, oyster shell, and sand. His men began digging the foundation on January 2, 1938, and by Thanksgiving Bass had used ninety-three thousand blocks to build the only true home Richardson would ever know. It was unlike any house in Texas. Designed by the Dallas architect O'Neil Ford, the home was an ultramodern concrete structure modeled on Le Corbusier's famed ship houses. Long and low, built to withstand the island's frequent hurricanes, its sole flourish was a dramatic exterior spiral staircase extending down from the master bedroom suite. The core of the home was a combination living room and porch, divided by aluminum and glass doors that receded into wall pockets. Open spaces were lined with terra cotta and mahogany.

St. Joe's became Richardson's very private world, off-limits to outsiders and rarely photographed, reachable only by boat or airplane. He staffed the island with Negro servants and a wrangler for his cattle and, along with the pilots and chauffeurs he accumulated in later years, this group became what amounted to his immediate family. "I'll always remember Sid sitting back in the 'staff room' behind the kitchen, playing dominoes with the black houseman's son, Joe the chauffeur, maybe a boatman from Rockport," recalls the Old Friend. "The pilots would be there, and Raymond the cowboy, who was shy and who would usually wander off. There was a poker table in the living room, but no, Sid preferred playing in the staff room."

It was a man's world, the days spent hunting and fishing, the nights playing cards. Richardson never married, and no one can remember a single woman who anyone considered his companion, leading to rumors about his sexual proclivities. In fact, the Old Friend confides, the only females in Richardson's life were what he gently calls "ladies of the evening" who could be flown in from Houston or New Orleans for a night's entertainment and spirited off the island by dawn.

By 1938 women weren't the only ones Richardson longed to be rid of. His shotgun marriage to the overbearing Charles Marsh was showing strains. Marsh had used his share of the Keystone Field proceeds to build a palatial English-style estate named Longlea on a thousand acres in the Northern Virginia hunt country. He had left his wife and children for his mistress, a tall, languid stunner named Alice Glass, and the two had taken to mentoring the Austin area's new young congressman, Lyndon Johnson, during long weekends Johnson spent at Longlea. Marsh lavished money and advice on Johnson even as Glass, as LBJ's biographers have shown, took Johnson quietly into her bed.

Then, in 1938, Marsh encountered a sudden, fierce financial reversal. What happened was never disclosed publicly. But from a single mention in a letter to Richardson—contained in Marsh's papers at the Johnson Presidential Library—it appears that the Internal Revenue Service served Marsh with a request for $1.2 million in overdue taxes. The high-living Marsh was habitually short of cash and was forever pestering Richardson's office for advances; most of his assets were tied up in his ongoing divorce dispute. The IRS claim led to a three-year fight during which Marsh was forced to repay much of the money. To raise it, he ended up selling all his Texas newspapers, including the one in Austin, and attempted to sell his share of the Keystone Field.

This effort, in fact, figures in the opening scene of Robert Caro's three-volume biography of Lyndon Johnson. In the anecdote Caro relates, Marsh and Brown & Root's cofounder, George Brown, met with Johnson during the summer of 1940 at the Greenbriar Resort in West Virginia. As two of Johnson's closest backers, they were attempting to find a way to augment Johnson's meager legislative salary. According to Caro, Marsh offered to sign over his partnership with Richardson. Johnson declined, believing such a sweetheart deal, if ever disclosed, would kill his future in politics. Marsh, it appears, then sold out to Richardson instead, in October 1941. Though the precise amounts remain unclear, Richardson paid Marsh roughly $260,000 plus several large loans. With one stroke of a pen, Richardson took outright control of the Keystone Field, nearly doubling his existing oil reserves.*

By the time he parted ways with Charles Marsh, Richardson had taken in Perry Bass as his junior partner; for the next twenty years the two owned everything they found 75/25. Already Richardson had leveraged his Keystone profits into discoveries in several other West Texas counties, including a series of strikes on the famed Slaughter Ranch. But Richardson wasn't satisfied. In 1939, following Bass's work on St. Joe's, "Sid told him, 'Goddamnit, Bass, you're a geologist, go find me an oil field,'" says the Old Friend. Like many Texas oilmen, the two began scouting the new frontiers in Louisiana, which was opening its coastal wetlands to drilling. Bass set up shop at New Orleans's Roosevelt Hotel in anticipation of a massive auction of state lands. Because Richardson was so frugal, Bass worked on the cheap. When Richardson refused to hire enough seismographic crews to map an area's geology, Bass chartered a plane and mapped the telltale paths competing crews left as their boats passed through the marshes below New Orleans; in this way he could see what lands other oil companies were studying.

According to family lore, Richardson employed a characteristic sleight

---

*In his 1989 biography of John Connally, *The Lone Star,* author James Reston Jr. gave another version of the Richardson-Marsh split. Reston quotes an unidentified "observer" of the deal who asserted that Richardson and Marsh had an agreement in which either partner could buy out the other in the event the second partner was unable to fulfill his financial responsibilities. According to this account, Richardson demanded $3.7 million from Marsh to develop leases around the Keystone Field; when Marsh couldn't produce the cash, Richardson forced him to sell out. "When Sid got the chance," the observer is quoted saying, "he screwed the guy who got him into business."

The Reston account makes no mention of Marsh's tax problems. Moreover, as Marsh's private papers show, Richardson and Marsh continued an amiable correspondence for years afterward. If Richardson really did "screw" Charles Marsh, and there's no documentation to support this assertion, Marsh apparently held no grudge. In later years Marsh regained his solvency and purchased a string of small eastern newspapers. He died in 1964.

of hand during the state's two-day land auction. On the first day, realizing his competitors were watching him closely, Richardson placed a huge bid on a parcel he didn't actually want. The next day, assuming Richardson knew something they didn't, Humble, Texaco, and other companies spent much of their money leasing parcels all around the ones Richardson had taken. As they did, Richardson swooped in and bought the distant parcels he actually sought at significantly lower costs. It was deep in these bayous below New Orleans where, over the next several years, Richardson struck the second- and third-largest fields of his career, at Cox Bay and Pointe a la Hache.

But the greatest oil field Richardson and Bass ever discovered lay beneath ground they already controlled. Richardson had found his first oil in the Keystone Field at thirty-five hundred feet. Over the next several years drillers in Winkler County had bored their way steadily deeper into the earth, until in early 1942, Richardson's friend, the newspaper publisher Amon Carter, announced he was drilling past nine thousand feet. At ninety-two hundred feet, Carter and his partners hit one of the largest oil sands ever found in Texas, the legendary Ellenburger Lime. Soon oilmen all over West Texas were scrambling to reach the Ellenburger. Richardson and Bass dropped well after well after well, and nearly all found vast amounts of oil. "The Ellenburger is just magical," marvels the Old Friend. "That was Sid's big pay. It made a big field, an elephant."

The Ellenburger production thrust Richardson into the top ranks of independent oilmen; his reserves approached one million barrels, at one point more than those of three major oil companies. "[Sid] drilled the deep wells and got a magnificent field," his friend Edgar Owen remembered years later. "Gulf Oil wanted to buy it. Sid and [Gulf's] Jay Adams were very close friends. Jay'd come over every day and dicker with Sid about buying the leases for the whole property.... [I remember one day] Sid said, 'Goddamnit, Jay, I've told you a dozen times I'm not going to sell it!' Jay said, 'Well, the company authorized me to pay eighty million dollars for it. Sid, that's more money than you'll ever know what to do with.' Sid said, 'No, I want to be *big* rich!' Jay said, 'You wouldn't know what to do with any more money.' Sid said, 'Oh yes, I'd just be like the rest of these rich bastards and have me some foundations.' [Jay] just gave up. Sid never seemed to take anything seriously. But he *always* knew what he was doing."

Richardson, like H. L. Hunt, became the rare independent to purchase his own refinery and make gasoline. Clint Murchison had picked up a refinery in Texas City, south of Houston, and didn't want it. The two old friends were negotiating its sale one day on Matagorda when the lawyers noticed they had gone missing. An aide found Richardson and Murchison in a barn, arguing over a debt—a very old debt. Murchison had reminded Richardson of a day when they were teenagers, packing peaches at the Richardson orchard outside Athens. Richardson had shortchanged Murchison by $4, and now Murchison was demanding repayment, plus interest. At prevailing rates, Murchison calculated, Richardson's four-dollar debt now came to thirty-five thousand dollars. Apparently Murchison won the argument, because Richardson ended up with the refinery.

III.

Though it made stabs at the southern gentility Dallas and Fort Worth achieved, Houston was always a rougher city, obsessively devoted to the dollar, where public displays of ostentation were not only tolerated but prized. The first of these began to appear in the 1930s, as newly wealthy oilmen began to build their dream homes. One of the more elaborate belonged to Clifford Mooers, a wildcatter who erected a colonial-style mansion as a gift for his second wife, a Cuban debutante, on a hundred-acre tract on Buffalo Bayou. Concerned about flooding, the Mooers home featured eighteen-inch concrete walls reinforced with steel, along with aluminum plates that slid over every door and window; the mansion was designed to remain airtight in the face of floodwaters twenty-four feet deep. On the grounds Mooers built a lake lined with tons of California sand, and a private zoo featuring dwarf Australian barking deer and a family of penguins. Despite their air-conditioned pen, the penguins did not survive their first summer.*[4]

---

*The home is today the Lakewood Country Club.

Roy Cullen's partner, Big Jim West, opted for an Italianate mansion he built south of the city on Galveston Bay. George Strake moved his growing family into a tasteful Tudor in River Oaks, where many oil families were building mansions. Strake saved his money for his Colorado summer home, Glen Eyrie, a vast stone castle, complete with turrets, that had been built in 1906 by the railroad magnate who founded Colorado Springs, William Jackson Palmer. After Palmer's death the castle, which sat amid several hundred thousand acres of ranchland, canyons, and mountain lakes, passed through the hands of several owners, none of whom could afford its maintenance costs. It had actually been vacant for thirteen years when Strake acquired it in 1938.*

The greatest of the new Houston mansions was Roy Cullen's. It was the "big white house" he had dreamed of as a boy, set on six acres at the head of River Oaks Boulevard just below the River Oaks Country Club. A bit chagrined to be building such a home during the Depression, Cullen justified the cost as a "civic enterprise" that would put hundreds of men to work. He put much of his energy into the adjoining gardens, sending horticultural experts traveling through the Deep South in search of the very best strains of azaleas and camelias; he bought so many flowers in Louisiana that, until his sellers protested, state officials briefly threatened an embargo.

Life at the Cullen mansion, however, was marred by loss. While the three youngest daughters were celebrated debutantes educated at eastern colleges, the eldest daughter, Lillie, was erratic, a headstrong young woman who yearned to be an actress. In the seven scrapbrooks devoted to Cullen's life and career at the family office in Houston, she is mentioned in only one clipping, the very first, a 1927 newspaper photo announcing her debut as a dancer in a play to be staged in San Antonio. Four years later, in 1931, Lillie traveled with an aunt to Los Angeles and, to her father's dismay, stayed. She married an Italian man, a would-be actor and self-styled baron named Paolo di Portanova. She gave birth to two sons, later divorced, and cut off

*Strake kept Glen Eyrie until 1951, when it was sold to the Navigators, a religious group then affiliated with the evangelist Billy Graham. The group uses the property as a summer camp and retreat to this day.

all contact with her family. No one but her father, who hired private detectives to trace her, knew of her whereabouts. "Gampa hated it; they tried
every possible way to approach her," Cullen's granddaughter Beth Robertson remembers. "In time, you know, she became a recluse. I think, everyone
thinks, that Lillie was sick, like manic depressive."

Of his five children, Cullen was closest to his strapping son, Roy Gustave, who after college followed him into the oil fields, drilling wells for
Quintana. Cullen was deeply proud of the young man, known inside the
family as Sonny, who he felt had the makings of a top-notch oilman; Roy
Gustave and his wife had a son, Roy Jr., and lived in their own River Oaks
home. Then, in May 1936, when he was thirty-three, he was summoned to
a Quintana drill site in the Rio Grande Valley, where pipe had frozen in a
well near Edinburg. The derrick, which Cullen had rented, was erected on
an old-fashioned wooden base, braced by iron pipe. "Be careful about that
rig, Sonny," Cullen warned. "It's not as solid as our rigs."

Roy Gustave drove south, and when he arrived he found the crew still
struggling to free the pipe. They were down seven thousand feet, and had
resorted to using a pulley system to lift and twist the mile-and-a-quarter
length of pipe stuck in the ground. Each time the pipe was wenched up and
twisted, the derrick shuddered. After a toolie went to check in with Cullen
in Houston, he returned with a warning. "Your dad says you are to come
down off that derrick!" he shouted. "He says it's an old rig and he doesn't
want you to take any chances."

"Okay! I'm coming down!" he yelled.

Just then it appeared the pipe jostled, as if freed. Roy checked a pressure
gauge, then yelled at the driller, Chester Kraft, "I think she's coming loose,
Cheety! Pull her wide open!" With each rotation of the pulley, the pipe
seemed to rise a few inches, but it was an illusion. In fact, as the family later
learned, the derrick was sinking into the mud. When Kraft switched the
lifting gear on full, the derrick began to vibrate violently. One of the iron
bracings on its legs popped loose. "Look out, Cheety!" Roy hollered. "She's
coming down!"

As the derrick began to collapse, Kraft jumped. Roy Gustave had no
chance. The derrick, more than a ton of twisting steel, collapsed atop him.

Moments later, Kraft found him pinned beneath the wreckage, unconscious. Cullen hurried down from Houston. His son lingered for two days before dying, never regaining consciousness. "My grandpa had a good friend, Sheriff Fisher, down in Calhoun County," Roy Gustave's son, Roy Jr., remembers today. "He had lost a son in a car wreck. I remember after Dad's death, Sheriff Fisher came into the office and, you know, he and Gampa looked at each other, and both of 'em started crying, not saying a word, these two tough old guys. Finally they blew their noses and went back to work. That was my grandfather."

IV.

Of the Big Four, H. L. Hunt was the only one who had yet to cash in. Not for Hunt the airplanes and private islands the others enjoyed. His sole extravagance was the beautiful Mayfield Estate in Tyler, where he rose every morning before dawn, packed his brown-bag lunch, and drove into the East Texas oil fields. Once the initial rush of drilling finished around 1934, Hunt was able to return home for dinner as he had in El Dorado; afterward, everyone gathered around the piano and sang songs. Of the children, Hunt was closest to his son Hassie, who had grown into the spitting image of his father; they once won the "look-alike" award at a local banquet. When Hassie was home from boarding school, he spent every spare minute with Hunt in the oil fields. In 1933 the oldest child, Margaret, began attending Mary Baldwin College in Virginia, and the following summer was named queen of the Tyler Rose Festival. A governess named Toogie taught French to the boys, Bunker, Herbert, and Lamar. "The time we had in Tyler," Hunt recalled years later, "may have been some of the best days we had as a family."

When he wasn't in Tyler, Hunt could usually be found in Dallas with his second family at their home on Versailles Avenue. Frania Tye was almost thirty now, still beautiful and still, she would always insist, under the impression she was married to Major Franklyn Hunt. When the Dad Joiner deal made headlines, she said years later, she did ask Hunt if he was the man in the papers; Hunt smiled and said no, H. L. Hunt was his uncle. Frania led a lonely life when her husband was away; for company she enticed her sister down from Buffalo. In 1932 they opened a beauty shop. Frania kept the books. Two years later, in the spring of 1934, she became pregnant for the fourth time.

That's when the trouble began. Somehow Frania learned the truth. How it happened would be the subject of courtroom debate decades later. Frania insisted a girlfriend told her. A Hunt attorney would claim that the confrontation came after someone pointed out Margaret Hunt to her at a party. Whatever the case, Frania confronted Hunt at the home on Versailles Avenue. Hunt confessed. Overwhelmed, Frania fled to her parents' home in Buffalo, though she couldn't bring herself to tell them the news. When she returned to Dallas, she and Hunt reached a compromise. She demanded and received permission to live openly, as "Mrs. H. L. Hunt." Hunt had only one condition: That she be Mrs. H. L. Hunt of Great Neck, Long Island. Which is how in mid-1934, Frania found herself moving her three children into a large white home thirty miles east of Manhattan. Hunt paid for everything. He visited every few months, telling Lyda he was going to New York on business. In October 1934, Frania gave birth to a fourth child, a son, Hugh Hunt.

For all the cloak and dagger, Frania wasn't a complete secret. Hunt's brother Sherman, knew her, though he seems to have believed she was a girlfriend. A few more Hunt Oil men made her acquaintance, not quite believing Hunt's explanation that she was a friend's sister. It is a measure of how close he was to the teenage Hassie that at some point, apparently while Frania was living in Shreveport, Hunt took his son to meet her and explained everything. For the time being, Hassie was the only person who knew the secret.

In June 1935, a year after moving Frania to Long Island, Hunt decided to diversify his holdings by buying a gold mine in Nevada. Hassie, who was then eighteen, and nineteen-year-old Margaret went along and spent six weeks gambling and relaxing in Reno. On the train back to Texas, Margaret found herself sitting alone with her brother, who wore an expression, Margaret recalled years later, "that clearly said that our vacation had come to an end."

"Margaret," Hassie suddenly said, "Dad has another family besides us."

Margaret didn't understand.

"There is a woman, Frania Tye," Hassie went on. "Dad lives with her. Like he lives with Mother and us. As if she's his wife, which she thinks she is. But he told me she's not. They have four kids. A lot of the times he's away he's with his other family."

Margaret thought her brother was playing a prank.

"I'm not joking, Margaret. Dad took me to see them. It's true."

Margaret was speechless. "One kid could be an accident," Hassie continued. "But not four. That family was planned. I could see that by the way Dad treated them, and her."

"Why would he take you to see them?"

"I was with him.... I guess he wanted to see them, so he brought me along. It was convenient."

"Did he explain anything?"

"You know he wouldn't."

"What about Mother? Does she know?"

"I'm sure she doesn't."

"How are you sure?"

"The way Dad took me there," Hassie said. "Like it was a secret between the two of us."

Margaret was devastated. Her head spun, trying to get her mind around something that was unimaginable. The only thing she decided was that their mother must never know. As the train crossed Nevada, she found herself staring out the train window as it began to rain.[5]

V.

By 1936 Hunt had all but finished drilling the Joiner leases. There were four hundred wells in all now, almost every one a major producer; one list showed Hunt to be the thirteenth-busiest driller in East Texas in those early years, more active than one or two of the majors. That October he finally bought out his minority partner, Pete Lake, for $1 million in cash, four of the Joiner leases, a drilling rig, and a Buick; once Lake was dispensed with, every dollar flowed into Hunt's accounts. At that point, one of his biographers, Harry Hurt, estimates that Hunt was worth about $20 million, roughly $250 million in today's dollars.

Hunt began to get organized, incorporating himself as a new Hunt Oil Company. As his venture into gold mining showed, he was determined to expand his holdings beyond East Texas. He hired more land men, then dispatched them into West Texas, Louisiana, and southern Mississippi, wherever there was an oil play. He had already incorporated his pipeline, as well as several other oil and gas-gathering lines, into a company called Panola Pipeline. When he purchased a small refinery near Tyler—it sold gasoline locally—he called it Parade Oil.

There were two final companies Hunt had in mind, and to name them he turned to his family. As Margaret remembered the conversation years later, they were sitting at home one evening when Hunt brought it up.

"Panola Pipeline and Parade Oil have been rewarding," Hunt said. "What other six-letter words beginning with P would you suggest?"

Hassie smiled. "Paltry? Putrid?"

"This is a business conversation," Lyda snapped. "No levity."

Hassie thought a moment. "Placid?"

Hunt mulled it a moment. "Placid.... Placid Oil.... Good. More?"

Margaret had been reading Booth Tarkington. "What about Penrod?" she asked.

"Perfect," Hunt said. "You might want to name your new company Penrod Drilling."

Margaret blanched. "Our new company?"⁶

The last two businesses Hunt formed were to be owned by the children. He transferred several choice properties into the first, Placid Oil, and placed the company under the control of trusts he formed in the children's names. At the same time, he gathered the eleven or twelve drilling rigs he had picked up over the years and formed them into a new contract-drilling company, Penrod Drilling, which was to be owned by the children outright. In later years the two companies would become the most important Hunt businesses of all.

The only serious shadow in the Hunt family's charmed years in Tyler fell at 3:05 P.M. on March 18, 1937, when one of Panola's pipelines exploded beneath the schoolhouse in New London, in northwest Rusk County, killing 294 people, most of them children. The New London disaster, to this day the worst loss of life in an American school, was not the Hunt company's fault; plumbers had attempted to illegally tap into the line to heat the school. The carnage brought out the best in Hunt, who sent scores of workers in to help clear the rubble. Hunt himself prowled the hospitals and funeral parlors, peeling off hundred-dollar bills and jamming them into the fists of brokenhearted parents.

As his horizons moved beyond East Texas, Hunt pondered moving his family once more, to Dallas. It made sense, he told Lyda. His banks and attorneys were there. He took Margaret along to look at houses. The one they chose was set on ten acres on White Rock Lake, eight miles north of downtown and several miles from Clint Murchison's spread. A large fourteen-room house if not quite a mansion, it was a replica of George Washington's home in Virginia, down to the glassed-in cupola and weathervane at its apex. Everyone called it Mount Vernon. The family moved in on January 1, 1938, prompting the *Dallas News* to note that "quite the nicest family has come to Mt. Vernon of Dallas to stay."

Hunt loved the lakeview; it reminded him of Lake Village back in Arkansas. Like Murchison's, Hunt's land had a rural feel, dotted with pecan trees and woods full of squirrels and songbirds; Hunt bought a half-dozen deer to make it feel even more like the countryside. He and Lyda took adjacent second-floor bedrooms. Margaret, who was now working as her father's assistant in Hunt Oil's new offices in the downtown Cotton Exchange Building, thought the house was impossibly isolated and fretted she would never meet a suitor; she did, however, and was soon engaged to a Hunt Oil accountant named Al Hill. The youngest boys, Bunker, Herbert, and Lamar, shared the master bedroom and used the second-floor laundry chute as a slide; Lyda had them enrolled at Dallas Country Day. The house came with five safes, which no one knew what to do with. Lyda used one to store canned goods. Six-year-old Lamar, a sports nut, took another to store his footballs and baseballs.

By the time he moved his first family to Dallas, Hunt had taken in his son Hassie as a partner. By 1938, when he turned twenty-one, Hassie was already shaping up to be the kind of stellar oilman his father knew he would be. At first he took a job leasing Penrod rigs, but Hassie's passion was finding oil. Having learned the basics at his father's side, he struck out on his own, drilling a series of wells in Southern Mississippi and hitting strike after strike in what became known as the Tinsley Field. Unlike Texas, Mississippi had no prorationing, so Hassie could pump all the oil he wanted. By his twenty-fourth birthday his wells were bringing in an estimated four million dollars a year. Hunt said more than once that Hassie had an unmatched gift, something mystical, for finding oil.

But Hassie had problems. He had been an unusual child, dyslexic and prone to temperamental fits. As a young man, his behavior grew stranger. Some of it was subtle; he laughed too loud, and at the wrong things. At a drill site, he might pick up a rock and sling it within feet of another man's head. Out in a duck blind, he would shoot a single duck just as a flock of geese hove into view, laughing as his fellow hunters fumed. Hunt could be hard on him, criticizing his sloppy paperwork and carping about how much he paid for land. Hassie responded by competing openly with his father, whether in buying acreage or in the memory games they played after dinner; each would study a deck of cards, then see who could remember the most. Once or twice these contests became so intense that Hunt and Hassie ended up in heated wrestling matches. Hunt believed Hassie was simply headstrong and needed "action," but some things couldn't be explained away so easily, as when Hassie showed up in the office barefoot with his shoes hanging by laces around his neck.[7] Or the time he walked past an automobile showroom, saw a car he wanted, then fired a rock through the window to get it. In time Hunt decided to assign a man to watch him full-time.

Once resettled in Dallas, Hunt began to miss Frania Tye, who remained in Long Island raising their four children. Though he was only able to see her every few months, Hunt's desire for Frania, as evidenced in poems he mailed her, remained intense. "Running now a little late / But hoping to keep our date," he wrote before a Christmas visit in 1937. "So strong is the urging / My engine is surging."

In the fall of 1939 Hunt moved Frania and their family to Houston. Maybe it was Frania's desire to continue living as "Mrs. H. L. Hunt" or maybe

Hunt had grown overconfident after more than a decade of successful secrecy, but instead of tucking Frania away in a discreet corner of the city, Hunt hired a New York architect to design and build her a large home smack in the middle of the city's finest neighborhood, right there on River Oaks Boulevard, one block down and just across the street from the Cullen mansion. As if to claim her rights as Hunt's wife, Frania applied to the River Oaks Country Club, just behind the Cullen spread, as "Mrs. H. L. Hunt, femme sole."

Given the insular world of the Big Rich, it was only a matter of time before someone learned the truth. As Frania told the story years later, it happened on the evening a prominent hostess named Ruby Matthews agreed to host an introductory tea party for her at her home. Many of River Oaks's glittering oil wives were to attend; Frania had even invited two of her girlfriends from New Orleans. Then, just as the guests began to arrive, the phone rang. On the line was a woman. She wouldn't give her name. "I know you're not Mrs. H. L. Hunt," she said. "But if you give me five thousand dollars, I won't tell what I know."

Frania later claimed she received as many as fifty similar phone calls in the ensuing weeks. She pleaded with Hunt to do something. He sidestepped the issue, sending a letter instead. "I know you were very much distressed," he wrote her, "and I am sorry, but don't let them get your goat. The best of them could never be friends of ours."

As the calls mounted, Frania became overwrought. No matter how she pleaded, Hunt would not come to her rescue. She felt abandoned. Finally she reached a breaking point. Diverting her anguish into anger at Hunt, she packed up the children and drove to Dallas, where she checked into a suite at the Adolphus and telephoned Hunt at his office.

"These are your children as well as mine," she told him. "Come take them off my hands." Then she drove back to Houston, leaving the four children, aged fourteen to five, at the hotel.

In River Oaks she returned to find the phone ringing. Hunt beseeched her to retrieve the children. She reluctantly did so, but the situation was fast approaching a climax. Not long after, Frania returned to Dallas. This time she demanded to see Lyda. She wanted a resolution. She wanted this to end. Hunt was almost out of options. More talk wouldn't help. Then he thought

of Margaret, now twenty-four, who had just married Al Hill. She was pregnant, and was busy overseeing construction of their home. Hunt dropped by
the building site on Vassar Drive and asked her to come with him for a drive.
After a few minutes in the car, he said, "You know about Frania Tye."

"Yes," Margaret said. "I know about Frania."

"She is here in Dallas," Hunt said. "She wants to meet you."

Margaret wondered whether it was Frania or her father who wanted this
meeting. "She is at the Stoneleigh Hotel," Hunt continued. "She wants me
to marry her, which I have told her I cannot and would not do." For the
first time Hunt described how he had built Frania the house in River Oaks.
"She's been threatening to call Mom," he said.

Margaret fought back her anger. She understood what her father was
doing: he obviously hoped that a plea from his pregnant daughter had a
better chance of dissuading Frania. They drove to the hotel. When they
entered the suite, Hunt said to Frania, "This is Margaret." Margaret accepted
Frania's hand with difficulty. For several minutes they made awkward small
talk, chatting about the weather.

Finally, as Margaret remembered the talk decades later, she said, "Frania,
we're chatting about trivia when in fact there is something important we
want to get out in the open. I understand you want to call my mother. Please
don't do that. You would hurt her severely. Mother is an innocent bystander
in all this." Lyda had high blood pressure, Margaret explained. The shock
could kill her.

"I don't want to hurt your mother, Margaret," Frania said.

"Then please don't call her."

"But what am I going to do? The children need to be with their father.
They love him."

Margaret shot a glance at her father, who averted his eyes.

"Frania," Margaret said, "I can only suggest that you continue your life as
you and Daddy have been doing, discreetly. I'm sure that you have been provided for, as well as the children, and that you can count on that for the rest
of your life. But he cannot be married to two people at the same time."

"But he's already married to me," Frania said.

"Now, Fran," Hunt interjected, "we have discussed that and you know that we did not take out a marriage license, that what we did was simply symbolic of a man and a woman coming together emotionally but not legally, as you understood at the time."

Something went out of Frania then. She sagged back into her chair. Margaret felt sorry for her. She walked over, leaned in close, and said, "Please don't call my mother." Then she left.[8] Once again Frania returned to Houston, distraught. Hunt pleaded with her to stay put. But she couldn't. After fourteen years in limbo, she needed closure. Margaret was again at the building site when her father came by. "Frania's in town again," he said. "She's going to move to Dallas." Margaret couldn't believe it. "Why? You just built her a house in Houston!" she said. "She must be crazy! No, she's not crazy. She's shrewd. Obviously her point is to make you feel threatened that you'll be pointed at as a bigamist. That's what's going to happen."

"I've offered Fran a million dollars," Hunt said. "She screamed at me that she wouldn't sell her children. I'm not trying to buy the children. I'm trying to support them. What else can I do?"

The next morning Margaret was at Mount Vernon when she found a note on her windshield. "Please see me before you leave," it read. "Mother." She found Lyda in the morning room, staring out the window. It took a full minute before Lyda could find the words. "I just had a telephone call from a Miss Frania Tye," she said. "She said that she and your father were married in 1925...and have four children."

"They were never married, Mother."

"You know about this?"

"Yes."

"Why didn't you tell me?"

"Why would I?"

Margaret stayed with her mother all day. At lunch Lyda said, "Those poor children." Before Margaret could respond, Lyda said, "Daddy always said that his genes were so outstanding that he wanted to leave a lot of them to the world. I am certain that he does not imagine there is anything the matter with this. He is so naive." She startled Margaret by suggesting she adopt Frania's children.

Three days later Frania called again. Lyda mentioned adoption. Frania refused, but softened. She ended up apologizing, saying that Lyda must be an extraordinary woman, and returned to Houston. "I arrived at the conclusion that Mrs. Hunt was the finest woman I had ever met," Frania said later. "I decided I would do everything to leave the family alone regardless of what [Hunt] would do."

Afterward, Lyda told Margaret they must never speak of Frania Tye again. "I don't forgive him like you do," Margaret replied.

"You must," Lyda said. "Do not dwell on something about which you can do nothing. It does not help anything. Or anyone. It makes it worse. Let us have our say now, but then after today never mention this again." It was, in fact, the last time Margaret ever heard her mother speak of Frania Tye. She felt certain her parents never spoke of it.[9]

After her talks with Lyda, Frania gave up attempting to reconcile with Hunt. On his urging, she moved once again, this time to Los Angeles, where she took rooms at the Santa Monica Club and enrolled the children in private schools. Friends urged her to sue Hunt for bigamy. Frania refused, saying it would irreparably harm the children. In December 1941 Hunt had trusts created for their children, mostly oil and gas leases, that initially threw off six thousand dollars in cash each year. The epistle led to a series of discussions between Frania, her attorneys, and Hunt at the Adolphus, aimed at a final resolution. The talks ended in a sixty-two-page agreement, signed on January 24, 1942, in which Hunt agreed to pay Frania three hundred thousand dollars plus two thousand dollars a month for the rest of her life. In return, she signed a statement swearing they had never legally married.

Twelve days later Frania married a Hunt Oil employee named John W. Lee. Lee, whose principal duties involved handicapping horse races for his boss, was tall, handsome, malleable, and willing to marry a beautiful, rich woman. After the war, they settled outside Atlanta, comfortably outside the Hunt family's orbit. For the time being.[10]

VI.

The story of H. L. Hunt, bigamist, should end there. But it doesn't. By the time he packed off Frania to her new life in Georgia, Hunt had already taken up with a new woman, a petite, twenty-five-year-old Hunt Oil secretary named Ruth Ray. Blessed with deep green eyes and a pneumatic figure that men noticed—"a real baby-doll type," one acquaintance remembered—Ray was a Depression refugee from dust-bowl Oklahoma, a religious girl who kept a can on her desk for tithes.[11] After dropping out of college she found work as a legal secretary in Shreveport, and when her boss joined the firm's largest client, Hunt Oil, she went along. As he told the story years later, Hunt noticed her one day outside the offices, waiting for a bus. He offered to take her for a drive in the country. Hunt being Hunt, she was soon pregnant.

This time he took no chances. Without any explanation to her co-workers, who guessed the truth, Ruth vanished from Shreveport. Hunt spirited her to an apartment in New York City, where Ruth mailed out wedding announcements saying she was marrying an army officer named Raymond Wright. The women at Hunt Oil weren't fooled. They wagered that the mysterious Mr. Wright would soon go missing in some far-off combat action.[12] In April 1943 Ruth gave birth to Hunt's twelfth child, a son they named Ray Lee Wright. Hunt was smitten with Ruth and their little boy, and couldn't bear to be away from them. For this, his third family, he decided it was worth the risk to keep them close.

He must have been emboldened by Lyda's reaction to Frania Tye, because the small bungalow he bought for Ruth was tucked away on the far side of White Rock Lake, a ten-minute walk from Mount Vernon. Unlike Frania, Ruth knew everything, including her place. She was perfectly happy to be H. L. Hunt's kept woman, living quietly and, as the years wore on, giving him child after child after child.

SEVEN

Birth of the Ultraconservatives

Virtually every Radical Right movement of the postwar era
has been propped up by Texas oil millionaires.
—THE NATION, 1962

I.

O ne of the most important, and most overlooked, legacies of Texas Oil has been its contribution to the growth of right-wing policies and politicians, especially in their most radical guises. In the decades after the East Texas strike, the state's oil millionaires would channel tens of millions of dollars into new conservative causes, bankrolling everything from mainstream Republican thinktanks to Senator Joseph McCarthy's red-baiting campaigns of the 1950s to extremist groups that openly espoused racism and anti-Semitism; later, oil money helped bankroll the birth of the religious right. In a very real sense, the influence of Texas conservatives in America today—in fact, the entire "Texanization" of right-wing politics that brought figures such as George W. Bush and Tom DeLay to national prominence—can be traced to forces set into motion by restive Texas oilmen during the 1930s.

Modern Texas conservatism sprang from the intersection of two disparate events: Franklin Roosevelt's New Deal and the Depression-era oil discoveries, especially those in East Texas. The New Deal outraged many Texas oilmen and East Texas gave them the money to fight it. Each oilman had his own pet peeve, but in general conservative fury was fueled by a fear of what is known today as "big government," the New Deal's introduction of a modern welfare state and deep-seated southern racism. "Nowhere in the United States, not even on Wall Street or the Republican epicenters in Michigan and Pennsylvania, did I find such a perfervid hatred for Mr. Roosevelt as in Texas," the author John Gunther wrote after touring the state in 1944. "[There] I met men who had been unfalteringly convinced that if FDR won again, 'it would mean that the Mexicans and niggers will take us over.'"

Politically, if not always socially, the Big Four oilmen moved easily into the society of oligarchs who controlled the state. Before oil the greatest Texas fortunes were made in ranching and East Texas lumber, where success depended on exploiting the labor of blacks, Latinos, and poor whites—the same formula necessary to succeed in the state's other industries such as sulfur mining and farming. The men who ran Texas oversaw a hierarchical, plantation-style culture, ruled by a southern aristocracy dedicated to harvesting the earth while keeping its workers subservient and poorly educated.

Even before making their fortunes, the Big Four enjoyed the trappings of Deep South culture—none more so than Roy Cullen. Though raised in San Antonio, Cullen identified deeply with his South Carolina–born mother. As a boy he dreamed of living in a "big white house...a great, spreading mansion with white porticoes and columns along the front," and his home in River Oaks was just that, complete with Negro servants. Clint Murchison and Sid Richardson hailed from East Texas, the westernmost bastion of the Old South, and Murchison was what Cullen only wished he had been: a child of southern privilege.

Texas oilmen shared a deep loathing of taxes, labor organizers, and anyone who looked to change their ways. Roosevelt was the first president since Reconstruction to try, at least indirectly. By the mid-1930s taxes were rising. Homeowners received protection against foreclosures, which angered real estate and banking interests. New labor standards and the growth of unions drove up wages, and thus the cost of doing business. Poor families received jobs from the federal Works Progress Administration, sucking the power from political bosses. Farm programs helped millions of families but upset the fragile relationship with landlords. Scores of new federal programs trampled territory long reserved for counties, cities, and states. Worse, the Roosevelts made public shows of helping blacks and other minorities, which didn't sit well with southerners who could still be surprisingly candid in their support of white supremacy. Everywhere Texas businessmen looked, it seemed, the federal government was poking its nose into their affairs. For many, there was no distinction between socialism and the New Deal—the "Jew Deal," as Texas racists termed it. Roy Cullen, for one, termed it "creeping socialism."

Like most southern states at the time, Texas had only one fully functioning political party, the Democrats, and the New Deal provoked a schism in its ranks whose repercussions are felt to this day. On one side were what the political theorist Michael Lind has called "modernists," ardent New Deal supporters represented in Washington by Sam Rayburn and his protégé, Lyndon Johnson, elected in 1937. On the other side were "traditionalists" riding the new tide of oil money, men who by the 1960s would inherit a new name: ultraconservatives.

Ironically, the man who can be considered the grandfather of ultra-conservative Texas oilmen, the man who triggered their entrance onto the nation's political stage in 1935, was no longer wealthy. In fact, he was bankrupt. John Henry Kirby, one of Roy Cullen's early backers, personified the transformation of Texas fortunes from older businesses—in his case lumber—to oil. The first of the great East Texas lumber barons, Kirby had also been the state's first industrial millionaire, compiling a fortune in backcountry sawmills long before finding oil. He is remembered today for Houston's Kirby Drive, for naming scores of East Texas towns that began as lumber camps—Kirbyville for one, and Bessmae, for his daughter—and for cofounding the Houston Natural Gas Company in 1925, a corporation that achieved notoriety decades later under a new name: Enron. In the years before World War I, John Henry Kirby all but *owned* East Texas.

Born near Tyler in 1860, Kirby had been a restless country lawyer in the 1880s when he defended a group of eastern lumber companies in a lawsuit. Intrigued, Kirby put together a group of Boston and New York investors and spent the next twenty years buying timberlands. In 1901 he merged these interests and took control, creating the giant Kirby Lumber Company—at one point Kirby controlled more pine acreage than any other man in the world—and the Houston Oil Company, which held Kirby Lumber's oil rights; a hundred years later, *Texas Monthly* termed it the most spectacular Texas business deal of the twentieth century.

By the 1920s Kirby had emerged as Texas's leading businessman, president of the National Manufacturers Association, a frequent appointee to presidential commissions, and an adviser to Warren Harding, Calvin Coolidge, and Herbert Hoover. He maintained suites at the Waldorf-Astoria in New York, a mansion called Dixie Pines at Saranac Lake, New York, and in 1928 built one of Houston's finest homes, a three-story brick mansion west of downtown. As glorious as his career had been, Kirby's empire crumbled during the Depression, leading to a bankruptcy filing in May 1933; Kirby retained only the ceremonial chairmanship of his companies and a minimal salary. His ruin left Kirby, by then seventy-three years old, deeply embittered, and much of his animus was directed at the Roosevelts. For years Kirby had served as president of an antitax group called the Southern Tariff Congress, and in the early 1930s he hired its most effective publicist, a fast-talking rooster named Vance Muse, to establish a series of anti–New Deal lobbying organizations.

By 1935, thanks to Muse, the Kirby Building in downtown Houston was home to a warren of shadowy, interconnected ultraconservative groups, all devoted to promoting white supremacy, fighting labor unions and communism, and, above all, defeating Roosevelt's reelection in 1936. The Kirby groups were little more than the Ku Klux Klan in pinstripes, a kind of corporate Klan: the Texas Tax Relief Committee, the Texas Election Managers Association, the Sentinels of the Republic, and the Order of American Patriots. In August 1935 Kirby and Muse unveiled their most ambitious group yet, the Southern Committee to Uphold the Constitution, or SCUC, which was immediately viewed as what it was, a southern counterpart to the northern Liberty League, a group of reactionary anti–New Deal millionaires funded in large part by the Du Pont family. In press conferences at Austin, Houston, and Washington, Kirby, with Muse at his back, announced plans for the SCUC to mount a challenge to Roosevelt's reelection the following year. "We plan to dictate the nomination," Kirby proclaimed. Democratic leaders snickered. The party's national chairman, James A. Farley, told reporters Kirby could hold SCUC meetings "in a phone booth."

In fact, though bankrupt, Kirby maintained influence in the oil and broader business communities, and many of his friends not only shared his ultraconservative views but were prepared to take action to defeat Roosevelt. Each had a cause. One of Kirby's most active allies was Maco Stewart of Galveston, an attorney who, after making a fortune in real estate titles, had seen his wealth mushroom when Humble found oil on land he owned south of Houston. Years before it became fashionable, Stewart's interest was fighting communism, especially in U.S. churches, a topic he began researching after witnessing a Socialist rally in New York's Union Square. Convinced that communism's goal was the "utter destruction" of America, Stewart "attended [radical] meetings, watched their parades, hired men to mingle with them, and trained investigators to get the 'low down' on all subversive activities," according to a privately published biographical pamphlet.

One of the first Texas oilmen to promote ultraconservative causes, Stewart in 1931 formed a group called America First to publicize his fears of the Red Menace, mostly in letters to newspapers and talks he gave around the state. The same year, much as Kirby hired Vance Muse, Stewart retained an oil field character named Lewis Valentine Ulrey to coordinate his personal anti-Communist drive. A onetime Indiana state legislator, Ulrey was a self-taught geologist who had wandered Louisana, Texas, and Mexico for two decades before, by his own telling, suffering a nervous breakdown and "washing up" penniless on a Galveston beach. Ulrey believed that Russian communism was part of an international Jewish conspiracy that had infiltrated the highest levels of American churches, universities, and the Roosevelt administration. He acted as a political tutor for Maco Stewart and his son, and by 1935 formed an intellectual alliance with Vance Muse.

As radical as these men were, the most extreme of Kirby's circle was George W. Armstrong, a Fort Worth oilman who owned Texas Steel, which made oil field supplies as well as concrete supports for Texas highways. A rabid racist and anti-Semite, Armstrong had been a top organizer for the Ku Klux Klan in the 1920s and during the 1930s and 1940s emerged as one of the country's leading purveyors of anti-Semitic hate literature—a fact that would lead to multiple investigations by the FBI and the Anti-Defamation League. An unsuccessful candidate for Texas governor in 1932, Armstrong secretly churned out a new book or pamphlet every year or two and hired a young man who handed them out in hotel lobbies and bus stations. His 1938 *Reign of the Elders* was a straightforward endorsement of the notorious anti-Semitic hoax, *The Protocols of the Elders of Zion;* in it, Armstrong characterized the "Jew Deal" as evidence that Roosevelt was controlled by an international Jewish conspiracy headed by the Rothschilds.

Though careful to moderate his public statements about Jews and blacks, Kirby was less guarded in private. In a letter to Armstrong, he characterized *Reign of the Elders* as "the most gripping comment on current events that I have read from any source. It is the greatest contribution to the current political literature of America that has been made.... This book has exalted my admiration for your patriotism and for the wholesomeness of your political philosophy."*

Armstrong joined the band of oilmen and southern businessmen who gathered beneath Kirby's anti-Roosevelt banner. In fact, though no historian has portrayed it this way, the SCUC was almost purely a creature of nouveau riche Texas oilmen. Its letterhead constituted a Who's Who of Lone Star oilmen, including Roy Cullen, Big Jim West, George Strake, and Clint Murchison— at least some of whom, such as Strake and Murchison, were not explicitly reactionary but in all likelihood joined Kirby's group as a personal favor. None, however, fully understood the vital role money would play in politics, and their contributions to Kirby's SCUC were meager. In fact, Kirby gathered almost all his funding, about ninety thousand dollars, from northern businessmen and angry Liberty Leaguers, including members of the Du Pont family. By the end of 1935, with the presidential election eleven months away and Roosevelt's approval ratings down sharply, Kirby's confidence was growing.

"We have fine prospects for rescuing the Democracy from the hands of the Socialists who are now in charge of the Party machinery," he wrote George Armstrong on January 2, 1936. "At the present writing it appears that we will be able to defeat the renomination of Mr. Roosevelt." Still, Kirby noted, "the New Dealers are well organized. They are supplied with

---

*The bond between Kirby and Armstrong was strong. When Armstrong went bankrupt in 1923, it was Kirby who stepped in with nine hundred thousand dollars to buy his various companies and return them to Armstrong's supervision. After Kirby's bankruptcy, Armstrong repaid the favor by buying Kirby's East Texas ranch and returning it to Kirby. Late in life, Armstrong would characterize Kirby as "the greatest man I have ever known."

enormous cash from the public treasury.... It will take wisdom, courage and great activity to dislodge them."

In forming the SCUC, Kirby had hoped to forge an anti-Roosevelt alliance behind a presidential ticket of Louisiana's fiery governor, Huey Long, and the fifty-one-year-old governor of Georgia, Eugene Talmadge. Talks between Kirby, Long, and Talmadge had begun that summer, but abruptly ended with Long's assassination in September 1935.[1] With Long dead, Kirby placed all his chips on the wary Talmadge. The Georgia governor, while making clear to reporters he wasn't yet a formal candidate, accepted Kirby's invitation to deliver the keynote address at the SCUC's first convention, held on January 29, 1936, at the Dempsey Hotel in Macon, Georgia. Kirby downplayed speculation of a presidential nomination, saying the SCUC was a movement, not a party.

On the convention's first day, the three thousand or so "delegates" barely filled half the auditorium, but what they lacked in size they made up in enthusiasm. As a band played "Dixie" and "Put on Your Old Gray Bonnet," farmers in dusty overalls and wives in faded calico dresses stomped their feet, clapped, and hollered encouragement to the speakers. "Give 'em hell!" they yelled. "Pour it on!" The tone of the gathering, which drew front-page coverage in the New York Times, was blatantly racist. On each seat attendees found a magazine called Georgia Women's World—actually produced by Vance Muse in Houston—that featured a photo of Eleanor Roosevelt being escorted by black ROTC officers at Howard University. The lead editorial assailed FDR's recent Jackson Day speech. "Andrew Jackson," it read, "didn't appoint a Negro Assistant Attorney General...a Negro confidential clerk in the White House...and when Andrew Jackson got to be President he didn't put in Republicans, Socialists, Communists and Negroes to tell him how to run these good old United States."

Standing beneath a Confederate battle flag, introductory speakers denounced Roosevelt as a "nigger-loving Communist," New Dealers as "social vermin," and the NAACP as "the worst communist organization in the United States"; one termed a federal anti-lynching bill an "infamous tyranny" and "total outrage." Kirby introduced Governor Talmadge, and in remarks carried live on the CBS radio network, termed him "a plumed knight on an errand for the Republic, refusing to bend his knee to dictatorship or barter the sovereign rights of a great people for Federal Gold." Talmadge did not disappoint, calling on southerners to launch a holy war to drive the "Communists" out of Washington. "Give it to 'em, Gene!" a man in the balcony bellowed.

In the wake of the convention, Kirby took out full-page ads in the New York newspapers to publicize his initiative. But the SCUC's effort to defeat Roosevelt was stillborn. The killing blow came when a liberal Alabama senator, Hugo Black, took offense at the photos of Mrs. Roosevelt—the "nigger photos," they came to be called—and summoned Kirby and Vance Muse before a Senate committee in April 1936. Kirby tersely answered questions about the groups headquartered at the Kirby Building, but it was Muse who made an impression, flippantly batting away questions about "the nigger photos."

"Can you describe the pictures?" Senator Black asked.

"Yes," Muse sighed, "but it is nauseating for me to do it.... I am a Southerner and I am for white supremacy.... It was a picture of Mrs. Roosevelt going to some nigger meeting with two escorts, niggers on each arm."

"You circulated them without anybody forcing you to circulate them?"

"No, sir," Muse replied, "except my conscience...and my granddaddy, who wore this kind of uniform right here." He pounded his chest. Asked what he meant by uniform, he said, "Why, my suit of Confederate gray."

For all his wisecracks, it was Muse who sank the SCUC ship, admitting that the southern committee had in fact been funded by northern industrialists. It was the end of the Southern Committee to Uphold the Constitution. But Kirby did not give up. By June he had created another group, the Jeffersonian Democrats, that managed to hold an anti-Roosevelt "convention" in Detroit in August. A Texas oilman or two showed up—Big Jim West joined, as did Maco Stewart's wild-eyed assistant, Lewis Ulrey—but the meeting was a farce, a motley bunch of fifty or so onetime governors and congressmen whose speeches drew derisive coverage. "Unhappy Has-Beens," *Time* dubbed them.

Thus ended Texas Oil's first foray into presidential politics. Roosevelt's victory in November 1936 disheartened Kirby, who afterward withdrew from political life and spent his last years—he died in 1940—puttering around his East Texas farm. But Kirby left behind the foundation of a movement. At its core were nouveau riche oilmen: Maco Stewart and his son Maco Jr., Big Jim

West, Marrs McLean of Beaumont, and others eager for a political fight. West bought newspapers and a radio station in Austin and Dallas, and was transforming them into ultraconservative organs when he, too, died in 1940.*

Though laughingstocks in the national arena, Kirby and his successors proved juggernauts within Texas. In his definitive study of Texas conservatism, *The Establishment in Texas Politics,* George Norris Green pinpoints 1938 as the year oil-backed ultraconservatives took control of the state's political structure. That year two outspoken pro-Roosevelt congressmen, including the fiery progressive Maury Maverick of San Antonio, were defeated in elections. Of far greater import was the seizure of the governor's office by an oil-and-business-backed flour salesman named W. Lee "Pappy" O'Daniel. Already famous thanks to his own radio show, *The Hillbilly Flour Hour,* O'Daniel was a clown and proud of it; his campaign stops featured a Hillbilly band playing his own homespun songs, such as "The Boy Who Never Gets Too Big to Comb His Mother's Hair." While his opponents promised Social Security benefits and industrialization, O'Daniel won with a simpler platform: the Ten Commandments.

Pappy O'Daniel's victory initiated two decades of ultraconservative rule in Texas. As governor, O'Daniel became Texas Oil's reliable partner, freezing wellhead taxes and backing oil industry lobbyists' takeover of the Railroad Commission. His administration was dominated by ultraconservatives, many of them oilmen, including his key financial backer, Maco Stewart, whose anti-Semitic adviser Lewis Ulrey corresponded with O'Daniel, and Jim West, who was nominated for highway commissioner before moderates in the legislature blocked the appointment. Oil's influence, as West's defeat showed, was not unlimited. Texas voters had a long, strong progressive streak, and many state legislators, especially those from districts with few oil wells, were notoriously independent and frequently blocked conservative initiatives. During the 1940s and 1950s centrists and even liberals continued to be elected to Congress, but those who stayed in Washington long did so only by tending to the interests of Texas Oil. The most visible example was Sam Rayburn, the Speaker of the

---

*After their father's death, West's two sons eventually sold the Austin radio station to a freshly minted Austin-area congressman named Lyndon Johnson, for whom it became the basis of a substantial personal fortune.

House, who in 1944 retained his seat only after a rare contested primary in his North Texas district; his opponent was backed by anti-Roosevelt oilmen, including Roy Cullen, who channeled ten thousand dollars toward Rayburn's defeat. The rest of his career Rayburn held his nose as he backed the oilmen's initiatives. "All they do," he once complained, "is hate."

In the decade following O'Daniel's election, a period that saw O'Daniel replaced by two more conservative, oil-friendly governors, Texas spawned initiatives that for the first time spread Lone Star ultraconservatism beyond the state's borders. One was spearheaded by a shadowy group called Christian American, founded in 1936 by Vance Muse and Lewis Ulrey and funded by Maco Stewart's son, Maco Jr. Its newspaper, the *Christian American*, issued regular broadsides against Negroes, liberals, unions, Communists, and the "International Jewish conspiracy." The organization's influence remained meager until 1941, when Muse decided to concentrate its energies on lobbying against the spread of labor unions; using thinly veiled language that equated labor power with Negro power, and with the help of Pappy O'Daniel's antilabor crusades in Texas, Christian American brochures and legislative lobbying were credited with the passing of union-limiting laws in a dozen southern and southwestern states by 1944. The Muse-Ulrey group, George Norris Green wrote, "did more than any other organization to awaken the South to the dangers of a unionized work force."[2]

Another ultraconservative initiative was led by an ambitious Southeast Texas congressman named Martin Dies, who in 1937 cosponsored formation of the House Un-American Activities Committee. The committee, which achieved lasting notoriety during the early 1950s, was formed after congressional investigations into Nazi and Communist front groups during the 1930s. Neither Dies, in his autobiography, nor any history of the committee explores the influence of Texas oilmen on the decision to form the committee. But John Henry Kirby and Maco Stewart were friends and longtime financial supporters of Dies, who was widely viewed as a tool of business and oil interests in the Beaumont area.* In congressional hearings and his own

---

*Dies's papers indicate he corresponded regularly with Kirby and Stewart; Stewart, in fact, wrote Dies a letter from his deathbed at the Mayo Clinic.

books, Dies spent the 1940s crusading endlessly against Communist influence in American politics, churches, and schools. For the most part, he was ignored. But Dies's activities laid the groundwork for the anti-Communist drives of the 1950s, in which Texas oilmen would play key supporting roles.

After John Henry Kirby's death the man who emerged as the standard-bearer for Texas ultraconservatives, who could legitimately be viewed as Kirby's intellectual successor, was none other than Roy Cullen, who in his mid-fifties began to channel his energy away from oil and toward politics. Cullen backed the entire spectrum of Texas ultraconservative causes, from Christian American's antilabor drives to Martin Dies's anti-communism to just about any outcry against Franklin Roosevelt and the New Deal. Roosevelt had frightened him from the beginning; Cullen's first step into national politics had been a single anti-Roosevelt letter he published in the *Houston Post* in 1932. As the New Deal's tendrils spread during the 1930s, so did Cullen's ire. Roosevelt's 1938 attempt to "pack" the Supreme Court was a turning point for Cullen, who fired off congratulatory telegrams to all 196 congressmen who voted against it. At the same time, in his first political foray outside Texas, Cullen took out a full-page advertisement in the *Louisville Courier-Journal* to excoriate one of its backers, the Kentucky senator Alben Barkley. Meanwhile, though their names were seldom associated in the newspapers, Cullen was quietly one of Pappy O'Daniel's most important financial backers.

At first Cullen only dabbled in politics, donating to favored candidates, firing off an angry letter to a congressman or the Houston newspapers, and funding conservative lectures around Houston. In 1938 he sponsored a Houston speech by Elizabeth Dilling, a nationally known author of anti-Communist and anti-Semitic literature who, after being linked to several pro-Nazi groups, was eventually tried (and acquitted) for sedition.[3] Kirby adored Cullen. Before his death he wired the younger man to suggest he run for office. "The vigor with which you are attacking national conditions is heartening to all of us old-fashioned Americans," Kirby wrote. Cullen demurred. "I can do more good helping other candidates—doing what I can to see that the good men get into office and the bad ones are kept out," he replied. Cullen was equally close with Maco Stewart and others in Kirby's circle; Lewis Ulrey described him as one of Houston's "leading anti-radicals."

Cullen's political awakening, such as it was, paralleled his emergence as a public figure in Houston. It had begun simply enough, with a fund-raising visit from the president of the new University of Houston in 1936; the two-year-old school was more an idea than a reality, its few classrooms in temporary buildings beside a high school. The visit, coming just months after the death of his son, struck a chord in Cullen, who wanted something to mark Roy Gustave's passing and thought Houston could use a university for the children of working men. At a time when none of the Big Four had thought much about philanthropy, Cullen stunned Houston by donating the entire $260,000 necessary to build the University of Houston's first building. He added another $90,000 and a second building soon after, and in time became the university's guiding patron. Among the few ideas he blocked was a proposal to rename the school "Cullen University." But he was thrilled when its ROTC marching team was named the Cullen Rifles.

Cullen's donations were front-page news in Houston. In 1939, seemingly emboldened by the acclaim that followed, he initiated correspondence with presidential candidates looking to unseat Roosevelt, including Wendell Willkie. The two, in fact, engaged in a set-to during a campaign stop Willkie made in Houston. When Cullen told reporters of a letter he had written Willkie denouncing his foreign policy, they asked Willkie about it. Willkie fibbed, saying he had never heard of Cullen. Cullen then released their exchange of letters, proving he did. Willkie thus became the first, but by no means the last, national politician to roll his eyes at the advice he was obliged to take from a man with a fifth-grade education. "You know the Good Lord put all this oil in the ground," Willkie quipped, "then someone comes along who hasn't been a success at anything else, and takes it out of the ground. The minute he does that he considers himself an expert on everything from politics to petticoats."

Undeterred, Cullen embarked on a series of speeches during 1941, mostly at schools his children had attended, during which he denounced the growth of the federal government. "A government of bureaus," he told the South Texas School of Law and Commerce, "leads to national socialism."

His concern about "big government" grew after Pearl Harbor, as Franklin Roosevelt oversaw a series of emergency measures that greatly increased White House control over the economy. During a visit to Washington in 1942, Cullen read of Roosevelt's Labor Day speech to Congress calling for a further increase in executive powers.

Infuriated, Cullen purchased a full-page advertisement in the Washington Times-Herald in which he quoted Senator Robert A. Taft's attack on the speech: "A deliberate effort to discredit and nullify Congress...to induce the American people to accept the rule of a man-on-horseback—a dictator!" Cullen's ad concluded with its own broadside against congressmen who supported the president. "The people of Texas do not want a dictator," it read. "And if you fail to pass proper laws to control a possible dictatorship—while our brave boys are fighting to preserve democracy—then you should resign at once and permit patriotic men to take your place, so that our children may enjoy the blessings of freedom!" Cullen, however, was just getting warmed up.

In 1944 Texas ultraconservatives, whipped to a near-frenzy by Roosevelt's wartime price controls on oil and other commodities, mounted their most serious challenge to Roosevelt to date. They attempted to seize control of the Texas Democratic Party, vowing to withhold the state's electoral votes in the event of Roosevelt's nomination for a fourth term, but were narrowly defeated at the state convention in September. Undeterred, ultraconservatives formed a third party, the Texas Regulars, whose membership was dominated by oilmen, including lobbyists from all the major oil companies active in Texas, as well as independents like Maco Stewart Jr. and Arch Rowan of Fort Worth. In the weeks leading to the November election, the Regulars mounted an elaborately financed anti-Roosevelt campaign of more than thirty statewide radio programs and front-page advertisements in newspapers across the state. The message was broadly antilabor, antigovernment, and openly white supremacist; one of the party's planks actually called for "restoration of the supremacy of the white race."

The Regulars were trounced that November—Roosevelt won the state handily—in large part because they refused to support the Republican candidate, Thomas Dewey. But the party's greatest failing, and the enduring metaphor of its aims, was its inability to field a candidate of its own; the men of Texas Oil, it appeared, had little to offer the American people beyond hatred. Their defeat was no surprise to the man who a decade later would admit to being the Regulars' largest single financial backer, Roy Cullen. While privately admitting they had no chance to stop Roosevelt, Cullen had hoped the Regulars could draw attention to the outrage of southern conservatives; he bought many of the group's advertisements and gave a speech or two on its behalf. For his trouble he received his first death threats. When a reporter asked whether he intended to keep fighting, Cullen snapped, "Just say we're starting now to work on the 1948 campaign!"

II.

Not all Texas oilmen were conservative; sometimes it just seemed that way. A handful, such as J. R. Parten of Madisonville, who took posts in the Roosevelt administration during World War II, could actually be called liberal. But the second major camp of politically minded oilmen channeled money and services to Washington not for ideological reasons, but for practical reasons, for access. By far the most successful of these was Sid Richardson, who gained entry to the White House at a time his peers were still dipping their toes in the political pool, a position he managed to sustain, to his great reward, through twenty-five years and three presidencies. In the process, Richardson helped transform the role of money—in his case, paper bags stuffed with cash—in the American political process.

Richardson's political involvement began when he was still a poor man, on a Saturday afternoon, March 11, 1933. Hobnobbing with friends and ogling cattle at Fort Worth's Fat Live Stock Show, he was introduced to a visiting VIP, twenty-three-year-old Elliott Roosevelt, Franklin and Eleanor Roosevelt's second son. A handsome young man, intellectually shallow but determined to escape his father's shadow, Elliott was the closest thing the Roosevelts had to a black sheep. One week earlier, on Saturday, March 4, he had attended his father's inaugural in Washington. Then, just four days later, he suddenly disappeared from the White House, abandoning his wife and young son.

According to her biographers, Eleanor Roosevelt was distraught, the more so because she had no idea where her son had gone. The answer came late the next day, when the Associated Press located Elliott in Little Rock, Arkansas, where he announced he and a friend had embarked on a cross-country driving trip, intending to visit Texas before heading on to Los Angeles. When the reporter asked if he might buy a ranch while in Texas, Roosevelt replied, "It takes money to buy a ranch. I haven't that money. I'm looking for a job."

In Texas he would find both. From Little Rock Roosevelt headed into Dallas, then to Fort Worth, where he attended the lifestock show. There, Roosevelt wrote later, he was introduced to local oilmen:

> I met Charles Roesser [sic], whose wells were earning him some money, and Sid Richardson, who had none, since the holes he was drilling seemed fated to be dry. Both men, along with another, Clint Murchison, whom I met later, were to show a certain interest in my career while Father was in the White House. I was vaguely aware that I was being sized up as a prospect. A real courtship was due to follow.

Though he later authored three books on his family, Elliott Roosevelt never disclosed precisely what the courtship by Richardson, Murchison, and Roeser—"my three oilmen," he called them—entailed. For the moment, he was preoccupied with a genuine courtship, that of a Fort Worth debutante named Ruth Goggins. Four months after his visit to Fort Worth, in July 1934, after his wife filed for and completed a Nevada divorce, Elliott and Ruth Goggins married. In early 1935, after Elliott took an executive job with Southwest Broadcasting, the couple purchased a 250-acre ranch seven miles southwest of Fort Worth. Later Elliott joined the Hearst chain of radio stations. It was clear from the outset, however, that he burned to work for himself.

The Roosevelts became high-profile members of Fort Worth society. Sid Richardson became one of Elliott's closest friends, a bond that strengthened after Richardson struck the Keystone Field in 1935. A turning point in their relationship, and in Richardson's political career, came in May 1937, when the president announced he was coming to Texas to visit Elliott and do some tarpon fishing in the Gulf of Mexico. Roosevelt's trip was historically significant because it marked the president's introduction to the young Lyndon Johnson. Historians, however, have overlooked a forgotten controversy involving Roosevelt's introduction to Texas Oil.

The president cruised to Galveston on the presidential yacht *Potomac*, where Elliott, with Sid Richardson and Clint Murchison in tow, met him for several days of fishing. On Wednesday, May 5, the *Potomac* anchored off Matagorda Island; the president dined ashore at Murchison's compound. Roosevelt's fishing was interrupted the next day by a disaster: the destruction of the German zeppelin, *Hindenburg*, in New Jersey. Friday the fishing resumed. At lunchtime the *Potomac* anchored off St. Joe's—its buildings still under construction by Perry Bass's crews—and everyone prepared to disembark.

"We went from the *Potomac* to the island by a small boat," one of the guides, Barney Farley, recalled in his 2002 memoir, *Fishing Yesterday's Gulf Coast*, "but when we got there, we were faced with the problem of how to get the president, in his wheelchair, from the boat to the island. Sid didn't have any way to unload him. He didn't have a dock for that. But he did have a cattle chute. So we pulled alongside the cattle chute and Sid explained to the president that he was going to roll him down. Mr. Roosevelt exploded, "What in the world—Sid, do you mean you're going to roll me down that bull chute?" Sid replied in a good-natured Texas drawl, "Why, Mr. President, you're the biggest bull that ever went down that chute!"[4]

On Saturday Roosevelt again lunched at Murchison's spread on Matagorda, then headed into Galveston, where he met Lyndon Johnson and boarded a train to Fort Worth. That night Richardson and Murchison, along with the crème of Fort Worth society, attended a barbecue in Roosevelt's honor at Elliott's ranch. On the surface it was a typical presidential trip, a chance for Roosevelt to visit his son and rub elbows with the new Texas millionaires. Not until eight years later, when Elliott Roosevelt underwent a five-day grilling by Internal Revenue Service attorneys examining his taxes, was it hinted that something more was afoot. The problem was Murchison.

Seven months earlier, in November 1936, a federal grand jury had indicted one of Murchison's largest subsidiaries, the pipeline that took East Texas crude down to the ports around Houston, for running hot oil. Murchison himself escaped charges; neither of his biographers even mention the episode. What Elliott Roosevelt's IRS interrogators suggested, and this was the story that leaked to columnist Westbrook Pegler in November 1945, was that a secret deal had been struck between Roosevelt, Richardson, and Murchison, apparently, the IRS believed, in a late-night meeting following the barbecue. The following week Murchison's subsidiary suddenly changed its plea from not guilty to no contest and paid a nominal fine of $17,500. The mainstream press didn't make the connection, but oil-industry periodicals did. "Roosevelt's Hosts Company in Texas Fined for Hot Oil Law Violations," read the *National Petroleum News* headline. Eight years later an IRS attorney pressed Elliott during a deposition whether he had interceded with the White House to go easy on not only Murchison but on Richardson and Charles Roeser, who the IRS suggested had been under investigation for hot oil violations.

IRS Attorney: There were rumors, and I don't know whether they are founded on fact, to the effect that Richardson and Roeser were both involved in the Hot Oil conspiracy and that you had interceded with the Federal officials and other officials and a member of the Cabinet in charge of the Government oil lands, in their behalf.

Roosevelt: I don't know anything about these rumors. I never knew that Mr. Richardson and Mr. Roeser were involved and I most emphatically state that I never interceded in their behalf with [anyone].

The inference was that President Roosevelt had persuaded federal prosecutors to go easy on Murchison. If so, the IRS attorneys clearly believed, the favor had been returned by Sid Richardson. Richardson gave two depositions in Elliott's tax case. As he told the IRS, he developed a new appreciation for Elliott in the wake of the president's visit to Texas. "Up to the time I had an opportunity to visit with Elliott on the island," he said, "I thought that he was more or less wild in his ideas, and after I had a chance to visit with him there I changed my opinion and thought the boy had something that I hadn't even had a chance to see in him before, and I thought it was worthwhile."

Two weeks after the president's visit, Richardson lunched with Elliott and listened as he sang the praises of several Texas radio stations he would love to buy—that is, if he had the money. "He told me one of them was a wonderful buy, not trying to sell me on it but merely discussing the situation," Richardson said. "[So] a few days after that I called [Elliott] myself and suggested to him that I loan him the money and to see if he could buy the station."

Which is how Sid Richardson, barely two years out of poverty, found himself bankrolling the president's son in the radio business. In July 1938, a year after President Roosevelt's visit, Richardson loaned Elliott twenty-five thousand dollars, which Elliott used to buy a Fort Worth station, KFJZ. Several months later came two more loans, totalling fifteen thousand dollars, which Elliott used to buy additional radio stations that soon formed his own network, the Texas State Network. Richardson became a major shareholder in Elliott's business.

Whether or not Richardson's dealings with Elliott Roosevelt had anything to do with Murchison's indictment, the relationship gave Richardson immediate entry to the White House. Just two weeks after agreeing to invest in Elliott's radio network, Richardson was invited to chat with the president in Washington. He used the occasion to complain about a proposal, backed by Treasury Secretary Robert Morgenthau, to eliminate the oil industry's all-important 27.5 percent depletion allowance, a tax loophole that allowed oilmen to write off nearly a third of costs associated with dry holes. "Mr. Richardson did ask me on several occasions whether there wasn't something I could do to stop Mr. Morgenthau from trying to ruin every individual oil operator in the country by eliminating the [allowance]," Elliott testified in 1945. "He was always complaining to me during the entire time of our acquaintance about that particular feature."

Though neither man ever described their meetings, Richardson and the president got along well, for Richardson was invited to the White House several more times. Richardson characterized his advice to the president as a counterpoint to the opinions of major oil companies, whom Roosevelt did not trust. Whatever his counsel, the oil depletion allowance stayed put. Richardson didn't get everything he wanted, however. On the only occasion he mentioned Richardson publicly, in his 1975 book A Rendezvous with Destiny, Elliott recounted a 1940 trip Richardson made to the presidential retreat in Warm Springs, Georgia. With war looming, the White House was poised to block oil shipments to Italy, one of Richardson's major customers. During a two-day visit, Richardson refused to stop supplying Italy with oil. It did him no good. Not long after, shipments were indeed blocked. Years later Elliott made clear he felt Richardson had used him. "I became the tool of the oil lobby in Washington," Roosevelt wrote.

Richardson's relationship with Elliott ended during the war, when Elliott took an army assignment in North Africa and his radio network foundered; Richardson's loans were never repaid, and he ended up owning one of Elliott's radio stations the rest of his life.* By all accounts, however, Richardson remained friendly with Elliott's father. In mid-1941, concerned with the nation's oil supply in the event of war, the president asked Richardson and Charles Roeser to examine the situation. Their report was to be delivered during a Thanksgiving dinner at Warm Springs, but at the last moment Roosevelt was called back to Washington.

Two weeks later, on Sunday, December 7, 1941, Richardson, Perry Bass, and Bass's new wife, Nancy Lee, trudged into the house on St. Joseph's Island after a morning of quail hunting to find the servants glued to the radio: Japanese airplanes were bombing Pearl Harbor. St. Joe had no phone at the time, but four days later, on Wednesday, December 11, a boat arrived from Freeport with an urgent message from the White House.† When Richardson managed to return the call, "the president's message," Perry Bass recalled, "in essence was, 'Sid, I want you to have lunch with me next Sunday and tell me what shape we are in for petroleum. I don't trust those major company bastards.'"

The next day Richardson was on a train heading east. One of his favorite sayings was that it was better to be lucky than smart, and when the train stopped in the North Texas town of Denison, Richardson had a chance

---

*According to his IRS testimony, Elliott thought so highly of Richardson that he named him godfather of one of his children. He doesn't say which one. In an interview with the author, Elliott's son, Tony Roosevelt, says he believes the story is true, though he, too, isn't sure which child was involved.

†In a foreword he wrote for a 1993 history of Aransas County, Texas, where St. Joseph's Island is located, Perry Bass dated this incident to the afternoon of December 7, 1941. But in his 1945 testimony to the IRS, Richardson put the call the following Wednesday.

encounter that would shape his life in later years. His attorney, William Kittrell, spied an army general he had met arranging for soldiers to appear at the state fair. His name was Dwight Eisenhower, and he couldn't find a seat on the crowded train. Kittrell invited Eisenhower to sit in Richardson's drawing room. The three men ended up talking most of the way to Washington. "I thought he was a pretty good hand," Richardson told the *Washington Post* in 1954. "The funny thing was, I didn't pay any attention to the name. Later on, Bill [Kittrell] said to me, 'Remember that fellow that shared your drawing room? He's the fellow that's in command over there in Europe.'"

III.

In the presidential election year of 1940, most Washington politicians, much like the rest of the country, hadn't the slightest idea how very rich Texas oilmen had become. Not one of the Big Four had received significant publicity, even in Texas itself, and national publications remained largely ignorant of the state's new riches. Typical of Washington's view was a memo Harold Ickes forwarded to President Roosevelt in August 1935, summarizing what Ickes termed an "interesting document... handed to me by a man in the oil game who thought it ought to reach you." The paper was a list of Texas oilmen who had given that year to the Democratic National Committee.

Of Herman Brown, the Brown & Root cofounder and millionaire oil investor, the memo noted, "Do not know of him." Sid Richardson—this was a month before he hit the Keystone Field—was described as "in debt and borrows money to develop leases from Charles Marsh." Only two names appeared promising: Clint Murchison and Dudley Golding. "Golding & Murchison," the memo concluded, "came into the East Texas Field several months after it started with no money and a little over three years [later] sold out... for about five million dollars."

Rarely in history can one name the single man who identified and singlehandedly developed a source of political power as vast as that afforded by the new Texas fortunes. In this case you can. His name was Lyndon Johnson, and he would ride the wave he discovered in 1940 all the way to the White House. No one, not even his most heralded biographer, Robert Caro, has been able to say where or how Johnson had his epiphany. In fact, it appears it simply arose from the small coterie of businessmen who had backed Johnson in his Texas elections. Chief among these was Sid Richardson's partner Charles Marsh, who introduced Johnson to his neighbor in Austin, Herman Brown.

It began in the fall of 1940, when Johnson, then a second-term Democratic congressman, used Texas Oil's cash to start his march to the apex of American power. His vehicle was the Democratic Congressional Committee, charged with raising money for campaigns nationwide. All that summer Johnson had pestered Sam Rayburn to allow him to lead the committee, but Democratic leaders judged him too young and inexperienced for such a high post. Finally, in September, with Democratic coffers almost empty, Johnson got the job.

Johnson left the White House that morning and hit the phones. His first call went out to his primary backers, George and Herman Brown of Brown & Root. Individual contributions were capped at five thousand dollars, but the Browns had six five-thousand-dollar checks on Johnson's desk by Friday afternoon; that single donation of thirty thousand dollars was more money than the Congressional Committee had received all that year from its principal benefactor, the Democratic National Committee. Charles Marsh prevailed upon Sid Richardson to donate five thousand dollars that same Monday, in the name of Perry Bass; Marsh's publishing partner chipped in another five thousand dollars. Richardson, in turn, reached out to Clint Murchison, who sent five thousand dollars on Wednesday. By Saturday, October 19, five days after his first phone call, Johnson was able to walk into Democratic National Headquarters with a check for a jawdropping forty-five thousand dollars. "We have sent them more money in the last three days," he wrote an acquaintance a week later, "than Congressmen have received from any committee in the last eight years."

You could almost see the lightbulb going on over Johnson's head. As Robert Caro wrote, "A new source of political money, potentially vast, had been tapped in America, and Lyndon Johnson had been put in charge of it." It would take years for the relationship between LBJ and Texas Oil to develop, but when it did, all of America would notice.

EIGHT

War and Peace

I.

On Sunday morning, December 7, 1941, H. L. Hunt, Lyda, and most of their family were in Denton County, north of Dallas, for a well test. Hassie was scanning the car radio for a weather report when he heard the news flash. It was war, a horrible thing, and though few said it at the time, the Big Four sat atop the liquid assets that would win it.

The families did their part. Clint Murchison's son John signed up the very next day. He ended up flying P-38s in Burma and China. His brother Clint Jr. entered officer-training school. Hassie Hunt ignored his father's pleas and signed up, too; he was sent to Washington as an oil "adviser" to the Chinese government. Perry Bass enlisted in the navy and began building shipyards in Florida. In Houston, Roy Cullen chaired bond drives. A few years earlier Cullen had reluctantly assumed chairmanship of the Houston Symphony Orchestra, which took to serenading him with one of his favorite songs, "Old Black Joe." For one bond drive, Cullen arranged a fund-raiser bizarre even by Texas standards, a professional wrestling match staged during a symphony concert, body slams and grunts to the gentle strains of Mozart.

The Big Four's real contribution to the effort, though, was not their energy but their oil. A single destroyer burned an average of three thousand gallons of oil every hour. One tank required ten thousand gallons of gasoline to drive one hundred miles. A single four-engine bomber used up to four hundred gallons of high-octane jet fuel per hour. The utility, and to some extent the size, of an army was directly linked to its oil supply. Even before

Pearl Harbor, that made Texas oil a critical natural resource. Roy Cullen's Tom O'Connor Field produced crude that made ideal jet fuel; when war came, the field was cordoned off and all approaches patrolled by military police.

The day the Japanese attacked Hawaii, 90 percent of Texas oil reached the East Coast in tanker ships sailing from the ports around Houston; the rest was shipped in rail cars. It was oil, much of it from East Texas, that heated Boston, New York, and Philadelphia and kept their automobiles running. That made the sea-lanes from Texas around Florida the nation's most critical transportation route. Even before fighting began, the White House worried about protecting them, and rightly so. World War II arrived on the continent just a month after Pearl Harbor, when a German U-boat sank the tanker *Norness* off Long Island on the night of January 14, 1942. In the next five months Nazi subs sank 171 ships between Florida and New York, another 62 in the Gulf of Mexico. Texas oil was still arriving on the East Coast; unfortunately, it was washing up on its beaches. Fuel deliveries to the East Coast fell by 90 percent. Consumer rationing was instituted. Gas shortages struck Washington, New York, and Boston. At the White House, Harold Ickes's people estimated it would take a train of seven thousand rail cars—fifty miles long—to replace the fuel that was being lost every day.

The obvious solution was a pipeline; as one Texas oilman quipped, "you can't sink a pipeline." Unfortunately, no one had ever attempted a pipeline as long as would be needed: 1,254 miles, the distance between Longview, in the heart of East Texas, to the nearest eastern refineries outside Philadelphia. Ickes had urged construction of a pipeline as early as December 1940. Now it would be built—and fast. There would actually be two, one for oil, dubbed the Big Inch, and a second smaller pipe for oil products such as kerosene, the Little Inch. Their size dwarfed anything in existence. Most American pipelines were eight inches in diameter or smaller; the Big Inch alone would be twenty-four inches around, the Little Inch twenty inches; together they would carry more than five times as much oil as any pipeline in the world. The job of getting them built fell to the liberal Texas oilman J. R. Parten, who had gone to work for the government petroleum bureau.

The Inch pipelines, however, did not enjoy unanimous support. A number of competing proposals sprang up that spring. As a stopgap measure, the government studied ways to get oil across Florida, which would at least avoid the shipping lanes south of Miami. A Florida group pushed for an eighty-million-dollar canal; tanks of oil could be floated across in barges. Parten thought of something cheaper. He remembered one of Clint Murchison's East Texas pipelines that had been used to funnel hot oil during the Depression but had fallen into disuse. Parten telephoned Murchison's office and asked whether the pipe could be dug up and reassembled across northern Florida. Murchison's men were at work in no time. Their new pipeline would ship its first oil in June 1943.

The issue of a Texas-to-East-Coast pipeline, however, remained up in the air. In June 1942 a House committee recommended building a 580-mile pipeline from the Tinsley oil field in Mississippi to Charleston or Savannah. To Parten's dismay, the bill passed. This new initiative, Parten realized, represented a dangerous alliance between several opponents to the Inch lines, not only politicians from South Carolina and Florida, but a man Parten had first met twenty years earlier when he was running an Arkansas gambling house: H. L. Hunt. It was Hunt and his son Hassie who controlled most of the Tinsley Field, an area Parten knew was already in decline; a pipeline linking it to East Coast ports would allow the Hunts to reach market, but it wouldn't make a dent in demand.

Parten discovered that the Tinsley pipeline was the first stage in what Hunt and his partners envisioned to be an even longer pipeline originating in Wichita Falls. Parten had already vetoed both ideas, even after one of Hunt's partners warned he would face "a great deal of political heat." The matter came to a head at a hearing in late June, where those testifying on behalf of Hunt's pipeline included, of all people, Hassie Hunt, whose job advising the Chinese government consisted mostly of laying around the Mayflower Hotel and chasing girls. The Hunts and their allies had their say, but in the end an army report favoring the Texas-to-Philadelphia route won out.

To build it, Parten hired a Texaco pipeline man named Bert Hull. Hull quickly assembled an army of fifteen thousand grizzled roughnecks. They worked in four groups, one each on eastern and western sections of both pipelines. It was a monumental task. Thousands of miles of American countryside had to be surveyed and cleared; tunnels had to be bored beneath the Mississippi and two hundred other rivers, streams, and lakes; almost three thousand miles of four-foot trenches had to be dug. Hull's crews laid the first pipe on August 3, 1942. Massive trucks delivered forty-foot sections of pipe to the crews around the clock; each section weighed two tons, and each had to be welded, sunk in the ground, and buried. On a good day, nine miles of pipe disappeared into the earth.

That winter, floods washed away entire sections of pipe waiting to be laid at the Mississippi and Arkansas Rivers. At the Mississippi, Hull's men managed to pull tons of pipe from the river mud and have it back in place in just two days. The Arkansas River damage took longer, forcing crews to stow seven miles of pipe in the streets of Little Rock. By January 1943, after barely six months of work, Hull's men had finished the initial 531 miles of the Big Inch. In February, as crews furiously worked to finish the rest, the first shipment of Texas oil flowed through the line to Norris City, Illinois, where more than a thousand railroad cars waited to take it east. Six months later the entire line was completed, and the first oil flowed from East Texas into the Sun Oil refineries at Marcus Hook, Pennsylvania. It was the world's longest pipeline, an extraordinary engineering achievement.

When the war was finally won, American oil was among the heroes. The Allies, it was said, "floated to victory on a sea of oil," and by and large it was Texas Oil, a good deal of it owned by the Big Four families. Between 1941 and 1945 the Axis powers produced an estimated 276 million barrels; in the same time span, Texas produced more than 500 million, 100 million from Hunt's East Texas fields alone. As Axis leaders acknowledged, they couldn't compete with the Allies' supply of aviation fuel and gasoline. "This is a war of engines and octanes," Joseph Stalin said in a toast to Winston Churchill in Moscow. "I drink to the American auto industry and the American oil industry."

If the Inch lines helped win the war for the Allies, they would win the peace for Texas Oil. When the federal government began auctioning off sixteen billion dollars of wartime factories and industrial assets in 1945, the two pipelines became the focus of intense speculation. Everyone had an idea what should be done with them. The most innovative suggested converting them for use transporting a product that for decades oilmen had simply thrown away: natural gas. A methane-rich vapor found in most oil and coal deposits, gas had been identified as early as 1683. Gas lighting had been used in

British factories since the early 1800s, and while some American cities began using gas streetlamps after the Civil War, it had never been widely accepted as an alternative to traditional heating oils. Well into the 1940s American oilmen burned off more natural gas—a process called "flaring"—than the entire country saw fit to use. An oil scout cruising West Texas one night in 1945 wrote that it felt like driving through a city—thousands of gas flares lighting the night sky for miles around. One town, Denver City, out near the New Mexico border, was wreathed with so many flares it didn't use streetlights for years.[1]

But, as happened after World War I, American energy demand soared after World War II, bringing yet another round of gasoline shortages. A new oil-drilling boom was soon under way. But by early 1946, when the Inch lines were put up for auction, there were already widespread calls, in Congress and elsewhere, that the nation's electrical utilities convert at least some of their plants to natural gas. Consumer advocates, realizing that gas could be much cheaper than oil, saw the future of home-heating and appliances. Texas Oil saw the future of its profits. The specter of one of its waste products heating all of Boston and New York and Philadelphia, and having it pumped directly there via an existing pipeline—it was the kind of treasure not seen since Spindletop or the East Texas field. Every major oilman in Texas began looking for ways to get his hands on the Inch lines.

Hunt joined one bidding syndicate, Murchison and Richardson another. Their bids, along with fourteen others, arrived in Washington that summer. The government needed more time to value the pipelines, so a second round of bidding was set for February 1947. In the interim, a new company, Tennessee Gas, was allowed to lease the lines, proving they could safely transport gas. When the final offers were opened, the winner, with a bid of $143 million, was a Houston group led by Lyndon Johnson's main financial backers, George and Herman Brown. They called their company Texas Eastern, and their goal was bringing Texas natural gas to the cities of the Northeast.

Within weeks utilities in Philadelphia, New York, and Boston announced widespread conversions to natural gas. The first Texas gas flowed through the new Texas Eastern–owned Inch lines into Philadelphia's Tilghman Street Gas Plant in September 1948. Even as other southwestern groups announced plans to build competing pipelines, Texas Eastern moved into New York. On

August 17, 1949, the mayor of New York City, William O'Dwyer, stood in a Consolidated Edison plant on the north shore of Staten Island. When a worker nodded his head, the mayor twisted a valve, releasing the first whiffs of Texas natural gas into a storage tank, where it was soon put to use warming apartments on Park Avenue and cooking meat loafs in Sheepshead Bay.

It was the beginning of a new era. Across America thousands of homeowners switched to cleaner, cheaper gas furnaces and appliances. Between 1945 and 1951, gas sales nationwide doubled. Everywhere, new pipelines were laid; one of the largest, the so-called Bigger Inch line built by El Paso Natural Gas, brought Texas gas to the burgeoning suburbs of Southern California, which quickly became the nation's single largest gas market. For Texas oilmen profits were stupendous. The men who bought Texas Eastern saw their stock, bought for $150,000, leap in value to almost $10 million in less than a year. H. L. Hunt's fields alone supplied a full quarter of the gas Texas Eastern brought to northeastern cities. Sid Richardson's Winkler County fields became the foundation of El Paso Natural Gas's supply; Richardson's gas lit up much of Hollywood and the Sunset Strip. For Hunt and Richardson, and for scores of other oilmen, it was found money.

And not a moment too soon.

II.

At mid-century America was the acknowledged king of oil-producing countries, pumping 63 percent of the world supply in 1941, and Texas was its prince, leading all states in production year after year. At the height of war in 1943 came the first hint of a serious challenge to these crowns. It rose in a land of desert nomads most Texans knew only from the tales of Sinbad: Arabia. Harold Ickes heard about it first. As the nation's wartime petroleum czar, Ickes spent the war worrying that America would run out of oil. In his search for more, he had his men study every corner of the globe, and as they did, they read of a smattering of small discoveries in the Persian Gulf, in some kind of sheikhdom called Bahrain in 1932, in something called Kuwait six years later. British crews had actually managed to fill an entire

tanker with Arab oil in 1939, and three years later, as Erwin Rommel's Pan-
zers rumbled into Egypt, Ickes decided it was time to find out how much lay
in Rommel's path. This called for a mission of the utmost secrecy, for which
Ickes selected a most unusual Texas oilman, Everette DeGolyer.

DeGolyer wasn't an actual Texan—he was raised in Oklahoma—and he
wasn't an actual oilman. He was a geophysicist, probably the world's best.
Short and rumpled, with an enormous head and roving intellect—in later
years he purchased the *Saturday Review*—DeGolyer made his name while
still an undergraduate at the University of Oklahoma in 1910, bringing in a
gusher that inaugurated the golden age of Mexican oil. He had gone on to
cofound the Amerada oil company, where he pioneered the use of the seismograph; colleagues kidded DeGolyer that he was "crazy with dynamite."
He had moved to Dallas in 1936 to start his own geophysical consulting
firm, and by the time Ickes loaded him onto a cargo plane for the Persian
Gulf in 1943, he was among the oil industry's most respected voices.

Hopscotching from Miami through the Caribbean, then onto Brazil and
Africa, DeGolyer arrived in the Middle East to find a land bleaker than
anything he had seen in West Texas. "In fact," he wrote his wife midway
through a tour of Saudi Arabia, Kuwait, and Bahrain, "Texas is a garden
compared to some places we have been." But the geology, DeGolyer discovered, was a thing of beauty. The more he studied it, the more excited
he became. The petty sheikhdoms of the Persian Gulf, he realized between
desert meals with menacing Bedouin, sat atop an ocean of oil that dwarfed
anything the world had ever seen. By the time he returned to Washington,
DeGolyer was certain he had glimpsed the future. A conservative man, he
estimated the Persian Gulf's reserves at twenty-five billion barrels—a guess
that further studies drove as high as three hundred billion barrels. "The oil
in this region," a man who accompanied DeGolyer wrote, "is the greatest
single prize in all history."

Great for the world; not so great for Texas. Projections by DeGolyer and
others triggered tense talks between the American and British governments
over just who would control Persian Gulf oil. At a White House meeting in
February 1944, President Roosevelt showed Lord Halifax a map he had drawn
of the Middle East. Persian oil, Roosevelt said, could remain British. The two

countries would share anything found in Kuwait and Iraq. But Saudi Arabia, he continued, would be the exclusive province of major American oil companies. The Anglo-American Petroleum Agreement, signed six months later, essentially codified Roosevelt's plan. As soon as the Nazis and Japanese surrendered, the majors would begin drilling.

Independent oilmen, especially in Texas, howled. Why should the majors control the Middle East? They wanted in, too. When the new agreement came up for a Senate vote in 1945, Texas producers led the drive against it, forcing the White House to withdraw. By the time the treaty was reintroduced, Texans had realized the problem wasn't the *control* of Middle Eastern oil but the *competition*. If the Persian Gulf held half as much oil as Everette DeGolyer thought, it would drive prices down sharply and, if imported, swamp Texas producers. In 1947 protests against the treaty erupted across the state. Even Texas schoolteachers threatened to go on strike if it passed. Again the treaty was withdrawn, but by then it was all but irrelevant.

Once the war ended, there was no stopping the tsunami of Middle Eastern oil. Massive new fields were discovered, one every few months during 1947 and 1948; much of the oil initially went to fuel the rebuilding of wartorn Europe, but for oilmen who saw the future, the tidal wave was growing, and it was heading straight for Texas.

III.

Compared to the decades that bracketed it, the 1940s were comparatively quiet years for the Big Four families, a time of steady corporate and family growth. The war itself was very good for Texas, especially the men of Texas Oil. Dallas and Houston boomed. The demand for jet fuel and all manner of chemicals led to the building of dozens of new refineries and chemical plants along the coast south and east of Houston—"an unbroken ninety-mile line of refineries," in one startled writer's words, "like a single throbbing factory." During the war a windfall-profits tax forced oilmen to plow every last cent of extra profit into the search for new reserves. The demand for oil skyrocketed, and when the fighting ended the postwar thirst for natural gas turned a waste product into Texas Oil's new profit engine.

Sid Richardson made a second fortune in gas, piping out billions of cubic feet of it along with millions of barrels of new oil from the amazing Ellenburger Lime in West Texas. In his new offices on the sixth floor of the Fort Worth National Bank Building, Richardson took the first steps toward forming a corporate organization, dividing his empire into halves, Texas and Louisiana. It was a munificent company. Out at the sprawling "Sid Richardson Oil" work camps in Winkler County, south of New Orleans, and at a new gas field near Eola, Louisiana, the supervisors all received Christmas gifts, while their children could look forward to something from "Uncle Sid" upon graduating high school. Once he returned from the war, Perry Bass kept things running smoothly, allowing Richardson to take long months at St. Joe's, or at one of several resorts in Arizona where he spent the winters, betting on the ponies, playing cards, and berating Bass when the need arose.

Unlike Clint Murchison, Richardson never seriously diversified. One of his few forays outside oil came in 1948, after he was outbid for the Inch lines. An oilman named Frank Andrews suggested he try to buy another soon-to-be-auctioned war asset, a factory in Odessa that was the world's largest producer of carbon black, a sootlike oil by-product used in making tires. Richardson put pressure on Lyndon Johnson to intervene on his behalf and Johnson came through. For the bargain price of $4.3 million, Richardson took control of not only the plant but 447 other buildings on 426 acres of land, employee housing, and fifty miles of adjoining pipeline. Johnson then pressed the White House to have carbon black classified as a "critical material" for national security, driving up its price, and even arranged for the government to stockpile its excess inventories at the Odessa facility. When Richardson cut several deals with his competitors, the Department of Justice opened an antitrust investigation—until one of Johnson's men intervened once more, killing the probe in its cradle.

Once the purchase went through, Frank Andrews suggested that Richardson use the massive plant to leverage a better price for the natural gas he was selling El Paso Natural Gas. Richardson was getting two cents per million cubic feet. Andrews suggested he might be able to pry four cents out of El Paso by threatening to fuel the plant with a competitor's gas. When the negotiation concluded, Richardson telephoned Andrews and, in his teasing way, hollered, "You're a liar, Andrews! I didn't get four cents." Before Andrews could respond, Richardson delivered the punch line. "I got five!" he crowed.

H. L. Hunt spent the war years drilling hundreds of new wells across Texas, Louisiana, and Mississippi, then branched east into Florida before turning west, snapping up leases in the Rockies and Canada. Hunt liked to boast that Hunt Oil had produced more oil during the war than all the Axis powers combined, and while it wasn't true, it was close. He bought dozens of new rigs, eventually sixty-five or more, and worked them around the clock, completing as many as three hundred new wells a year. Oil was harder to find outside Texas, but Hunt's new son-in-law, Al Hill, urged him to hire his first geologists, and under their guidance Hunt Oil piled on massive new reserves. By 1946 the company was bringing in an estimated one million dollars a week in free cash flow. Hunt purchased a new refinery outside Tuscaloosa, Alabama, and began marketing his Parade brand gasoline in five southern states.

Everything was run out of Hunt Oil's new offices in the Mercantile Bank Building in downtown Dallas, where Hunt relocated in 1945. The headquarters staff remained small in those years, rarely more than a few dozen secretaries and bookkeepers. Working out of a corner office behind a large mahogany desk, Hunt was an unremarkable manager for his day, taciturn and impatient, never one to lavish compliments on his people; his only eccentricities were a penchant for singing gospel songs and a habit of taking off his shoes and socks and propping his bare feet on his desktop. Outside the office, Hunt was just another balding Texan in his fifties in a gray suit and fedora; as a man with three separate families now, it was no accident he avoided all publicity, refusing to join local boards or give to Dallas museums or charities. For the most part, Dallas returned the favor. In 1941, when Hunt applied to the exclusive Brook Hollow Country Club, he was turned down. A number of oilmen suffered the same fate. It would take many years, in fact, before Dallas became entirely comfortable with oil money.

Hunt's only hobby in those years, the one pastime that generated an adrenaline rush to rival an oil discovery, was gambling. He frequented underground joints around Dallas, playing craps as well as his beloved poker. He kept a regular game at the Baker Hotel, and for the first time began matching wits with some of the country's best-known players, including the legendary Indiana gambler Ray Ryan. The two engaged in games whose stakes sometimes topped two hundred thousand dollars. Hunt, like both Clint Murchison and Sid Richardson, stopped carrying cash altogether; when he needed money for a game, he dipped into a box he kept in an office closet, or another he kept at First National.

During the mid-1940s the one thing that took Hunt away from his daily routines was the situation with his son Hassie. It was unbearably sad. After he testified against the Inch lines in Congress in mid-1942, Hassie had begun to unravel. Later there would be stories that he grew increasingly paranoid, believing the Roosevelts were attempting to kill him, or perhaps his father's enemies. The turning point came in 1943, when the army transferred him from his government job in Washington to infantry training in Louisiana. There Hassie had a breakdown of some sort, washing up in an army mental ward near New Orleans. Hunt learned the news from one of his son's drilling partners in Mississippi, who said it was sunstroke. Hunt hurried to New Orleans. When he telephoned Margaret his tone was grave. "Hassie does not seem right," Hunt said. "Too many jokes, too much hilarity for a boy weakened by sunstroke."

On his return to Dallas, Hunt took Lyda to talk things over with Margaret. The army was now saying Hassie would be medically discharged. Though no one knew what it was, he clearly had some sort of mental problem. "They are recommending he be sent to the Menninger Clinic to be evaluated," Hunt said. "Is that okay?"

Lyda couldn't speak. She turned to Margaret, who said "Yes."[2]

Hunt took his son to a clinic outside Andover, Massachusetts, where the doctors suggested electric-shock treatment. Hassie refused. According to Harry Hurt, Hunt tried his own treatment. Believing Hassie just needed more "action," he brought in a string of young women for Hassie to sleep with. It had no effect. Hunt, unwilling to shut his eldest son away in a mental ward, decided to take him back to Texas. "Great!" Hassie said when he heard the news. "I'm going back to the oil fields where I belong." Instead,

Hunt installed him at the family's pecan farm outside Tyler, hoping a period of relaxation would help him recover. It didn't. If anything, Hassie's behavior grew worse.

He grew increasingly confused, unable to tell the difference between his father's secretaries and his own sisters. The slightest disagreement could provoke violent tantrums. At times Hassie challenged his father to fistfights. Over breakfast at Mount Vernon he picked up a grapefruit and hurled it at his mother, breaking her glasses. Hunt spent long periods doing little more than telephoning and dictating correspondence to hospitals across the country, searching for a doctor who could cure his son. "It was killing Daddy, making him old," Margaret recalled years later. "Hassie was his favorite, his most adored child. Daddy could not, would not fail him."

When the war ended, Hunt took Hassie to more clinics for treatment, at least one of which diagnosed him with schizophrenia. Doctor after doctor advised that he be institutionalized. Hunt wouldn't hear of it. Hassie consented to electric-shock treatments at one clinic; the treatments, however, had no effect. Finally, at a facility in Hartford, Connecticut, a doctor urged Hunt to consider a new procedure called a prefontal lobotomy. Developed in Europe, the lobotomy involved inserting a scalpel behind the forehead to physically sever the prefontal lobes from the rest of the brain. No one understood why or how it worked, and results were uneven. Some patients calmed. Others became zombies. Tears welled in Hunt's eyes when he told Margaret what had to be done. "It's the only thing there is left," he said.

The operation was performed in Philadelphia in 1946. It transformed Hassie from the energetic young wildcatter his family remembered into a quiet, lethargic man who lost all interest in business. He turned pensive and introspective; about all he enjoyed doing was reading and taking long walks. Hunt had Hassie placed in a New York psychiatric hospital for a time, but on Lyda's urgings eventually brought him home to Dallas. They purchased a house for him adjacent to Mount Vernon, where he lived in seclusion. For the rest of his life Hunt was plagued with doubts over what he had done. During the 1950s, when doctors began to produce drugs to treat schizophrenia, his guilt began to mount. "Daddy was certain he had made a terrible, tragic mistake," Margaret recalled. Margaret, in turn, blamed the family fortune. If they hadn't had the money for the lobotomy, she reasoned, perhaps they could have somehow found another treatment.

No one outside the Hunt family saw much of Hassie after that. For years Hunt kept the office next to his vacant, for his son; he never came to work. Instead, from time to time, friends would spot a thin, ghostly presense, a latter-day Boo Radley, walking the shores of White Rock Lake. The sightings went on for years, then a decade, then another and another. In the end, Hassie Hunt outlived his parents, finally passing away quietly, at the age of eighty-eight, in 2005.

IV.

In the 1940s most Americans still thought of millionaires as stuffy, tuxedoed Monopoly Men in Fifth Avenue mansions. While Hunt and Roy Cullen, sober men who wore suits each morning to a downtown office, conformed to the expectations of how businessmen should look and act, Clint Murchison shattered the mold. Years before Bill Gates made it acceptable for executives to doff their jackets, Murchison became one of the nation's first—maybe *the* first—casual billionaire.

While he embraced ostentation in private life, Murchison abhorred it in business. The downtown Dallas building he moved into after the war, a bland two-story affair he picked up with an insurance company, didn't even announce his presense; its only adornment was a metal plate that read 1201 MAIN. For Murchison, every day was casual Friday. If he was seeing people, he wore khakis and an open-neck shirt; he kept a tie in his office closet, but the only one he wore on a daily basis he used as a belt. If his schedule was free, his taste ran to shorts, a battered straw hat, and sandals; he didn't care for socks, and the ones he wore were often mismatched. His only formality was reserved for his secretary, Ernestine Van Buren. When she was needed, he didn't holler, always gliding to his office door to murmur, "Mrs. Van Buren."

The relaxed atmosphere at 1201 Main belied the sophistication of Murchison's growing basket of investments. "Money is like manure," Murchison liked to say. "You gotta spread it around for things to grow." The diversification he began in the late 1930s picked up steam during the 1940s. Figuring that returning soldiers would trigger a baby boom, he invested in textbooks by buying the New York publisher Henry Holt & Co. in 1945. He tried to buy one of his favorite magazines, *Popular Mechanics*, but couldn't, picking up *Field & Stream* instead. Betting all those new families would take to vacationing, he bought a Seattle steam line and a pair of bus companies. Still, he never lost touch with oil. In 1948 he had Southern Union spin off its exploration arm, Delhi Oil, into a separate company traded on the New York Stock Exchange; it was the only publicly traded company Murchison managed, taking a slot as Delhi's chairman.* Delhi began as a small outfit headquartered above a corset shop in a run-down building a few blocks from 1201 Main, but its oil finders proved top notch. Ranging across the Rocky Mountains and Canadian West, they unearthed massive quantities of natural gas in the next decade, turning Delhi's handful of executives, plied with stock at Murchison's insistence, into millionaires.

He did it all with other people's money, gobbling up tens of millions in loans from eager banks in Dallas and New York. As his reputation grew, 1201 Main became a magnet for every speculator and investment banker in the region, all arriving with deals for Murchison to consider. He would flip through a prospectus quickly, then toss it to one of a half-dozen bright young men he had hired. The group would argue an investment's pros and cons until Murchison made a decision or, if he needed more time, threw up his hands and announced, "I'm going fishing." They made million-dollar decisions sitting at picnic tables gnawing at the barbecue ribs Murchison loved; his aides knew to always take along money, for their boss, like Sid Richardson, never kept a cent in his pocket. One of the few times Ernestine Van Buren saw him handle actual money was when a bum wandered into 1201 Main and asked for a handout. Murchison found a ten-dollar bill and gave it to him, which brought the bum back for more. He smelled, though, and in the ensuing weeks Van Buren watched, smiling, as Clint Murchison,

---

*Not long after, Murchison divested the last of his Southern Union stock, his last link to the company.

one of the country's richest men, leaned out his office window and dangled a ten-dollar bill down to his new friend on the end of a long string.

For the most part, that was the extent of the Big Four's philanthropy. H. L. Hunt regularly had the representatives of Dallas charities escorted from his office. In later years an unnamed Dallas oilman told of trying to interest Sid Richardson in charity. "Sid, why don't you give Dallas a children's hospital?" the oilman suggested during a cross-country flight. Richardson replied, "Now if I do that, why everyone in the world will come around asking me for money and I just don't want to be bothered." His peers didn't seem to hold this attitude against him. "Why, Sid has no more civic responsibility than a coyote," an oilman told the same writer. "But he's a nice guy."

The most prominent philanthropist among Texas oilmen was probably George Strake, who gave millions to the Catholic church and was generally recognized as the most decorated American layman; in 1946 Pope Pius XII personally decorated Strake a Knight, Grand Cross, of the Order of St. Sylvester, the oldest and most prized of papal orders. Of the Big Four, only Roy Cullen, with his givings to the University of Houston, had done any serious thinking about charity. Cullen turned sixty-four in 1945; his last two unmarried daughters wed returning servicemen, nice, clean-cut young men, both of whom Cullen took on as assistants. In the coming years they managed to find the odd pocket of oil in the Rockies, but by and large Cullen had discovered all the oil he was to find. As the years wore on, the men coming to his office wanted to talk as much about philanthropy and politics as oil. Cullen endowed a pair of buildings at the Gonzales Warm Springs Foundation for infantile paralysis—after receiving assurances it had nothing to do with Franklin Roosevelt's Warm Springs in Georgia—and enjoyed it.

"Lillie and I are pretty selfish about our giving," Cullen said in 1947. They liked seeing the fruits of their gifts, unlike wealthy couples who donated posthumously. "Honey, we've got the children taken care of," Cullen told Lillie one evening as they sat in their living room in River Oaks, watching the sun set over the mansion's reflecting pool. "There is a lot of money coming in from the wells that we don't need." Cullen suggested they make a large donation to the university.

"But what about the hospitals?" Lillie asked. Houston's medical facilities were aging, run-down, and poorly staffed. Several civic leaders were making noises about improving them. Cullen thought she had something. The next morning he called a man at Memorial Hospital and told him he wanted to write a check for one million dollars. The man nearly choked. Cullen's next call went to Hermann Hospital. It, too, received a check for what he called "a million plus." Houston had a tiny, struggling Methodist hospital as well. Cullen's third call went to its fund-raiser. The man scurried to Cullen's office and left, dumbfounded, with a check for one million dollars. The next day, as word of Cullen's donations spread through the medical community, a man from the Episcopal diocese—which didn't yet have a hospital—appeared in Cullen's office with a set of construction blueprints under his arm. He left an hour later with a check for "one million plus." Later Cullen added another million-plus to the Baylor University medical school.

Cullen's charitable spree was front-page news in Houston; in later years, when the city emerged as a national center for heart surgery and cancer research, his gifts that week would be remembered as its genesis. Afterward Cullen found himself deluged with requests for money from every fund-raiser in Houston. He and Lillie began discussing whether to establish some sort of foundation to handle all the solicitations. Cullen thought it might need more than just another few million dollars. He broke the news on the evening of March 27, 1947, when the Texas Hospital Association honored him at a banquet in Houston's Music Hall. "I'm going to let you in on a little secret," Cullen told the crowd. "Mrs. Cullen and I are now having our attorneys draw up papers to create a foundation in which we will put oil properties estimated to produce eventually some thirty to forty million barrels of oil, worth eighty million dollars or more."

There was an audible gasp from the audience; no one in Texas—no one in the entire South—had ever given away such a sum. Cullen told the crowd he hoped the money would be given to local hospitals and the University of Houston. Two days later, the Houston Chamber of Commerce's entire board of directors came to his office en masse to thank him. "Thanks for your kindness," Cullen told them. "But since I made that announcement, Lillie and I have been thinking this thing over, and we've changed our minds a little bit."

The chamber of commerce men exchanged nervous glances.

Cullen cleared his throat. "We've decided," he went on, "after looking over our property, that we can double that figure."

It took a moment for the enormity of Cullen's remark to sink in: $160 million; in today's dollars, nearly $1.7 billion. It was, at the time, the single largest gift ever made by a living American. Once the paperwork was complete, the new Cullen Foundation became the largest charitable foundation in the South and, after the Ford and Rockefeller Foundations, third largest in the nation. Its gifts were limited only to projects in Texas. Cullen spent the rest of his life plowing money into the foundation and the University of Houston; after its football team won a match against Baylor University in 1953, Cullen was so happy, he donated another $225,000. He would eventually give away 93 percent of his fortune.

News of Cullen's gift made national headlines. Overnight, he found himself inundated with two hundred thousand pieces of mail, much of it from people seeking money. A man in England addressed his letter to "Hugh Roy Cullen, Texas Oil King, Somewhere in America." Two thousand miles to the east, news of the donation prompted brows to furrow in the newsrooms of New York and Washington. Someone gave away how much? In Texas? What the hell was going on down there?

NINE

The New World

Could we have kippers for breakfast?
Mummy dear, mummy dear
They gotta have 'em in Texas
'Cause everyone's a millionaire
SUPERTRAMP, "BREAKFAST IN AMERICA"

I.

In the first years after World War II, H. L. Hunt, Roy Cullen, Sid Richardson, and Clint Murchison had emerged as a handful of the richest men in America—*and no one knew it.* It wasn't just that few people understood how wealthy they were. Beyond the insular world of Texas Oil, almost no one knew they existed. By 1948, despite Hunt's historic dealings in East Texas, Cullen's philanthropy, and Richardson's dinners with the Roosevelts, the Big Four had garnered precisely three references in the nation's newspaper of record, the *New York Times.* Cullen earned the only *Times* headline, when he announced his new foundation. It identified him as a "Former Texas Oil Field Laborer."

Their anonymity was largely a function of how the press worked during the 1930s and 1940s. By and large the Texans' few public mentions came in the oil-industry press, whose arcane regurgitation of drilling data was as indecipherable to the average newsman as Urdu. National publications, focused on the news of Washington and New York, tended to rediscover mid-America only when undistracted by world events, and the events of the previous twenty years had long diverted their newsrooms. The Depression, Hitler, Pearl Harbor, the war—for two decades there had been far more important matters to cover than the goings-on of inaccessible Texas businessmen. It wasn't until the smoke cleared over the battlefields in Europe that the press slowly awakened to the profound changes the intervening years had wrought at home.

Even then, many writers who passed through Texas in 1945 and 1946 failed to notice the new wealth coursing through the state. A notable book of 1947, *Inside U.S.A.*, was a massive, thousand-page survey of American life written by the correspondent John Gunther, who in the words of Arthur Schlesinger set out to discover "a new America, hardly known to the world—or to itself." Gunther wrote chapters on every state—three on Texas—and while he noticed its new industries, he had almost no sense how rich its oilmen had grown. In fact, Gunther's theme was the opposite, that "Texas is probably the richest 'colony' on earth...and has been badly fleeced by outsiders"; this was boilerplate dating to the 1920s. Studying those who "owned" Texas, he ignored Hunt, Murchison, and Richardson in favor of conventional business-men such as Houston banker and real estate magnate Jesse Jones, though Gunther did note that Houston's "richest citizen, and reputedly the wealthi-est man in the state, is an oil operator named Roy Cullen."

The first hint of impending change came on a cool, windy afternoon in February 1948, when Hunt, wearing an off-the-rack tan gabardine suit and gray fedora, emerged from the Mercantile National Bank Building onto the sidewalk along Commerce Street. He was on his way to a card game at the Baker Hotel. When he reached the corner at Ervay Street, across from the Neiman Marcus department store, he stopped for a red light. Suddenly a man rushed up, lifted a camera, and snapped a photo. Before Hunt could react, the man disappeared into the crowds. Hunt headed on, thinking the man was probably taking a picture of the building behind him.

In fact, the photographer belonged to a team sent by *Life* magazine, whose Dallas bureau chief, Allene Pohlvogt, had been mulling a story on Hunt for months. Both *Life* and its sister publication, *Fortune*, were preparing layouts on the Texas economic boom and had caught wind of just how wealthy Hunt and his peers had become. Still, just about everyone in Dallas—and around the country—was floored by the headline in the April 5, 1948, issue of *Life*: Under the photo of Hunt, it read: "Is this the richest man in the U.S.?"

Fortune's April issue went a step further, painting a portrait of oil million-aires popping up daily across the Southwest, especially in Texas. Its article carried an early mention of the term that would come to define the Big Four and their brethren: The Big Rich. Hunt, *Fortune* said, was "the biggest of The Big Rich, and thus also probably the richest individual in the U.S." Both magazines underestimated Hunt's net worth at $237 million at a time that it was probably closer to $600 million. Both misspelled his first name as "Haralson."

It's difficult to overstate the impact of the *Life* and *Fortune* spreads, which triggered a seismic shift in the way America viewed Texas, especially its oilmen. The story of Texas Oil, in fact, can be divided into halves, the anonymity of the pre-1948 years, and everything after. Until April 1948 Texas had been known mostly for cowboys, cattle, and braggodocio.* On the morning after, you could almost see editors up and down the Eastern Seaboard scratching their heads. America's richest man? In Texas? And there were more? It was as if a curtain had been lifted and a new band of actors, mysterious Texas millionaires, had wandered onto the national stage. Who were they? What did they want? For the Big Four, nothing would ever be the same.

The *Life* and *Fortune* articles sent dozens of eastern writers scurrying into Texas for the first time, many eager to advance the stereotype of eccentric, nouveau riche Texas zillionaires tossing hundred-dollar bills like confetti. From 1948 until well into the 1950s their articles crowded every newspaper and magazine of the day, from *Time* and *Colliers* to glossy spreads in *Holiday*, including piece after piece in the *New York Times*. In short order books began to appear: *The Lusty Texans of Dallas* in 1951, *Houston: Land of the Big Rich* the same year, followed by a a collection of Texas primers such as *The Super-Americans* by *The New Yorker*'s John Bainbridge.

At least initially, the Big Four were too canny to engage with snooping reporters. Hunt gave only a single interview, to the *Dallas Morning News*, then scrambled for cover. Still, to the chagrin of those in the state who prized discretion and taste, many writers found exactly the kind of Texan they were looking for. His name was Glenn McCarthy.

---

*To be fair, oil had been associated with Texas in the public mind since Spindletop. Several novels and minor films had been issued about Texas oilmen in the 1920s. As late as 1940, the movie *Boomtown*, starring Clark Gable as a wheeler-dealer Texan, achieved wide notice. Tellingly, *Boomtown* was set during the Ranger-Breckinridge booms of the early 1920s. Hollywood, like the rest of America, had yet to learn of the state's new wealth.

II.

The stereotype of the raw, hard-living, bourbon-swilling, fistfighting, cash-tossing, damn-the-torpedoes Texas oil millionaire did not exist before Glenn McCarthy rocketed into the national imagination in the late 1940s. Yet McCarthy was all that and more. Little remembered today, it was McCarthy, and his quixotic dreams, who, more than H. L. Hunt or Roy Cullen or his wealthier peers, introduced Americans to the changes oil had brought to Texas. The distilled essence of swaggering Texas id, McCarthy rubbed elbows with Howard Hughes and Hollywood stars, drank and brawled his way from Buffalo Bayou to Sunset Boulevard, and, at the peak of his fame, adorned the cover of *Time*. No Texas oilman ever rose so high or fell so hard.

McCarthy's legend began, fittingly, near Spindletop, where he was born on Christmas Day 1907.* His father, an itinerant oil field worker and plumber, shuttled the family between Beaumont and Houston's rough Third Ward; at one point, Glenn was the Howard Hughes family's paper boy. A scrappy, sinewy teenager, he worked odd jobs in the oil fields and became a standout youth football player and amateur boxer. He dropped out of high school but leveraged his gridiron prowess into brief stays at Rice Institute (now Rice University) and Texas A&M, where he was expelled for hazing. By 1930, though nominally still enrolled at Rice, he was pumping gas at a Houston service station.

McCarthy was twenty-two then and handsome, with dark eyes, an Errol Flynn mustache and a single brown curl that fell rakishly over his forehead, a kid on the make with a temper to match, known for slugging just about anyone if he'd had enough to drink. His life changed one night that spring when a girl he knew stopped for gas and brought along a pretty sixteen-year-old named Faustine Lee, daughter of the wealthy oilman Thomas Lee, co-owner of Beaumont's Yount-Lee Oil Company. Glenn took Faustine to a dance, then another, at which point, within weeks of meeting, they eloped.

---

*Interestingly, several of McCarthy's high school and college transcripts list his date of birth as December 25, 1906, suggesting he may have lied about his age as he became older.

Her father was not happy. McCarthy told him not to worry. He refused to take a penny of Lee money. In fact, he swore to Lee he, too, would become an oilman. For now, however, the only oil McCarthy saw on a daily basis came in cans.

The newlyweds rented a one-room apartment and slept on a rollaway bed. Soon after, McCarthy quit his job after a disagreement with his boss. Money ran low. On a rainy morning in November 1931, Glenn and Faustine spread their entire net worth on the kitchen table. It came to $2.65. Sliding the money into his pocket, McCarthy took a streetcar downtown and began looking for work. By midday, rain-soaked and forlorn, he ran into a buddy, who lured him to the racetrack. With little to lose, he put three-quarters of his life's assets down on a $2 bet. In a bit of family lore all the McCarthys swear is true, he hit the Daily Double. McCarthy came home that night and dumped $700 in front of Faustine. Her response took him aback. "Glenn McCarthy," she said, "we don't need money *that* badly. You take it straight back." She assumed he had stolen it.

The only job McCarthy knew was pumping gas, so he set his sights on opening his own station. For days he counted the cars passing a busy nearby corner, at Main and McGowen Streets, and armed with his calculations he began paying daily visits to the Sinclair Oil office in suburban Angleton, where he pestered the man in charge, practically begging him to let him open a station. In time the man, impressed with McCarthy's tenacity, agreed, but the deal was harsh: Sinclair would take almost all profits from gasoline sales. McCarthy could keep anything he made fixing cars and selling tires and accessories.

As a gas station manager, McCarthy was by every account a force of nature. In return for wash and grease jobs, he would drive customers to their offices in the morning and pick them up in the afternoon with a greased and washed automobile. He worked twenty hours a day, and kept the store open the other four by hiring his father. He bought huge stacks of tires at a discount, sold them for twice what he paid, and moved used cars on the side. He claimed his station sold more gasoline than any other in Houston, and he might have been right. Sinclair rewarded him with a second station, on

Houston's West Side, and by early 1933 McCarthy was clearing fifteen hundred dollars a month.

His confidence bursting, McCarthy decided it was finally time to enter the oil business. A seismographic crew was analyzing a section of scrubland along the Houston–La Porte highway, and McCarthy, on a hunch, decided to option the land and drill it. He sold the gas stations, bought up several adjoining parcels, and hired a contractor. He and Faustine put most of their savings into the discovery well—and lost it all when the well churned up nothing but salt water. Undaunted, McCarthy invested his last thousand dollars in a friend's well. It turned out to be a small producer. McCarthy sold it for an eight-thousand-dollar profit.

Hustling back to his land on the La Porte highway, he began another well, only to have it ruined when heavy rains caused an adjacent creek to overflow, flooding the site. He sold some of the neighboring leases and managed to raise enough for one last wildcat, but this, too, came in dry. All told, McCarthy had tried three wells and had precisely nothing to show for it—except a painful education in oil. Out of cash, and determined not to borrow from Faustine's father, he decided to try his luck as a contractor, drilling wells for others.

With no rig of his own, all McCarthy could do was beg for a drilling job. Oilmen laughed him out of their offices. But McCarthy didn't give up. He offered to drill wells for next to nothing, and a few wildcatters, no doubt mindful who his father-in-law was, gave him a chance. His results were impressive; even when McCarthy found no oil, he drilled fast, worked hard, and came in under budget. His father-in-law was sufficiently impressed to introduce him to the cotton magnate M. D. Anderson, who needed someone to drill him a well on the edge of George Strake's field at Conroe. As he had done before, McCarthy begged, borrowed, and "covertly borrowed" most of his equipment, sending his men, all of whom worked for food and the promise of an eventual paycheck, sneaking into the woods at night to liberate pipes and barrels of drilling mud from nearby drill sites. When McCarthy couldn't pay his rig's rental fee, the owner secured an injunction preventing its use. McCarthy ignored the court order, but it worried Anderson, who brought in a thirty-year-old attorney named Leon Jaworski—later to achieve fame as a Watergate prosecutor—to handle things.

Jaworski drove deep into the pines to find the drill site, and what he saw when he arrived fired the McCarthy legend:

> I found [M. D. Anderson] in his usual spot in the rear of his Cadillac. He was ashen-faced, visibly shaken and sputtering disjointed sentences, from which I gathered he had just witnessed some horrifying spectacle. Finally I was able to piece together what had taken place, an incident he kept referring to as a "miracle." Only moments before, a platform some fifty feet above the ground had collapsed as Glenn, his brother Bill and another worker were standing on it. As the three men were falling, Glenn had grabbed a cross-iron, and the other two grabbed him. His brother clung to Glenn's waist and the third man to his leg.
>
> "Glenn not only held himself and the two others securely," Mr. Anderson said, sputtering, "he then maneuvered the lowering of all three to the ground." McCarthy had suffered friction burns on both hands, sliding down the derrick to safety. Otherwise they were all uninjured....
>
> As I began to question his story, McCarthy walked up, stoic, unexcited and austere. He looked a little like Barrymore's Hamlet. When Anderson introduced us, Glenn merely grunted. He made no reference whatever to his acrobatic landing and his manner discouraged any questions. But those involved, and the others who witnessed it, repeated the story until it became a part of McCarthy lore.[1]

McCarthy eventually hit oil for Anderson at Conroe, and he soon had a car, a larger apartment, and a reputation. "He was known as a man who could drill a well in half the time it would take a major oil company," a geologist named Michel Halbouty recalled, "a man inclined to raise his fists at every affront whether large, small or imaginary...but he could charm Lady Godiva off her horse." As good as he was at drilling, McCarthy, whose own father had grown poor working on other men's wells, knew he could never become wealthy unless he worked for himself. By 1935 he had enough money to try, and his father-in-law introduced him to the talented Halbouty, who had proven to have an uncanny nose for knowing where to find oil.

Halbouty leased a tiny but promising plot of land west of Beaumont, but there was a catch: the existing lease expired in twenty days. If McCarthy hadn't drilled by then, he couldn't drill at all. Somehow McCarthy managed to wheedle a rig on credit, but the following week it rained so hard, it took eleven days just to drag it to the drill site. By the final day they had the rig in place, but couldn't get the motor running to begin drilling. By dusk the owner's lawyers were standing in the mud alongside, ready to stop McCarthy at the stroke of midnight. At eleven o'clock, with the motor still dead, McCarthy hollered, "All right, boys, we're gonna start drilling this son-of-a-bitch by hand." At 11:45 McCarthy used a pair of chain tongs to inch the drill bit into the mud. He had a notary standing by to prove it. They called the well Longe No. 1, and a week later it came in strong, the discovery well of what became known as the West Beaumont Field. McCarthy had found the first oil of his own.

They had just started work on Longe No. 2 a few weeks later when, one afternoon resting in a Beaumont hotel, Halbouty spied smoke on the horizon: the well was on fire. The two men raced to the site but it was too late. Their rig and all their equipment had melted. If that weren't bad enough, nearby residents began pelting McCarthy with lawsuits, claiming the sooty smoke had killed their crops and cattle. McCarthy hired Leon Jaworski to defend him in court, and while McCarthy underwent an appendectomy, Jaworski appeared to be swaying the jury. McCarthy's inate flamboyance, however, nearly sank the case. When he arrived to testify, he sat in a wheelchair pushed by a Negro man—"his valet!" observers whispered—and attended by a buxom blond nurse. Even worse, he wore a silk bathrobe and a silk scarf knotted at the neck. Somehow Jaworski won the case, and saved McCarthy's career.

McCarthy went on to hit a string of strong producers outside Beaumont, earning him nearly two million dollars, enough to build his dream home, a seven-thousand-square-foot columned southern mansion just south of downtown Houston. He had a family now, four girls and a little boy on the way, and while never exactly faithful to Faustine, he was a good father, staging impromptu plays for the kids and wrestling with them. A lesser man might have been satisfied. But McCarthy wanted more.

In 1939 he took everything he had and plunged into a risky play around the town of Palacios, on the Gulf Coast southwest of Houston. Geologic charts suggested it was one of the most promising formations in years. McCarthy optioned 562 parcels in and around the town, then borrowed heavily to buy five new drilling rigs, which cost $1 million. He started five wells simultaneously, but the formation was laden with unstable natural gas. One after another, all five wells blew out. What gas McCarthy found he was unable to sell. By the time he gave up, he had lost $1.5 million and was heavily in debt.

Facing bankruptcy, McCarthy convened a meeting of his creditors. The CEO of one was named to oversee what remained of his business, while McCarthy went back to contract drilling in hopes of recouping his debts. A prewar drilling boom was under way, and a hapless Minnesota businessman named Frank Anderson, facing deadlines to drill thousand of acres he had amassed north of Galveston, hired McCarthy when no other reputable driller would take him on. McCarthy drilled Anderson's first well to nine thousand feet and came up dry. His contract fulfilled, McCarthy was about to pack up his rig when, after dipping his tongue in a final coring, he told his tool-pusher to drill another two hundred feet. They found oil. Anderson gave him a $100,000 bonus and a piece of the upside, and in the ensuing two years McCarthy drilled him another sixty wells in the area, earning a total of $1.5 million, enough to pay his creditors.

By 1942 McCarthy was back drilling his own wells. He kept to areas he knew, the swamplands and buggy moors south and east of Houston, and his crews hit gusher after gusher. At Anahuac, Chocolate Bayou, Angelton, Collins Lake, Coleto, North and South Stowell—overnight, it seemed, McCarthy became the hottest oil finder in Texas. Michel Halbouty, who went on to become one of Houston's most celebrated oilmen, called McCarthy "possibly the best practical oilman the country had produced, an improviser who could drill with junk [and who possessed] a knack for finding oil he couldn't explain because it came installed in his system like an antenna.

And he was a plunger, always willing to shoot the moon on his chances of finding oil."

All through the war years, with little fanfare, McCarthy opened new fields, extended old ones, and fattened his bank accounts. By 1945 he was a very wealthy man, his oil reserves worth $50 million, about $535 million today. He bought the twenty-two-story Shell Building in downtown Houston, then led a group that bought the Second National Bank. Like every successful oilman, he added a ranch, fifteen thousand acres of West Texas prairie outside Uvalde. He enjoyed nothing so much as racing around his land in one of his Cadillacs, a bottle of bourbon at his side, slowing down just long enough to blast his shotgun at a rattlesnake or dove.

Had McCarthy stopped there, he might have been recognized as a fifth member of the Big Four, lazing away his days playing cards like Sid Richardson, or sniping at politicians like Roy Cullen. How differently things would have turned out had he stuck to oil. But Glenn McCarthy was a man with dreams—vast, historic, world-altering dreams. Whether they sprang from his hardscrabble beginnings, a desire to eclipse his wealthy father-in-law, or something else altogether, McCarthy burned to eclipse the Hunts and Cullens and Richardsons atop the Texas pyramid, to own titanic refineries and office towers and continent-spanning pipelines. He wanted to create a legacy, a landmark, something no one else had ever done. And the more he thought, the more he planned, the more he mapped out what postwar America would need, the more McCarthy's dream crystallized into a single idea:

A hotel.

Not just any hotel: the world's finest hotel, built for Texas, a mammoth, ornate structure to eclipse the Waldorf-Astorias and Ritzes and Hiltons, the largest building between New York and Los Angeles, a glorious symbol not just of his own mushrooming power and confidence but that of Texas as well. It would be the Lone Star Taj Mahal, its Eiffel Tower, its London Bridge. And McCarthy, as its builder and owner, would emerge as the state's dominant impresario, using newfound fame and contacts to expand into every conceivable business, to conquer New York and Hollywood, even Dallas. He would be King of Texas.

This was Glenn McCarthy's dream. Figuring, naively, that it wouldn't be too difficult to get bank loans on his oil reserves, he announced the project in November 1944. The McCarthy Center, as he called it, would consist of the hotel, apartment buildings, a theater, and a shopping center. After a day or two of local headlines, Houston yawned. No one outside the oil business had heard of Glenn McCarthy. There was a war on. After a few weeks, the city forgot about it.

But McCarthy would not be deterred, not even when his bank, Republic of Dallas, passed on the financing; an officer suggested Republic might regard the project more favorably if McCarthy would repay the $12 million he already owed. Undaunted, McCarthy looked elsewhere. As the war drew to a close in the summer of 1945, a broker put him in touch with an officer at the Equitable Life Assurance Society of America. It was a fateful choice. Up and down Park Avenue scores of New York lenders, keenly aware of the massive loans Sid Richardson and Clint Murchison had taken, were flocking to find oilmen of their own to fund. America's third-largest insuror, The Equitable, was late to the oil game but determined to catch up. Its executives had opened their first oil-lending office in Dallas just weeks before, in May 1945. Its first loan was for $425,000. The second, to McCarthy, would be for a jaw-dropping $22 million, among the largest single loans Equitable's board had ever approved. That was only the beginning.

To ascertain McCarthy's creditworthiness, Equitable first commissioned an analysis of his business. In February 1946 a petroleum geologist it hired valued McCarthy's oil and gas reserves at forty-five million dollars. On the strength of that report, Equitable advanced McCarthy that first twenty-two million, all but three million of which went to pay off existing debts. But McCarthy knew more was coming, and in short order he began to put his plans into motion. He moved his offices into the Shell Building, taking up the entire sixth floor, and hired a public relations man, a Houston radio veteran named Fred Nahas; if McCarthy so much as burped, Nahas made sure it got into the papers. Texas reporters quickly discovered McCarthy was an entirely different animal from taciturn Roy Cullen or other press-shy Texas oilmen; eminently quotable, with his dashing mustache and good looks,

McCarthy was the kind of oilman writers loved: handsome and energetic, larger than life.

So frenetic were McCarthy's preparations that year, and so effective was Nahas at publicizing them, it sometimes seemed McCarthy was the only oilman in Houston in 1946. Their campaign began with a "coming out" spread in the *Chronicle*, a long valentine wrapped around a family portrait. That spring, three months after the Equitable loan closed, McCarthy took the unusual step of publicly announcing it, something no Texas oilman had ever done; he told reporters it was "the largest loan of its kind in Texas financial history." Ignoring the fact that much of the money had gone to repay old debts, McCarthy simultaneously announced a thirty-three-million-dollar drilling program in ten Texas counties, along with new natural-gas facilities and pipelines.

Every week seemed to bring a new McCarthy initiative. Out at the village of Winnie, southeast of Houston on the edges of the Anahuac field, he began construction on a four-million-dollar plant that was to convert oil into formaldehyde and other industrial chemicals. At the same time, he started scouting for a pipeline to buy. In May he announced plans to bid on the Big Inch and Little Inch pipelines. Two months later he followed through, offering eighty million dollars he didn't have. He was outbid. A week later McCarthy, an aviation nut since boyhood, was reported in talks to build a new airport in the Rio Grande Valley. In December he acquired a large position in the stock of Eastern Airlines, after which the chairman, McCarthy's new friend Eddie Rickenbacker, named him to the company's board of directors.

But the big event, the announcement that introduced much of Houston to the McCarthy whirlwind, was the hotel's groundbreaking. It was to dwarf anything ever built in Texas. They had decided to call it the Shamrock, after the Houston *Post* ran a naming contest, and everything about it was to be Irish-themed, the dining areas the Shamrock Room and the Emerald Room, the night club The Cork Club. McCarthy had hoped to break ground on St. Patrick's Day, but the master of ceremonies, the actor Pat O'Brien, couldn't make it till March 22.

That day at 4:30, three thousand people crowded the fifteen-acre building site to watch McCarthy, O'Brien, and a host of Texas dignitaries, including two former governors, give speeches from a flag-decked platform. "I'm a little bit choked up," McCarthy said after several speakers praised his vision. "To me this symbolizes the future greatness of Houston." O'Brien told the crowd, "The eyes of the other forty-seven states are upon Texas and this project." The crowd cheered lustily, and grew so riotous that the actual groundbreaking had to be eliminated when McCarthy and O'Brien were mobbed as they went for their shovels. The two were hustled into a limousine. It was a sign of things to come.

The next day's press coverage was breathless, even by Texas standards. "Speakers Hail Shamrock as Symbol of New Era" blared a *Chronicle* headline. The *Post* coverage swallowed the entire front page. The *Chronicle*, terming the project "perhaps the greatest ever undertaken by an individual in Texas," listed the Shamrock's features: "largest suburban theatre in the South, seating 1500...biggest food market, and biggest drug store in Texas, maybe anywhere...electronic gadgets and plastic materials throughout...radios in every room...a two-floor registration desk...a 10,919-square-foot main dining room, the Shamrock Room...dancing nightly to nationally-famed bands...a five-story parking garage with room for 5,750 automobiles...and all topped off with a blazing neon sign that can be seen from Sugar Land and Humble and which will fill that part of the city with soft green moonlight."

Yet even before the groundbreaking, there were naysayers. McCarthy had chosen to erect the hotel three miles south of downtown, on Main Street near his home. One morning at the Rice Hotel Jesse Jones took him aside and gently warned that business travelers wouldn't stay that far from downtown. McCarthy didn't listen. Houston was growing fast, and though no one said it at the time, the Shamrock's location constituted a bet on postwar suburbanization.

What no one understood was that McCarthy didn't have a fraction of the cash he needed to complete the Shamrock or any of his other projects. He was counting on the Equitable's eagerness to produce it. It was a sound bet. A second analysis of his reserves was already under way, and when completed in May 1947, it valued McCarthy's oil and gas at $61.9 million. A third analysis, in 1948, would boost the figure to $73.5 million. With that kind of collateral, Equitable's board felt secure pushing mounds of cash at McCarthy, loan after loan after loan, eventually reaching a total of $51.8 million. But not without strings. Equitable's board arm-twisted McCarthy into downsizing the Shamrock project, eliminating the apartment buildings, the shopping center, the grocery, the drugstore, and the theater, leaving only the hotel, which was fine by McCarthy. It was all he really wanted anyway.

Construction began, and proceeded fitfully through 1947 and 1948. The Shamrock was as big as McCarthy had hoped, a gray granite colossus eighteen stories high and almost as wide. Building it, however, was the easy part. McCarthy dreamed of making the Shamrock's opening, scheduled for St. Patrick's Day 1949, a national event, with coverage in *Time* and *Life* and newspapers around the world. He envisioned a Hollywood-style gala, complete with spotlights and movie stars. Unfortunately, Houston had no movie stars. So McCarthy hatched the idea of holding a simultaneous movie and hotel opening. To do that, however, it would be necessary to make an actual movie.

And so, in March 1948, one year before the opening, McCarthy flew to Los Angeles and announced the formation of Glenn McCarthy Productions, telling reporters he was embarking on a series of major film projects. His announcement, coming just as *Life* introduced America to H. L. Hunt and the strange new world of Texas millionaires, created a stir in film circles. "A spectacular entrance," one columnist termed it. "It looked as though McCarthy would turn Hollywood topsy-turvy with his ambitious production plans." McCarthy then embarked on a manic tour of Hollywood parties and nightspots, befriending a slew of stars, from Errol Flynn to John Wayne. Making a movie, he discovered, wasn't nearly as difficult as finding oil. McCarthy hired a director and a cameraman and had talent agents hire actors. He had a script ready to film, *The Green Promise*, the story of a young girl's struggle to save her family's farm. To McCarthy's delight, the child star Natalie Wood agreed to play the lead. Walter Brennan signed to play her father.

But the movie was only the beginning. The Shamrock's biggest star, McCarthy sensed, had to be Glenn McCarthy himself. If he was to be King of all Texans, he needed to assume the role. The first step was to dress the part. Out the door went his business suits. In their place came a look equal parts Houston and Hollywood: dark sunglasses, leopard-patterned ascots, gleaming leather jackets, a diamond on his pinkie the size of a dime. His deeds, McCarthy knew, had to match the look. He appeared at the Houston Fat Stock Show and outbid everyone for the champion steer two years running. Not satisfied with a regional victory, he proceeded to the International Livestock Exposition in Chicago, where he was so confident of purchasing the champion bull, he marched into the hall carrying banners that read, SOLD TO GLENN MCCARTHY. He was as good as his word, laying out a world-record $12,900 for a single cow, then set a second record by becoming the first individual to buy the champion steer and its runner-up.

The cattle-buying excursions brought McCarthy his first national headlines, but it was his love for flying that really got him noticed. In 1947 McCarthy purchased a P-38 fighter for four thousand dollars, spent fifty thousand dollars upgrading it, then entered it in the cross-country Bendix Trophy Races from Los Angeles to Cleveland. Though fifty planes entered the race, McCarthy's promotional skills secured him the advance headlines; he held a series of star-studded Hollywood launch parties, then prevailed upon the actress Joan Crawford to christen *The Flying Shamrock*, smashing a champagne bottle across its nose as flashbulbs burst. More headlines ensued when a fire erupted in one of the McCarthy plane's engines, forcing his pilot to parachute from twenty-five thousand feet; a day later the pilot was found wandering an Arizona Indian reservation. McCarthy got still more headlines when the race's winner, the celebrated flier Paul Mantz, scrambled through the finish-line crowds in Cleveland to claim a ten-thousand-dollar bet McCarthy had made him on the race.

The following year McCarthy hired Mantz to fly one of three separate planes he entered in the race. They finished first, second, and fourth, and as reporters mobbed McCarthy afterward, he announced his hopes that the Bendix races would come to Houston in 1949. They wouldn't, but it was the kind of proposal McCarthy had begun making regularly. Around Houston he had become a human tornado of postwar progress, donating land for a new airport, holding press conferences to announce charitable gifts, riding a horse in the rodeo parade, lobbying to bring the city a professional baseball or football team. Between appearances he purchased a steel factory in Detroit, Houston's largest radio station, and a string of small newspapers. Every appearance and every press release generated headlines, and headlines, McCarthy knew in his heart, would be the key to the Shamrock's success.

By January 1949, with the hotel's opening now three months away, McCarthy rivaled Roy Cullen as Houston's best-known millionaire. He even wangled a profile of his own in *Life*. "Brawny Glenn McCarthy Embodies City's Success," the headline read. Construction on the Shamrock was almost finished; it had cost twenty-one million dollars. *The Green Promise* was complete, and would premiere at a Houston theater the night after the hotel's opening; city fathers had agreed to stage a torchlit evening parade in its honor. In those frantic final weeks, no detail seemed too small for McCarthy's notice. When every major grass supplier in Texas refused to hand over the ten acres of San Augustine McCarthy needed to seed the Shamrock's lawn—they insisted it was needed for other customers—McCarthy simply bought one of the suppliers, trucked in a bulldozer, and planted the grass himself.

McCarthy spared no expense. The Shamrock pool, he claimed, was the world's largest. Its towels: the world's largest. The opening alone was costing him $1.5 million. He mailed gold-filigreed invitations to every movie star and CEO he could think of, then invited dignitaries from twenty-four foreign countries. The *Post* and the *Chronicle* planned special editions. Thousands of shamrocks were flown in from Ireland; McCarthy even prevailed upon a Dublin newspaper to publish a special edition of its own. Hundreds of reporters were coming, from as far away as London.

With six days to go, the hotel was still not complete. The top floors still needed painting. Everywhere the walls and ceilings were lined with wet plaster. McCarthy brought in massive fans to dry it. On Monday, three days before the gala, the Emerald Room was finally ready. But when McCarthy strode in to see it, he found a workman had just fallen through the damp plaster atop its dome, leaving a gaping hole. He ordered the entire dome rebuilt. Industrial heaters were rolled in to dry its plaster. Later that day, McCarthy was checking the Emerald Room's lighting when he discovered there were no "pin lights" in the ceiling to spotlight the entertainers. A telephone call turned up the fact that the nearest lights were in New York. At four the next morning McCarthy found a seller. The man said a shipment would take ten days; it would take three just to pack the lights. "Don't pack 'em," McCarthy barked. He had a plane on the way to New York at daylight. The lights were in Houston by nightfall. Installation was finished by dawn.

On Tuesday, with two days left, McCarthy flew to Los Angeles to fetch his friend Howard Hughes and finalize travel arrangements for the movie stars. The two returned to Houston the next morning on Hughes's million-dollar Boeing Stratoliner, the world's largest private plane. McCarthy was thinking of buying it. He and Hughes ate a pancake breakfast by the Shamrock's pool. "Keep the plane," Hughes said, rising to leave. "Fly it around. Let me know what you think."

That morning the stars began to arrive. McCarthy had chartered an entire fourteen-car Santa Fe train—the "Shamrock Special"—to bring them from Hollywood. A crowd of five thousand dominated by teenage bobby soxers ringed the train station and lined nearby rooftops for its arrival. Girls squealed when Dorothy Lamour emerged and kissed McCarthy on the cheek. Cheers erupted as other stars followed: Robert Ryan, Andy Devine, Alan Hale, Ward Bond, Kirk Douglas, Stan Laurel, Buddy Rodgers, Ruth Warwick, Robert Stack. Nearly fifty others arrived on an American Airlines charter that afternoon. Dozens of reporters trailed in their wake. Neither Houston nor Texas, as the newspapers reminded readers every morning, had ever seen anything like it.

Finally, the day arrived. McCarthy glided through the Shamrock's carpeted hallways that morning inspecting his handiwork. Even his critics had to admit his decor made a statement. The walls and carpets were shades of coral and lime. Columns were rose and pink. The cavernous lobby, paneled in acres of Honduran mahogany, was dominated by a life-size portrait of McCarthy himself. Outside, the pool shimmered with kelly-green water. Uniformed guards lined its edges; McCarthy explained that a cabal of Texas A&M students had threatened to dump maroon dye into it. "It's the greatest hotel the world has ever known," he told a reporter. The architect Frank

Lloyd Wright, attending a convention in Houston that week, took the tour as well, and emerged in a daze. Asked his opinion of the interior, Wright remarked that he'd always wondered what the inside of a jukebox looked like. "Tragic," he said.

Night approached. The staff appeared ready. The news vendors shifted uncomfortably in rented tuxedos. The security guards had actually been given elocution lessons, shown how to holster their Texas accents, and instructed to address guests with "Good evening, sir" instead of "Howdy." Buxom girls stood ready to hand out the little packets of Irish shamrocks as the guests arrived.

By dusk a crowd of three thousand had surrounded the hotel. Floodlights crisscrossed the night sky, just as McCarthy dreamed they would. At seven shrieks rose from the crowd as limousines began to arrive and disgorge the celebrities. McCarthy's new pal Errol Flynn waved. Lou Costello waddled in. Close behind came the rest: Ginger Rogers, Van Johnson, Edgar Bergen, Van Heflin, Sonja Henie, Earl Wilson, Eddie Rickenbacker. The Texans came pouring in as well, the governor and a string of politicians, oilmen, bankers, Amon Carter with a delegation from Fort Worth. Sid Richardson ambled in with his niece; only Howard Hughes failed to appear. All the men wore tuxedoes, the women in chiffon and tafetta trains and backless dresses and mink after mink after mink, diamonds dripping from every neck. The evening was a coming-out party, not just for McCarthy but for Texas Oil itself.

Inside, everyone crowded into the lobby for champagne; a cowboy actor, Don "Red" Berry, sipped his from the slipper of Beaumont oil heiress Ann Justice. By 7:30 the public areas were jam-packed. People couldn't move. Waiters gave up trying to wade into the crowd. Off in the corners, McCarthy's security men exchanged nervous glances. There were too many people. Two thousand had been invited. Three thousand had gotten inside. McCarthy, dressed in a white dinner jacket, did his best to navigate the throng, which was growing louder as the champagne began to dwindle. He had contracted with the National Broadcasting System to air Pat O'Brien's opening remarks to a nationwide radio audience at eight o'clock, live from the Emerald Room. The plan was to announce O'Brien's appearance at 7:45 over the hotel's public-address system, at which point the crowd would file to their tables in the ballrooms.

But when McCarthy went to make the announcement, the PA system didn't work. Some of his waiters tried to shout at people to take their seats, but no one moved. Then O'Brien couldn't be found. At the last minute McCarthy delayed his appearance and told the network to cancel the first part of the broadcast. They would go live with Dorothy Lamour when she took the stage at 8:30. But by 8:20 the PA system still wouldn't function. McCarthy and Lamour caucused and decided to begin the broadcast anyway, with or without an audience in the Emerald Room.

At precisely 8:30 radio listeners around the country heard Lamour welcome them to the gala opening of the Shamrock Hotel. Word quickly spread through the packed lobby that the show was beginning. Chaos ensued. Hundreds of tuxedoed men and fur-clad women began to push their way into the Emerald Room, knocking over chairs and a table or two. A cacophony of shouts, curses, and wolf whistles erupted as everyone attempted to find a seat. Amid the din, Lamour's radio audience heard little but crowd noise, interspersed with curses from the control room.

When a Chicago technician phoned the ballroom in exasperation, listeners heard his Houston counterpart's voice: "I can't hear you! I can't hear you!"

"They're fucking it up," the Chicago technician muttered—live, on the air.

Another technician chimed in: "All they're getting is swearing on the line!"

"What?"

"Swearing!"

Meanwhile, a stream of tipsy Texans began taking shortcuts across the Emerald Room's stage to find their tables. When a dumbfounded NBC producer told one matron she was interrupting the broadcast, the woman snatched the microphone from Lamour. "I don't give a damn about your broadcast!" she snapped. "I want my dinner table!"

After seven minutes NBC took the show off the air, citing technical difficulties. When the network returned live five minutes later, however, the

Emerald Room's microphones failed. Lamour, now joined by Van Heflin and a comic named Ed Gardner, tried in vain to get the crowd's attention, but it was no use. No one could hear them.

"People are milling around here, the PA system don't work," Gardner groused over the air. "Nothing is gonna get a laugh anyhow." In an attempt to engage the crowd, Gardner began shouting the names of stars in the audience: "Over there, Pat O'Brien, ladies and gentlemen—Pat O'Brien!" Nothing worked. "I've been in radio a long time," Gardner quipped, "but who has ever seen anything like this!"

In an act of comic desperation, Gardner began calling an imaginary horse race: "And a big crowd is here tonight at Santa Anita." Lamour pleaded for him to stop. "Now, there's a big crowd listening on the air," she begged. "Come on." After a half hour the NBC producers gave up and ended the broadcast. Lamour fled to her suite in tears. "I've been on the road with Bing Crosby and Bob Hope but never through anything like this," she moaned.

The dinner service was a comedy as tuxedoed waiters weaved through the crowd. There had been considerable speculation surrounding the menu, given a price tag one writer dubbed "astronomical": forty-two dollars a plate. Dinner turned out to be beef and a fruit cocktail dubbed "pineapple surprise." Some guests managed to take delivery of their food; others, notably the matron who received a pineapple surprise splashed onto her bare back, wished they hadn't. A ceremony featuring Pat O'Brien finally began around midnight, three hours behind schedule. Meanwhile, someone stole Mayor Oscar Holcombe's chair, forcing him and his wife to sit in a hallway for two solid hours. "It was the worst mob scene I ever witnessed," the mayor fumed afterward. "It was ridiculous." A reporter for *Time* magazine wrote that the party "combined the most exciting features of a subway rush, Halloween in a madhouse and a circus fire." The *Chronicle*'s society editor dubbed it "bedlam in diamonds."

Bedlam it was, but many regarded the Shamrock's gaudy, chaotic, diamond-strewn opening as an apt metaphor for the new Texas. It proved to be exactly the media event McCarthy yearned for, drawing coverage around the world; *Life* ran a five-page photo spread. "Shamrock Puts Eyes of Nation on Houston," read the *Chronicle* headline. Overnight the Shamrock became not only the dominant symbol of Houston, but of Texas. People in Dallas, unsurprisingly, hated everything about it, nicknaming it the "Damn-rock." Still, every reporter who wrote about Texas visited the Shamrock, until its fame overshadowed anything else in the state. Most Americans, a San Antonio columnist wrote, "think of Houston as a cluster of mud huts around the Shamrock Hotel, in the cellars of which people hide from the sticky climate, emerging at long intervals to scatter $1000 bills to the four winds."

The morning after the hotel's unveiling, McCarthy opened the doors for guests. Business was strong those first few weeks. Tourists from all over the world poured through the front doors, ogling McCarthy's opulent "Texas Riviera," as the gossip columnists quickly dubbed it; one Englishman told a reporter the only things he knew about Texas were the Shamrock and Roy Cullen, and he intended to see them both. Dinah Shore and Mel Torme sang in the Cork Club; Frank Sinatra was booked for January. ABC began broadcasting a weekly radio show in the Emerald Room, *Live from the Shamrock*. The pool became the showcase where Houston's youngest, most beautiful, and richest women came to see and be seen. "I like it here," one was heard muttering. "It's like you were somewhere else—not in Houston at all."

To an outsider the Shamrock appeared to be exactly the shining new symbol of Texas of McCarthy's dreams. Inside the New York offices of the Equitable Life Assurance Society, however, his lenders weren't so pleased. The hotel was losing money, and fast. Red ink was to be expected in any new venture, but McCarthy's spending was simply out of control. He was laying out three thousand dollars a year for a golf pro, never mind that he didn't have a course, plus a half million a year for a promotional magazine named *Preview*, which McCarthy labeled a "cowpuncher" alternative to *The New Yorker*.[2] On Easter the Shamrock hosted what it called the largest Easter egg hunt in history; hotel workers spent days hiding more than ten thousand eggs all over the property.

McCarthy unveiled his biggest show on July 4, a fireworks display he boasted would be the greatest ever seen on earth. By dusk fifty thousand people had lined the streets around the Shamrock, a crowd dwarfing the throng of opening night; traffic was backed up for miles on all three lanes of South Main, families in fishtailed Cadillacs and pickup trucks scanning the sky. At nightfall skyrockets of every hue arced up over the hotel, exploding in gushers of yellow, blue, and green. "Some of the fireworks seemed to be improvements on anything ever set off before," the Chronicle reported the next day. "Their startling 'whoosh' sounded like the passing of a jet fighter, their whistle like the falling of 1000-pound bombs, and their bursting was like all glory let loose." The massive show ended with a series of rockets that burst into a giant green shamrock over Houston.

Every day, it seemed, McCarthy had something new to show the press. In August he led reporters through his newest toy, the Boeing Stratocruiser he bought from Howard Hughes for $500,000. Adorned with a wet bar, two desks, and sleeping berths for eight, the interior was done in uncharacteristically tasteful shades of pastel blue and pink. It was McCarthy's third plane. He named it The Gooney Bird. Meanwhile, at Houston's Fat Stock Show, McCarthy broke his own American record by purchasing the champion steer, an eight-hundred-pound heifer, for $15,400.

In his new role as Texas kingmaker, he threw himself into efforts to attract a professional football team for Houston. He sponsored an exhibition game dubbed "The Shamrock Bowl" at Rice Stadium between teams of NFL all-stars that December—Bob Hope led the entertainment—and afterward unveiled plans to build a covered stadium that could seat 130,000 people. McCarthy flew to Philadelphia to crash a meeting of football owners and make his case for Houston. He managed to corner Commissioner Bert Bell but got nowhere. Afterward, people snickered at the very idea of a covered stadium, but McCarthy, as usual, was ahead of his time. Sixteen years later Houston built the Astrodome.

All this activity, and all the expense, had the Equitable on edge. On June 30, 1949, three months after the Shamrock's opening, its board was given a fourth updated report on McCarthy's oil reserves; the valuation came in at fifty-nine million dollars, down fifteen million from the previous report, but still enough—just—to cover his outstanding loans. For Equitable executives, the first sign of real trouble came not from the oil fields or the Shamrock but from the chemical plant McCarthy had begun building east of Houston in 1946. Its construction, budgeted at four million dollars, came in at eight million. The plant was to use a new technology called the

"Bloodworth Process" to strip liquid petroleum gases out of natural gas and convert them into methanol for use as an antifreeze additive. Unfortunately, the plant didn't work. Engineers spent months tinkering with equipment but could never figure out why. McCarthy spent millions replacing machinery before finally giving up.

Equitable had first realized the plant was in trouble the previous summer when McCarthy approached its executives for money to fix it. They politely declined. Undeterred, McCarthy had pried fifteen million dollars from the Metropolitan Life Insurance Company in return for a lien on all his chemical-related assets and a written guarantee that the plant would be up and running in a year. When the deadline arrived that August, however, the plant was still comatose. With no money coming in, McCarthy was unable to pay his quarterly interest payment to Met Life that September. Far worse, he told Equitable executives he would need to delay his next three debt payments to them as well.

Alarm bells were already ringing at Equitable's headquarters when McCarthy stunned the company by announcing he had found a new oil field at New Ulm, eighty miles west of Houston. It should have been welcome news. But developing the field took money McCarthy didn't have. Lender and borrower arrived at a face-off: Equitable insisted it wouldn't lend McCarthy another dime until he paid what he owed; he was already giving Equitable half his income. McCarthy insisted he couldn't pay unless allowed to develop New Ulm. In the meantime, he began paying many of the Shamrock's entertainment acts with shares in the new field.

McCarthy found relief from his mounting financial pressures in bottles of bourbon and night after night of drunken revelry. And when McCarthy drank, he fought. That spring a Houston radio announcer sued him for eighty-seven thousand dollars for punching him at a party. In June McCarthy engaged a twenty-six-year-old Hollywood "producer" in a wild early-morning melee inside the Cork Club. The young man, William Kent, was in the club on McCarthy's invitation and was thus surprised when McCarthy accused him of insulting behavior and punched him in the head. Kent threw McCarthy to the floor and was sitting on his chest when four burly Shamrock waiters attacked him and, along with McCarthy, began chasing him around the club, overturning tables in little explosions of expensive stemware. As Kent raced down a hallway to safety, he heard McCarthy shout, "One Irishman can beat up eight Englishmen any day!"

September brought the strangest fracas to date. McCarthy had wagered fifteen hundred dollars on a Texas A&M football game, but there was a mix-up; the bookie thought McCarthy bet on the loser, not the winner. McCarthy angrily summoned the man who arranged the bet, a gambler named Larry Rummens. The ensuing discussion ended when Rummens called McCarthy a liar, at which point the new King of Texas leaped onto his desktop and, as Rummens stood before him, kicked his guest in the chest, then launched himself onto Rummens and began pounding him with his fists. The incident led to a set of nasty headlines and a lawsuit in which Rummens claimed McCarthy held him hostage at the hotel for two days.

The newspapers made light of such incidents; fistfighting and carousing were seen as part of McCarthy's larger-than-life persona. Some of McCarthy's peers, however, tried to calm him down. Roy Cullen's grandson, also named Roy Cullen, remembers his grandfather taking McCarthy into the pantry of the Cullen mansion and gently admonishing him to take it easy. Good families, Cullen suggested, wouldn't be seen at the Shamrock if it were viewed as a haven for hooliganism. "My father viewed Glenn as a very bad guy," recalls George Strake Jr. "We wouldn't have anything to do with him."

But McCarthy wasn't listening to anyone—even the FBI. In October, on the same day Larry Rummens sued him for $210,000, McCarthy received an extortion note from a would-be kidnapper who threatened to take McCarthy's family hostage unless paid $50,000. The money was to be placed in a nearby culvert and against the FBI's advice, McCarthy strapped on a shoulder holster, slid in a .38-caliber snub-nosed revolver, walked to the culvert, and threw in a reply note stating he needed more time to raise the money. The note was never retrieved, but the next day police arrested a Shamrock janitor, a onetime deputy sheriff named Raymond "Good Buddy" Chambers, so named because he ended his sentences with "good buddy." Chambers was later convicted.

Through it all, McCarthy's finances continued to deteriorate. Though word of his loan defaults hadn't leaked, by Thanksgiving rumors of financial strain were growing by the day. McCarthy laughed off the questions, but signs of distress were evident. In a matter of weeks he sold the Shell Building, closed the chemical plant, then sold the Detroit steel plant—ten months after buying it. He tried to sell the New Ulm field to Howard Hughes but couldn't. In October speculation about McCarthy's future spread to Washington when he was seen sliding into a side door at the White House for what officials told reporters was a private meeting with President Truman. Three months later McCarthy stunned the financial press by confirming that he had asked the federal Reconstruction Financial Corporation—and the president—for a seventy-million-dollar loan package, which, if approved, would be the largest government loan ever granted a private businessman in peacetime.

Still McCarthy denied he was in trouble, even as rumors swept Houston that the Shamrock was poised to close. When the *Chronicle* reported "curbside gossip" that he was "on the threshold of the poorhouse," McCarthy flatly lied, claiming all his operations, including the Shamrock and the moldering chemical plant, were running at a profit. "I have no problem I will not be able to overcome within a very short time by gearing operations to my income," he told a reporter. His only problem, McCarthy claimed, was new competition from low-priced Middle Eastern oil, which was being imported at $1 a barrel at a time Texas oil cost $2.65. "Every time a barrel of foreign oil comes in America, a barrel less is produced in Texas," he groused. "Texas is taking most of the licking, but the rest of the country will feel it before long."

No one around Houston seriously believed McCarthy could go under. An oilman going bankrupt in Texas? He was too big, people said, the Shamrock too glamorous, the times too giddy to even consider such a thing. McCarthy's legend, in fact, continued to grow. Just three weeks after disclosure of his government-loan request, he reaped the ultimate American accolade: the cover of *Time* magazine. Beneath an oil derrick adorned with cowboy boots, a ten-gallon hat, and flexing its muscles Adonis-style, the headline read: "Texas' Glenn McCarthy: Since Spindletop a Jillion Jackpots." The package inside, including a map of Texas titled "Land of the Big Rich," marked the apex of two years of nonstop Texas hype in the national press. "The Lone

Star State," it noted without irony, "is one of the few places left in the world where millionaires hatch seasonally, like May flies."

Texans, for the most part, ate it up. For now.

III.

The media's coverage of Glenn McCarthy spawned a new cultural icon, the Lone Star playboy, the swinging oilman who romances starlets between trips on his airplane to see his next gusher. In the postwar years many of the stories fueling this caricature emanated from Hollywood, where McCarthy and other wealthy Texans, like the nouveau riche of every American generation, were drawn to the glamour and glitz and welcomed by money-hungry movie producers and a bevy of young actresses all too happy to take up with Texas sugar daddies. As a Hollywood columnist wrote in 1954, "The Texas jillionaires seem to gravitate to the motion pictures like a moth to a candle."

Among the first to arrive was Jack Wrather, a Dallas oil heir who, bored with life in the oil fields, moved to Los Angeles and married the actress Bonita Granville in 1947. Wrather used his fortune to produce seven movies in the next several years, then branched into television, eventually producing *Lassie* and *The Lone Ranger*. In time he purchased the Queen Mary entertainment complex in Long Beach, built the Disneyland Hotel in Anaheim, and became a founding member of Ronald Reagan's "Kitchen Cabinet" of political advisers.

By and large, though, the Texas playboys took home far more actresses than Oscars. The starlet Jane Withers wed an Odessa oilman and would-be producer named Bill Moss in 1947; after their divorce, Moss married the dancer Ann Miller. The Moss-Miller wedding, in La Jolla, California, brought together an increasingly common constellation of Hollywood stars and Texas oilmen, everyone from Ginger Rogers to Clint Murchison. Miller, however, couldn't cope with Moss and his hard-living oilman buddies, whose lives were a series of drinking binges wrapped up around evenings at night spots like Ciro's in Hollywood and the Cipango Club in Dallas. "I simply didn't have the energy or the patience to keep up with him, especially when we went out to parties that lasted two or three days," Miller wrote in her autobiography.

When Miller and Moss divorced, she married another Texas oilman, Arthur Cameron, who had purchased Louis B. Mayer's Benedict Canyon mansion and built one of the largest private estates in the desert outside Palm Springs. On their European honeymoon Cameron summoned a man from Harry Winston's, the New York jeweler, and bought Miller a 20-carat white diamond. Cameron, however, was a serial philanderer, and Miller left him several years later upon finding him engaged in an impromptu pool party with seventeen young ladies in bikinis.

The Houston oilman W. Howard Lee, meanwhile, romanced and married the actress Hedy Lamarr; the two lived for several years in River Oaks, until their divorce. Afterward Lee married the actress Gene Tierney. One of Clint Murchison's closest friends, the Dallas oilman E. E. "Buddy" Fogelson, wed Greer Garson at his New Mexico ranch in 1949. A six-time Academy Award nominee—she won the 1942 Oscar for best actress in *Mrs. Miniver*—Garson all but gave up films after her marriage and settled into a long, quiet life in Dallas with Fogelson, who died in 1987. Garson became a beloved Texas philanthropist, endowed the Greer Garson Theater at Southern Methodist University, and died in 1996.

Money, not romance, was at the heart of most Texas-Hollywood partnerships. One of Fort Worth's richest oilmen, W. A. "Monty" Moncrief, took to wintering in Palm Springs, where he began playing golf with a number of stars, including Bob Hope and Bing Crosby. When the two actors expressed an interest in investing in an oil well, Moncrief cut them into shares of a new West Texas field for a pittance. Following a single dry hole, after which Moncrief had to explain to his new friends that not every well actually struck oil, he hit a string of twenty straight producers. By Moncrief's estimate years later, Hope and Crosby walked away from their investments more than ten million dollars richer.

Actors weren't the only stars Texas oilmen took under their wing. D. H. Byrd and a group of oilmen were among the first sponsors of a piano prodigy named Van Cliburn. In 1951 Sid Richardson befriended a young evangelist who arrived in Fort Worth to deliver God's message. His name was Billy

Graham. Richardson, though never known for his piety, took a liking to Graham, introducing him to Murchison, Sam Rayburn, and Lyndon Johnson. When Graham attempted to deliver a radio broadcast from the Capitol building in Washington, he was told it was impossible; a single phone call from Richardson to Rayburn led to a special Act of Congress that allowed Graham to hold the first-ever religious service on the Capitol steps. Graham was so inspired by Richardson and other Texas oilmen he met that his fledgling film company's first two movies, Mr. Texas and Oiltown U.S.A., told the stories of hard-living Texans who found salvation in Christ. Oiltown U.S.A., which Graham debuted at a Hollywood premiere and aired at his crusades, featured a Houston oilman whose conversion, in the words of promotional materials, featured "the development and use of God-given natural resources by men who have built a great new empire."*

IV.

As colorful as the jet-setting Texas playboys were, the oilman eastern writers most wanted to profile proved elusive. His name was James Marion West Jr., though he was known in Houston as "Silver Dollar" Jim West. West was the son of Roy Cullen's onetime partner, Big Jim West, who died in 1940, leaving his family an estate of seventy million dollars, which doubled during the ensuing decade. Jim and his brother Wesley built homes two blocks down from the Cullen mansion in River Oaks, but there comparisons between the families ended.

Jim West was hands down the most flamboyant of all Houston oilmen. A roly-poly five feet nine and 210 pounds, he waddled the streets of downtown wearing cowboy boots, eight-inch-wide belt buckles adorned with oil derricks or dancing cattle, his trademark orange-flannel shirt, and a pistol, usually a .45, strapped to one hip. A smiling if sometimes cantankerous

---

*Richardson remained a quiet supporter of Graham's the rest of his life, at one point bankrolling a white-tie dinner at London's Claridge's hotel in which Graham preached the gospel to almost two hundred members of the British social elite.

prankster, he got his nickname, "Silver Dollar," for his habit of tossing dollar coins everywhere he went, at doormen and yardmen as tips, on a sidewalk for bums, at reporters for a grin. When a local union struck the downtown parking garage he owned, West broke up the picket line by stepping to his second-story office window and scattering silver dollars all over the sidewalk below. He kept the coins in massive racks in his basement; it was a Negro servant's daily duty to dust them.

George Strake Jr. remembers diving for silver dollars in West's pool as a teenager. "Except he always threw them into the deep end, and I got an earache, so my parents wouldn't let me go any more," Strake recalls. West never left home, in fact, without stuffing silver dollars into two large saddle pockets he had custom-sewn into all his trousers; he jingled when he walked. At hotels he took plastic "cartwheels" of dollars to hand the help; the Shamrock doormen got three wheels just for retrieving his Cadillac.

West had the usual oilman toys, a DC-3, a Twin Beech, and two converted trainers, plus several ranches—he liked to say he wasn't sure how many—and several thousand cattle. He wasn't happy about anyone telling him what to do. When he got into a quibble over his water bill, he drilled two artesian wells for his own supply. When he judged his electric bill too high, he bought a massive diesel power plant to provide his own electricity. He was especially ingenious in thwarting the Houston zoning commission, which outlawed tennis courts with backstops. He built a court anyway, with a deep slit dug beside it; at the press of a button, a backstop rose. If an inspector appeared, he pressed another button and it sank into the ground.

West's passion, though, was police work. He spent most nights either riding in patrol cars or racing to crime scenes in his own. Beside his bed he had an entire bank of radios and electronic equipment to monitor calls. He owned thirty cars, including eleven Cadillacs, stored in a six-car garage at home—it was connected to the main house by a 225-foot white-tiled tunnel—and the three-story garage he owned downtown. Each car carried a sawed-off shotgun, a Thompson submachine gun and, beneath the dashboard, four telephones, one wired to a police wavelength, another for the sheriff's department, and two for personal use; there were twenty-four more telephones at home and a dozen at the garage. Several of West's cars had

racks for tear-gas canisters, which he wasn't shy about using. When a crowd
of teenagers descended on his home one Halloween, he dispersed them with
a cloud of gas.

The Houston police loved him. At Christmas every captain on the force
received a crisp hundred-dollar bill, every lieutenant a fifty, every sergeant
twenty five silver dollars, and every patrolman ten. One cold winter West
ordered a gross of European sheepskin coats and handed them out to police-
men. When the Texas Rangers gave him an honorary commission and a
gold Ranger star to wear, he had it encircled by nine large diamonds and
wore it religiously. West made it to every national police conference and just
about every shooting in Houston; often he arrived before the police. His
nocturnal missions were seldom dull. Returning from a jaunt at 3:30 one
morning in May 1949, he fell asleep at the wheel on West Gray. His Cadillac
struck a parked car, careened across a lawn, and smashed through a concrete
mailbox before plowing through a hedge. Startled awake, West went to press
the clutch but pressed his siren instead, waking the neighborhood. A crowd
in bathrobes and slippers was soon on the scene, but no one was hurt.

One night in October 1954, during a time of tensions between Mayor
Roy Hofheinz and the police chief, West heard that Hofheinz was throwing a party for a hundred friends. West bet his best friend, Lieutenant A.
C. Martindale, a Coca-Cola that Hofheinz didn't have a hundred friends.
When West parked his Cadillac in front of the mayor's home, Hofheinz saw
them and assumed they were spying on him. West and Lieutenant Martindale drove away, but the mayor leaped into his son's car and gave chase,
following the pair just long enough to identify the car. Two days later the
episode hit the papers, making everyone look silly.

Not all West's adventures were so harmless. One night in 1952, he was
accompanying Lieutenant Martindale on patrol when they responded to
reports of a burglary downtown. A man had broken the front window of
a sporting goods store and taken a shotgun. West and Martindale spotted
the man on Dowling Street; the lieutenant jumped from the car and yelled
"Halt!" The burglar opened fire with the shotgun. When a police cruiser
reached the scene minutes later, the burglar was down and bleeding; so
was Lieutenant Martindale, shot in the shoulder and in the ankle. When a

ballistics report indicated that the bullets that struck Martindale had been fired by different guns, the Houston *Post* reported speculation that West had mistakenly shot his friend. Martindale defused the crisis, however, insisting that he had accidently shot himself. The *Post* took West to task in a front-page editorial anyway; the headline read, "Let's Disarm the Amateurs."

Many readers agreed. In fact, it was a sign of Houston's gradual maturation that, when a *Collier's* writer finally cornered him for an interview in 1953, Silver Dollar Jim West felt his fellow citizens were beginning to view him as an embarrassment; he knew he was the kind of caricature many Americans expected a Texas oilman to be, and he didn't give a damn. But what really galled him was mail. Publicity had triggered an avalanche of money-seeking letters to several top Texas oilmen—Roy Cullen got bags a day—and West couldn't stand it. Some of his mail was marked, "Any Millionaire, Houston."

"I just hate to get letters!" he told the man from *Collier's*, brandishing one envelope. "Man here wrote me from New Orleans and asked for five thousand dollars. Said he'd insure himself for ten and commit suicide in case he couldn't pay it back in a year. That's awful, ain't it? Trouble with the bunk you write, people think I'll send 'em money, or a maybe a Cadillac. Tell them I won't answer."

V.

By the time Glenn McCarthy appeared on the cover of *Time* in February 1950, ten months after the Shamrock's opening, he was already in deep financial trouble. Unable to make debt payments to either the Equitable or Metropolitan Life, he was facing the unimaginable: a foreclosure, loss of the Shamrock, loss of his dreams. All through the early months of 1950 he endured a series of make-or-break meetings with his creditors in Houston and New York. When Metropolitan ordered a new estimate of McCarthy's gas reserves, it was stunned to find they were down by almost two-thirds in several key fields. While McCarthy blamed the drop on foreign oil, Metropolitan's engineers deduced that the greatest damage to McCarthy's reserves had been administered by McCarthy himself. In his rush to raise cash, he had pumped so much natural gas so quickly that well pressures could not be maintained, rendering wells less productive and in some cases useless. Auditors uncovered telltale hints of financial chicanery. Wells had been completed and put on McCarthy's books when it was doubtful they would ever produce commercial amounts of oil or gas. The lesser wells—whether weak from birth or crippled by McCarthy's haste—had been abandoned in droves without telling the creditors.

Behind closed doors, executives at Equitable and Metropolitan Life pondered whether to foreclose. They ultimately decided against it, fearing the public-relations nightmare of crippling a Texas legend and recognizing that the Shamrock, if run properly, might yet turn a profit. In May 1950 McCarthy hammered out a deal in which Metropolitan ponied up an additional six million dollars in cash in exchange for liens on his New Ulm field. Equitable, however, was out of patience; its board decided the only way to bring McCarthy's spending under control was to take control itself—gradually and in secret, so as not to antagonize McCarthy or those in Texas who regarded him as a hero. Months of secret meetings culminated in an agreement signed January 15, 1951, in which McCarthy reluctantly ceded Equitable complete control over all his oil fields, the Shamrock and just about everything else he owned for ten years. McCarthy would still run things, but under the watchful eye of a comptroller Equitable brought in to oversee his finances.

It took three days for the deal to fall apart. On January 18, an updated analysis of McCarthy's overall reserves reached Equitable's board. It was catastrophic. Reserves were down a third. Equitable's board met and decided gradual control would no longer do. It needed immediate and total control. At a meeting in Houston on March 1, the insurer's executives ousted McCarthy as president of his companies—he remained chairman—and replaced him with a squat Equitable attorney named Warner H. Mendel. In effect, Mendel became McCarthy's new boss. Equitable needed McCarthy's continued presence if the Shamrock was to maintain its stature; any hint that Equitable had pushed him aside, an internal report noted, would lead Texans to view it "as the big, bad absentee Scrooge." Mendel's mandate, as stated in another report, was "to obtain complete control of the operations of McCarthy Oil & Gas and the Shamrock Hotel, while preserving sufficient anonymity to assure McCarthy's cooperation."

For Warner Mendel, it was the job from Hell. McCarthy had gone along with the change because he had no choice, and because he felt he had wrung from Equitable a pledge "to keep your damned hands off the hotel." But of course that was the whole point. Mendel moved to Houston, overseeing McCarthy Oil & Gas from 6 A.M. to 7 P.M. each day and running the hotel until 11 every night. When he began slashing costs, McCarthy resisted. It was no use. Mendel fired the golf pro, killed the magazine, and suggested that the next time McCarthy was in New York, he might inquire as to the market for used private planes.

The Shamrock spiraled into an internecine war between McCarthy and Mendel. When McCarthy proposed spending fourteen thousand dollars to fly in a planeload of Hollywood stars for the Shamrock's third-anniversary celebration, Mendel said no. Over howls of protest from McCarthy, Mendel cut back the July 4 fireworks display. Matters came to a head when Mendel questioned the personal expenses McCarthy was billing to the hotel. At that point, McCarthy, in the words of an internal Equitable report, "threatened to use physical violence upon Mendel, a man of slight stature." Mendel stood his ground. As the Equitable report concludes: "Violence did not ensue, and Mendel told McCarthy that 'there were to be no further illusions as to where the final control lay on matters relating to hotel operations.' "

McCarthy withdrew to his mansion, beaten. No matter how hard he tried, he realized he couldn't best the Equitable, not with his fists, and not in court. But he was damned if he was going to be a glorified greeter at his own hotel. He was still *Glenn McCarthy*. He was still a living legend. He was only forty-three, still a young man. If he could make a fortune once, McCarthy reasoned, he could do it again. But not in Texas. The big fields had all been found. And then it hit him: the Middle East. He would beat the oil importers at their own game.

Rumors that McCarthy was mounting some kind of mysterious Middle Eastern excursion were already swirling when he abruptly disappeared from Houston in early November 1951. "McCarthy on Mystery Oil Junket," read the *Houston Post*'s front-page story, which speculated that McCarthy might be heading to Egypt to hire the famous belly dancer, Samia Gamal, who had just married a Texas playboy named Sheppard King III. "Glenn has an addiction and attraction to dancers," Warner Mendel wryly noted, "[so] maybe he is." When a reporter tracked him down at New York's St. Moritz Hotel, McCarthy said he couldn't speak publicly, explaining, "I don't think the United States government would like me talking about its business." Another round of front-page stories ensued, now suggesting that McCarthy was engaged in a diplomatic mission or some kind of government espionage. No, no, no, the oilman told a second reporter the next day. It was personal business, and he couldn't talk about it.

A few days later reporters trailed McCarthy to Paris, where he was staying at the Hotel George V with Howard Hughes's onetime PR man, Johnny Meyer. Each morning a new round of speculation erupted in Houston: Where was McCarthy going? And what was he up to? The answer came when a wire-service reporter spotted him in Cairo, and McCarthy admitted he was there to look at the prospect of obtaining an oil concession in the Middle East, though he wouldn't say where. "You can rule out Iran and Saudi Arabia," his PR man whispered.

A week later McCarthy resurfaced, again in Cairo, grumping about an Egyptian law mandating that any Egyptian oil company had to be at least 51 percent owned by Egyptians. Reporters refused to believe there wasn't more to the trip, repeatedly asking McCarthy whether he had come to sign Samia Gamal. "I came out here to see if we could buy up some land and drill for oil," he snapped, "not to sign up any dancers." From there it was on to Rome, where he stopped by to see the pope.

Still McCarthy would not give up. All that winter he engaged in a frenzy of would-be deal-making. Returning to Houston, where he insisted he would prevail upon the Egyptian parliament to change its ownership laws, he announced plans to build the largest hotel in Latin America, a six-million-dollar resort in Guatemala City. Next he said he had purchased exclusive rights to broadcast television in Guatemala. A week after that he confirmed he was exploring a massive oil deal with the government of Venezuela. Then an entire chain of Shamrock hotels across America. To top it all off—and to have any hope of financing his fanciful schemes, not one of which ever saw the light of day— McCarthy announced he was starting a new oil company and intended to sell its shares to the public on the New York Stock Exchange.

This was too much for Warner Mendel and the Equitable's board of directors. They wanted McCarthy where he could do them the most good, shaking hands at the hotel and, most of all, sorting out the mess at his oil company. Mendel gave McCarthy an ultimatum: devote himself full-time to the Shamrock and his oil fields, or else. McCarthy refused. Mendel demanded that McCarthy hand over his stock, giving the Equitable outright ownership of his companies. McCarthy refused.

At that point, Mendel wrote the board that McCarthy had outlived his usefulness. He was doing more harm than good at the hotel, and Mendel couldn't see anyone buying McCarthy's oil properties—which appeared to be the only way Equitable could ever recoup its investment—if it meant dealing with McCarthy. Mendel wanted to foreclose, but before doing so he made McCarthy one final offer. In exchange for signing over all his stock, and a promise to refrain from interfering in the Shamrock's operations, McCarthy could keep a suite at the hotel and spend "a limited amount of money" on business entertainment. McCarthy would retain only "redemption rights," that is, the rights to regain ownership in the unlikely event his debt was ever repaid, which at the Shamrock's current rate of profitability, Warner Mendel estimated, might occur as early as 1977.

McCarthy signed. He had no choice. Everyone issued upbeat statements to the press, and the Houston newspapers, backing McCarthy to the end, ran stories emphasizing his tenuous chance at regaining ownership. Still, everyone knew it was over. A scant three years after its legendary opening, the Shamrock was no longer Glenn McCarthy's. For McCarthy, the only good news was that the national publications that celebrated his rise generally failed to notice his fall. In those last few months, in fact, there had been just one eastern writer snooping around the hotel. McCarthy had met her once or twice and forgotten about it. She said she was writing a book. Her name was Edna Ferber.

VI.

Probably the best-known female novelist of mid-century America, Edna Ferber was a creature of effete New York literary salons, a witty, headstrong member of the Algonquin Round Table and a lesbian, all of which made her an improbable chronicler of muscular, nouveau riche Texas. She had been famous since the mid-1920s, when one of her early novels, *So Big*, won the Pulitzer Prize. Ferber had been mulling over a novel about Texas since at least 1939, when she first toured the state. At the time, she found Texas too foreign, too outlandish, too *big* to easily grasp. There was a great American novel there, she decided, but not hers. "Let Michener write it," she told a friend.

All during the 1940s, however, Texas called to her. Though Ferber was always vague about details, her decision to actually tackle a Texas novel coincided with the 1948–49 boom in stories about its new millionaires. "Texas," she once explained, "was constantly leaping out at one from the pages of books, plays, magazines, newspapers.... The rest of the United States regarded it with a sort of fond consternation. It was the overgrown spoiled brat, it was Peck's Bad Boy of today.... This Texas represented a convulsion of nature, strange, dramatic, stupifying." She was especially curious about Glenn McCarthy. During one of several visits to Texas the two were introduced. Neither ever spoke meaningfully about their meeting.

There was no mistaking McCarthy, however, as the model for the tempestuous wildcatter Jett Rink who sprang from the pages of the book Ferber decided to call *Giant*. *Giant* is the story of an oil and ranching family clearly modeled on the Kleberg clan who owned the vast King Ranch. The Rink/McCarthy character brings all the main characters together in the book's early scenes for the massive opening of his "El Conquistador Hotel" in the sprawling city of "Hermosa." Ferber captured the Shamrock's gala in vivid detail, down to the malfunctioning PA system. Jett Rink first appears staggering onto the Conquistador's stage on page 49, and by page 50 has already knocked another character senseless with one punch. Everyone in the book appears to own a private airplane and boasts incessantly; there's even a visiting king and queen, a nod to a weekend the Duke of Windsor and his wife spent at Clint Murchison's Mexican ranch in 1950.

Serialized in *Ladies' Home Journal* beginning in the spring of 1952, *Giant* was released that summer to immense sales, quickly leaping to No. 1 on the *New York Times* Bestseller list. In a matter of weeks the book brought Ferber's rich caricatures of nouveau riche Texas and its "oilionaires" to a far broader audience than any magazine writer. Critics, at least in the East, ate it up. "*Giant* makes marvelous reading—wealth piled upon wealth, wonder on wonder in a stunning, splendiferous pyramid of ostentation," a *New York Times* reviewer wrote. "[Ferber] paints a memorable picture of that new American, *Texanicus vulgaris*, which is all warts and wampum."[3]

Texas critics weren't so kind; they began savaging the book even before it was published. *Giant*, a reviewer for the *Houston Press* wrote, "is the most gargantuan hunk of monsterous, ill informed, hokum-laden hokus-pokus ever turned out about Texas." "Ferber Goes Both Native and Berserk: Parody, Not Portrait, of Texas Life," read the headline atop the *Dallas Morning News* review. "For sheer embroidery of fact—an art of which Texans are rarely surpassed—Miss Ferber takes the cottonseed cake," reviewer Lon Tinkle wrote. "She has us all riding around in our own DC-6's. At first blush, you might call it, her book on ranch and oil empires of South Texas reads like a parody of that grand old melodrama, 'A Texas Steer.' But Miss Ferber has written a bum steer.... *Giant* is a triumphant parade of platitudes."

It was what everyone expected Texans to say. In the end, of course, all efforts to fight the myth of brawling, hard-drinking Texas oil millionaires proved fruitless, in large part because there was so much truth to it. Thanks to Glenn McCarthy and oilmen like him, *Giant* the book, and four years later the motion picture starring Elizabeth Taylor and a young actor named James Dean, completed the new picture of Texas that eastern writers had been painting since the day *Life* ambushed H. L. Hunt on a Dallas street corner in 1948. It was an image of frivolity fueled by condescension, a cartoon that irritated many Texans even though it remained essentially sunny and upbeat.

It wouldn't stay that way for long.

VII.

With the last of his dreams evaporating around him, Glenn McCarthy was in no mood to discuss Edna Ferber or her book, icily refusing to answer any reporter's question. Not till years later would he even acknowledge meeting

Ferber. "I thought she was a perfect lady," he once said. "A perfect whore, rat, sneak, thief."

By early 1953 McCarthy had given up all efforts to regain control of the Shamrock. He kept his office there, coldly staring at Warner Mendel in the hallways and cold-cocking a carpenter he suspected of being an Equitable spy. For the most part, though, he made himself scarce. In February 1953, after several scouting expeditions to Nicaragua, Argentina, and Guatemala, he announced he would finally launch a South American drilling program, on a million acres he leased in Bolivia. He tried to sound excited about the venture, taking along Houston reporters to watch him begin drilling, but it felt like exile. His first three wells found natural gas, but McCarthy was soon obliged to shut down production. Bolivia had no pipelines, and thus no way to transport the gas. McCarthy fumed and waited for one to be built.

The Equitable, meanwhile, slowly broke apart his empire. It sold his oil fields and pipelines piece by piece over the next few years. In August 1954 it handed over management of the Shamrock to Hilton Hotels. A year later Hilton bought out McCarthy's redemptive rights, severing his last ties to the Shamrock, which was renamed the Shamrock Hilton. Workers lowered the giant portrait of McCarthy in the lobby and replaced it with one of Conrad Hilton. It was the last most of Texas would see of McCarthy for a very long time. McCarthy himself, like a latter-day Butch Cassidy, disappeared into the wilds of Bolivia, vowing to return with a second fortune. For a long time the only indication he was even alive came in the odd stories that floated up from South America every few months, none of which, as with the case of an Ecuadoran doctor he attacked in a drunken rage on a flight to La Paz, suggested Glenn McCarthy was walking gently into the night.

TEN

"A Clumsy and Immeasurable Power"

It's funny about Texans.
They have to hate somebody, a whole lot of them.
—ALLA CLARY, SAM RAYBURN'S SECRETARY

I.

The mass media's discovery of ultrawealthy Texas oilmen in 1948, and the resulting caricature of flamboyant, jet-setting billionaires popularized in *Giant*, introduced the country to a new regional archetype—funny, silly, harmless Texans who rode ostriches, wooed Hollywood stars, and scattered silver dollars on the sidewalks of Houston and Dallas like so much pocket lint. It was as if America had acquired an exotic new animal for the national zoo, *Texas oilicus*. It didn't take long for those ogling the rough-hewn beast to realize it had teeth, and it intended to use them in ways no one had foreseen.

The venue would be national politics. In 1948 few outside the inner ranks of Washington fund-raisers appreciated what the new Texas millionaires were capable of, much less their intent. Insiders appeared to regard Sid Richardson and Clint Murchison as reliable Democratic donors but little more, cash-heavy martinets controlled by Lyndon Johnson. Roy Cullen's bankrolling of Texas ultraconservatives, and his penchant for firing off angry telegrams to half the Senate, had made him a curiosity, a humorless, self-important old man who deluded himself he was a power in politics. Hunt remained a cipher.

What few understood was that America was on the brink of a new school of political thought, modern conservatism, and the Big Rich would be a driving force in its spread. To most, the rise of postwar conservatism came out of nowhere, a notion enshrined by Lionel Trilling's observation in 1950 that "liberalism is not only the dominant, but even the sole, intellectual tradition" in the United States. America in 1950 had not a single leading politician who could be termed conservative by today's standards and, other than *Human Events*, an eight-page newsletter launched in 1944 with a readership of 127, not a single conservative media outlet of note. Until the late 1940s American conservatives, what few there were, remained relegated to the fringes, a motley collection of marginalized academics and midwestern businessmen whose interests ranged from World War II isolationism to the extremes of racism and anti-Semitism. To the extent it noticed, the mainstream press dismissed them as kooks and fanatics.

Despite this, in the first years after the war a number of groups, each with its own pet issue, began to coalesce into something approximating a consistent conservative philosophy. One consisted of former Communists and Socialists, many of them European emigres, who felt America wasn't doing enough to contain Russian and Chinese communism. Onetime Trotskyites such as James Burnham fanned the flames of anticommunism, which exploded into a national issue during the prosecution of the diplomat Alger Hiss in 1949. A second group, led by a pair of eccentric cape-wearing writers, Albert Jay Nock and Ayn Rand—one historian dubbed them "the caped crusaders"—railed against what's known today as "big government," the idea that the spread of federal power under the New Deal would lead to totalitarianism. Many leading Texas oilmen, including Cullen and Hunt, drew ideas and inspiration from these two groups but belonged squarely to a third, southern traditionalists opposed to almost all aspects of Rooseveltian thought, especially civil rights and the winnowing of states rights.

By 1950 a series of books had appeared laying out the philosophical elements that would comprise American conservatism: anti-Communist, anti-labor, pro-religion, pro-business, favoring limited government and opposing civil rights. Conservatives were not yet identified with a political party; given that the feeble Republicans hadn't won a presidential election since 1928, most conservatives, especially southerners, remained uneasy Democrats, as evidenced by Strom Thurmond's breakaway Dixiecrat candidacy of

1948. What conservatism needed was a leader, a prophet, and in 1951 it got one in the son of a Texas oilman, a Yale undergraduate named William F. Buckley, who attacked the trend toward atheism and collectivism in his book God and Man at Yale. Buckley's father had worked in the Mexican oil fields; as a boy, his babysitters included Mr. and Mrs. George Strake. In 1955, with money raised from his father and friends, including a Houston oilman named Lloyd Smith, Buckley would found The National Review, which became the crucible for conservative thought. Until 1955 the rising voices of conservatism remained far-flung and disjointed. Many of the loudest, though, were coming from Texas.

II.

When it finally dawned on the rest of America how very, very conservative Texas oilmen were, many asked why. The best reason was the simplest: the deep-tissue insecurity of the nouveau riche. As one oilman told a magazine writer in 1954: "We all made money fast. We were interested in nothing else. Then this Communist business suddenly burst upon us. Were we going to lose what we had gained?"[1]

Roy Cullen certainly thought so. In the summer of 1947, fresh off the wave of publicity that accompanied his donations to the University of Houston and local hospitals, Cullen began to flex his political muscles. His years of "intermittent fire," as he termed his telegrams to congressmen, were over; he turned sixty-six that year, and if he was to have any chance of halting the march of "creeping socialism," it had to be now. His resulting initiatives, though alternately ignored and ridiculed by historians, were to have a lasting impact on Houston, on Texas, and, to some small degree, insofar as they constituted an early effort to push the Republican Party to the right, the nation.

The first fight Cullen picked, however, was a purely local affair. Not by accident it brought him into direct conflict with his longtime rival, "Mr. Houston" himself, Jesse Jones, the banking magnate who had served in several positions during the Roosevelt administration. More than one

Texan viewed the subsequent tussle as Cullen's attempt to dethrone Jones and the so-called Suite 8-F crowd, a cadre of downtown power brokers so named for the hotel suite where they met. Oddly enough, it all began with the white camelias in Cullen's garden. One morning he noticed they had turned yellow. Sniffing the air, he guessed the cause: toxic smoke from a new paper mill on the edge of Houston. At his office, he could smell the mill's odors wafting in on an easterly wind. He was forced to shut his windows. "If I believed in Hell," Cullen complained to a lunch partner at the Houston Club, "I would say the odors of the paper mill and those that steam out of the cracks in Hell must be very similar."

When a new mill was proposed for a site east of his beloved University of Houston, Cullen declared war, firing off a series of angry letters to the newspapers. At one point, he threatened to leave Houston if the mills weren't stopped. Told that the mills' locations were a question of zoning, Cullen began a study of Houston's ordinances. What he discovered changed not only his philosophy on the mills, but the focus of his anger. The zoning board seemed to have complete control over who could build what where in Houston, and the more he studied, the more Cullen realized that the board's guiding hand was none other than Jesse Jones. Suddenly the mills were forgotten. What began as a dispute over noxious odors evolved, in Cullen's mind at least, into a debate over zoning laws and individual freedom.

Both headstrong, independent men, Cullen and Jones had been squabbling for years. After a disagreement in 1946, Cullen wrote Jones: "Our philosophies of life are so different. You build houses of mortar, stone and steel, while I build Man." Theirs was the rare public dispute between nouveau riche oilmen, few of whom had involved themselves in their communities, and the downtown businessmen who ran Dallas and Houston. The rancor grew when Cullen managed to push through a referendum on whether Houston should have *any* zoning. Jones, joined by the Suite 8-F crowd, fought hard for the status quo. Cullen gave no quarter. In one letter published in the *Chronicle*, he wrote that it had been "a pleasure to help build this city up to now, but Jesse Jones has been away (in Washington) the last twenty-five to thirty years and has...decided, with the influence of the press here, and the assistance of a bunch of New York Jews, to run our city."

In the 1948 referendum, Houston voters backed Cullen's position by a 2-to-1 margin. To this day, Houston is the only American city without major zoning laws, a distinction it owes largely to the efforts of one man, Roy Cullen.

The victory appeared to embolden Cullen. All through 1948 and into 1949, letters spewed from his typewriter, to the *Chronicle*, the *Post*, the *Press*, to Harry Truman and every congressman in the country. A public figure now, recognized as the Southwest's greatest living philanthropist, Cullen threw himself into public life with abandon, delivering speeches that set the tone for booming Houston to soon be recognized as a national center for ultraconservative views. In one speech, he called for the impeachment of "sixty percent" of sitting Supreme Court justices. In October 1948 he blasted a pending civil-rights law that sought to make lynching a federal crime, abolish poll taxes, and outlaw segregation in interstate commerce. "If [these bills] become law, totalitarian government will follow, and that means atheism," he wrote the local papers. "If enforced, [they] must result in a police state, and in a Communistic government, and in the end of our freedom and of our democratic form of government."[2] Sometimes he got carried away, as when he went on the University of Houston's radio station and called Secretary of State Dean Acheson "a homosexual," at which point the station abruptly went off the air.

One thing Cullen didn't announce, though it was clear to those few paying attention, was his plan for the Republican Party in Texas. The hapless Texas GOP was a second party in a one-party state, a cadre of nattering nobodies who hadn't won a single statewide election since 1874; in the 1914 elections, the Republicans actually garnered fewer votes than the Socialist Party. What the party stood for was anyone's guess; its reputation, as the political historian V. O. Key Jr. noted of Southern Republican parties in general, wavered "somewhat between an esoteric cult on the order of a lodge and a conspiracy for plunder."

Change had begun in the late 1930s when a number of Texas oilmen, most notably Cullen's friend Marrs McLean of Beaumont, joined the party to fight Roosevelt. Cullen had voted Republican off and on since 1938 but, as a practical man, kept his hand in Democratic politics because it remained the only politics that mattered in Texas. Yet as far back as 1938 Cullen had a vision, of a Republican Party shoved sharply to the right, a party of onetime conservative Democrats fed up with FDR and creeping socialism, a party to be reckoned with. He took to discussing his ideas with an energetic young oilman whose offices adjoined his own, Jack Porter, and found him in agreement.*

In December 1946 Cullen took Porter to Washington, where they met with the new Republican Speaker of the House, Joseph W. Martin of Massachusetts, who replaced Sam Rayburn when the Republicans took Congress that November. Cullen had been courting Martin for years, throwing him a fund-raiser in Houston that fall, and Martin was seemingly among the few in Washington who took Cullen seriously. Probably not expecting much, Martin encouraged the two in their plans to take over the Texas party. Porter got himself appointed head of a committee seeking a Republican challenger to Lyndon Johnson's senatorial bid in 1948, and when one couldn't be found, he ran himself.

Cullen publicly endorsed him. That November, backed by cash from Richardson, Murchison, and other oilmen, Johnson beat Porter easily, but Porter did far better than most observers expected. The senatorial bid instantly made Porter the best-known and most active Texas Republican, although insiders accepted that he was essentially a proxy for Cullen. In 1949 Porter began a far-reaching effort to recruit new Republicans, focusing on the resuscitation of the dormant Young Republicans organization, which became a power base for the Cullen-Porter forces.

As Jack Porter laid the groundwork for a takeover of the Texas GOP, Cullen decided to step up his efforts on the national level. In late 1949, livid over Truman's nomination of a liberal economist named Leland Olds to head the Federal Power Commission, Cullen and Porter announced a "grass roots campaign" directed at defeating the reelection efforts of the fifteen senators who had supported Olds or, as Cullen dubbed them, "the inner sanctum of the New Deal–Fair Deal politburo" in Washington. The resulting storm of

---

*Porter was already a leader among Texas independents, having spearheaded, along with Glenn McCarthy, their opposition to the Anglo-American treaty in 1944. Porter and McCarthy went on to cofound one of Texas Oil's largest lobbying arms, the Texas Independent Producers and Royalty Owners Association.

letters and telegrams to editors and congressmen had little obvious effect, at which point Cullen appears to have realized that words alone wouldn't change the status quo. In American politics, the Texans were slowly learning, arguments only mattered when they came clipped to a check.

It wasn't unknown for a major political contributor to donate money to elections in far-off states, but as 1950 dawned, Cullen took the practice to a new level. He and Porter drew up a list of every congressman up for reelection in 1950 and, where they felt a candidate wasn't sufficiently tough on "creeping socialism," began mailing checks to his opponent. Their money, in amounts from five hundred dollars to as much as ten thousand dollars, found its way into dozens of national campaigns. "I guess I'm supporting as many Democrats as Republicans," Cullen told a reporter. "A lot of 'em may not expect my support, and some of 'em may not want it, but if they believe in our American system of constitutional government and the free enterprise system, they're going to get it."[3]

Cullen's checks often came attached to a favorite book. One was *I Chose Freedom*, by a Russian refugee named Victor A. Kravchenko. After reading it, Cullen mailed a check for twenty-five thousand dollars to the publisher, Charles Scribner's Sons, asking that the book be distributed to libraries and schools across the country. Librarians received the unsolicited book with a note from Scribner's saying it had been donated "with the compliments of a believer in American freedom, who hopes this book will have the widest possible reading."[4]*

The most influential book on Cullen's thinking was *The Road Ahead*, a 1950 best seller written by John Flynn, a onetime liberal writer who argued that New Deal policies were transforming America into a totalitarian state. Cullen mailed thousands of copies of Flynn's book to libraries and universities, and in time the two men became friends. Yet money and books and telegrams alone weren't enough to spread the conservative cause, Cullen saw. He needed to make war on the battlefield of ideas. In late 1949 he and

---

*Cullen wasn't so wild about Kravchenko's second book, *I Chose Justice*, which detailed his fight against a libel suit filed against him by communists in Paris. "The story of the libel suit is fine," he wrote a Scribner's editor who mailed him galley proofs, "but I think the man is a socialist."

Jack Porter began scouting for a publication they could groom as a voice for Texas conservatives. That December they approached Ida Darden, the sister of John Henry Kirby's onetime mouthpiece, Vance Muse. Darden, who practiced a paler version of her brother's overt racism, had started a newspaper called the *Southern Conservative*, in Fort Worth; most of her seed money had come from Kirby's crony, the Klan organizer George Armstrong.

The year 1949 was a very bad year to consider going into business with the ultraracist Armstrong, who, much like Cullen and so many other Texas oilmen, was taking his first steps into the public arena. That summer, after years evading FBI probes of his anti-Semitic pamphlets and books, the eighty-four-year-old Armstrong had made a startling proposal to a small school in his native Mississippi, Jefferson Military College. Armstrong offered the school half the mineral rights on twenty-six thousand acres of his oil fields, a proposal he valued at fifty million dollars. His only conditions, Armstrong said, were that Jefferson exclude blacks and Jews and teach its students white supremacy. The school turned him down flat. The resulting furor made headlines around the country and did nothing to further the portrait of Texas oilmen as enlightened businessmen.

Cullen made the approach to Ida Darden in partnership with H. L. Hunt, the only time the two are known to have worked in tandem. But the talks foundered, either because of concerns over Armstrong's activities or, as Darden's papers suggest, difficulties structuring a legal partnership between Cullen and Hunt. Still Cullen did not give up.

Eighteen months later, in the summer of 1951, he received a call from the writer John Flynn. Flynn had launched a program on the nation's second-largest radio network, the Liberty Network of Dallas, which had leveraged its broadcasts of major-league baseball games into a news service carried in forty-three states. Its rapid growth, however, had plunged Liberty and its founder, the radio pioneer Gordon McLendon, into financial chaos. Flynn asked Cullen to come to Liberty's rescue. He assured Cullen that McLendon was a "good conservative" and that, if Cullen didn't help, a "radical element" might buy Liberty.

Cullen saw the opportunity. Fifty years before Fox News became a conservative media bellwether, he envisioned transforming Liberty into something similar. After a single meeting with McLendon, and without so much as examining Liberty's books, Cullen paid one million dollars for 50 percent control. He was named cochairman, announced plans to move Liberty to Houston, and threw himself into a new calling. "I recently purchased controlling interest in the Liberty Network to keep it out of bad hands," he wrote one of his newest pen pals, the presidential candidate Dwight Eisenhower. "This is the second largest network in the country, and soon will be probably the largest."

Red flags, however, rose immediately. A New York newspaper reported that a local station had broken off talks about joining Liberty over concerns about Cullen's political views. In November *The Nation* reported that Flynn had taken over Liberty's news operations; "his official mission," the magazine stated, "is to make sure the news is no longer given what he calls a 'leftist' slant."[5] In later years both Flynn and Gordon McLendon would deny that Flynn had done any such thing, but the rumors did nothing to attract the new advertisers Liberty badly needed. Liberty was forced out of business and declared bankruptcy just nine months after Cullen's cash infusion, in May 1952, effectively ending the oilman's attempts to create a national platform for ultraconservatism. In hindsight his defeat appears preordained. To succeed in politics, Cullen needed a support organization of some kind, but building one was something he was unwilling or incapable of doing.

It was a mistake H. L. Hunt would learn from.

III.

Much as publicity over his philanthropic efforts seemed to embolden Roy Cullen, his unofficial coronation as the nation's richest man tranformed H. L. Hunt. All the resulting attention—the bags of mail, the interview requests—seemed to confirm what Hunt had long believed, that he was a unique intellect, a superman, a figure whose ideas could save the nation from the mounting perils of communism. More than one of his aides sensed a new messianic quality in Hunt. Said one, "He thought he was a second Jesus Christ."[6]

If anything, Hunt was further to the right than Cullen; he believed deeply, in his bones, that communism and socialism were poised to over-run the world. He distrusted Jews, whose loyalty he questioned, and felt that anything that hurt American business, especially the oil industry, was anti-American. At first his political work paralleled Cullen's. He wrote letters to congressmen, gave the occasional speech around Dallas and brought in a series of radical right-wing speakers, including Frederick C. Schwartz of the Christian Anti-Communist Crusade. ("Karl Marx was a Jew," Schwartz told the Dallas rally. "Like most Jews, he was short and ugly, lazy and slovenly, and he had no desire to go out and work for a living. But he was also pos-sessed of a keen intelligence, a superior evil intelligence like most Jews are, and his mission was to destroy all of Christian civilization.")[7]

Unlike Cullen, who found his ideas in books and articles written by others, Hunt had original ideas and, fancying himself an author, decided to put them in print himself. In 1950 he published a pamphlet titled *A Word to Help the World*. The word he coined was *constructive*. Hunt detested liberals but never saw himself as conservative; there was nothing conservative about the way he built his business, he liked to say. "Constructive," Hunt wrote, was a "trademark for the wholesome in government." Hunt's new philosophy was broadly pro-business and anticommunist, in favor of eliminating federal controls on all aspects of individual "freedom" and favoring smaller govern-ment in all areas but defense; others would term Hunt's "constructivism" ultraconservative or reactionary.

In time, like Cullen, Hunt began to cast about for a broader platform from which to spread his ideas. In his pamphlet he called for "constructives" to form an "Educational Facts League," whose purpose would be "to secure an impartial presentation of all the news through all the news channels concerning issues of public interest." He envisioned an organization where ordinary Americans would be supplied the "facts" of political life, via a newsletter or maybe a radio program, which they could debate in small dis-cussion groups. He decided to call it Facts Forum, and announced its forma-tion in a speech in June 1951. One of his aides approached a Dallas-based FBI agent he knew about heading the group, and the man suggested he hire an agent named Dan Smoot, who was retiring. On a long drive through East Texas, Hunt gently interrogated Smoot, found him to be an ardent anti-Communist who idolized J. Edgar Hoover, and spoke of his goals for Facts Forum. By the time they returned to Dallas, Hunt was convinced Smoot was his man. "The philosophy of the New Deal," Smoot liked to say, "is also the basic philosophy of Communism, fascism and Nazism."

At the Mercantile Bank Buildling, Smoot was given Hassie Hunt's old corner office, an Oldsmobile, a secretary, a handful of files, and not much else; Hunt himself then disappeared on a three-week business trip. Unsure how to proceed, Smoot began telephoning names in the Facts Forum files, trying to organize discussion groups. Hunt had mentioned a radio broadcast, so he contacted a radio producer who arranged for him to host a fifteen-minute weekly discussion on the Dallas station WBAP. On his first broadcast, Smoot moderated a debate between two high school boys on the United Nations, "Should the U.S. get out of the U.N. and or get the U.N. out of the U.S.?" Returning from his business trip, Hunt was thrilled; on the spot, he decided to focus Facts Forum on radio broadcasts, eliminating the idea of discussion groups. After Smoot's seventh broadcast, Hunt informed him he had arranged for Facts Forum to go national, broadcasting a thirty-minute show from a Washington studio to a handful of syndicated stations. Smoot began commuting to Washington, replaced the students with congressmen, and moderated their debates through the end of 1951.

In January 1952, after the resignation of the secretary who arranged his guests, Smoot switched to a format that became his trademark. He began arguing both sides of an issue himself, first blandly presenting the far-left viewpoint, then energetically advancing Hunt's "constructive" alternative. Smoot's early broadcasts tended to focus on threats against democracy and religion, but in time they became overtly political, branching out to "debate" the Korean War and the fight against communism. "Democracy," Smoot observed on one broadcast, "is a political outgrowth of the teachings of Jesus Christ. Christianity is essential to the creation of our democracy. We in Facts Forum know that American democracy is still the most nearly perfect expression ever made by man in legal and political terms of a basic ideal of Christianity." Hunt was actually outraged by this statement; he considered America a republic, not a democracy, and lectured Smoot that democracy "was the handiwork of the devil himself," a watered-down version of communism. "In an ideal society," Hunt was prone to say, "the more taxes you pay, the more votes you get."[8]

Hunt himself never appeared on Facts Forum, preferring to act as Smoot's tutor and "research assistant," shoveling books and articles his way. In time Smoot began filming the shows, and Hunt found a television station or two to broadcast them. For nine more months Facts Forum remained a modest operation, its annual budget barely one hundred thousand dollars. Then, in the fall of 1952, everything changed. Hunt found a new intellectual partner, and their collaboration would transform Facts Forum into a national phenomenon. His name was Joseph McCarthy.

IV.

While one camp of Texas oilmen plunged into national politics for ideological reasons, a second camp—Sid Richardson, Clint Murchison, Brown & Root's George and Herman Brown—remained primarily interested in laying its hands on the levers of powers. Their man in Washington, as he had been since 1940, was Lyndon Johnson, a congressman with little interest in anything at that point beyond scaling the political ladder. Theirs was the perfect marriage: Richardson, Murchison & Co. had only one thing to offer, and Johnson, who was running for the Senate in 1948, needed only one thing to get elected. Cash.

He got it. Johnson narrowly defeated Coke Stevenson in the 1948 Democratic primary, a contest memorably chronicled in Robert Caro's 1990 book *Means of Ascent*, before besting Jack Porter that November. Both victories were fueled with vast quantities of illegal cash, much of it from Richardson and the Brown brothers, but also from Murchison, Amon Carter, and the Taylor independent Harris Melasky. Gathering it took a series of secret missions run by a half-dozen Johnson aides led by a sharp young attorney named John Connally. Whenever money was needed, typically for radio advertising, Connally or one of his men would charter a private plane to Austin, Dallas, Fort Worth, Houston or, in Richardson's case, the airstrip on St. Joseph's Island, returning with paper bags and briefcases packed with ten, twenty-five, or in some cases as much as fifty thousand dollars in cash. There was so much of it Connally couldn't keep track of it all, at one point misplacing forty thousand dollars he had hidden at his home. The money was never found; Connally guessed he had left it in a suit he sent to the cleaners.[9]

Texas had never seen anything like it; no state had. "This was the beginning of modern politics," Connally recalled. "It was the dawn of a whole new era."[10] The oilmen thought so, too and were dumbfounded when confronted by an honest politician who wouldn't take their cash. During a Washington dinner for Sam Rayburn in 1949, Sid Richardson beckoned the Speaker to join him in the men's room. Several minutes later, Rayburn emerged sputtering mad, only to encounter Creekmore Fath, chairman of the Democrats' national finance committee. "Creekmore!" Rayburn snapped. "You go in there and talk to him."

Fath pushed into the men's room and found Richardson standing alone, confused. "I don't know what gets into Sam sometimes," Richardson complained. He reached into a pants pocket and withdraw a wad of five thousand dollars in cash. From a second pocket he fished out another five thousand. He handed it all to Fath.

"How am I going to list this contribution?" Fath asked.

"I don't know," Richardson said. "Use the Bass boys."[11]

Richardson, however, found more subtle ways to woo Rayburn. On a visit to the Speaker's farm with his attorney, William Kittrell, Rayburn complained about his skinny cattle. "Goddamnit, Sam," Richardson said, "what you need is a decent bull!"

"Well, I've got my eye on one bull," Rayburn muttered.

"Listen," Richardson said, "I've got this scrawny little old bull that's no use to me. Goddamnit, Sam, I want you to have it."

"Don't do it, Sid."

"Well, I'm going to."

Rayburn shrugged. Afterward Richardson took Kittrell to a livestock dealer and scribbled out a twenty-thousand-dollar check for the man's best bull. He had it delivered to Rayburn with a note apologizing for handing over such a pitiful specimen, figuring that Rayburn wouldn't know the difference. "If you ever tell this to Rayburn," Richardson warned Kittrell, "I'll de-nut you."[12] In time Richardson realized he needed a more polished approach, and in 1951 he hired John Connally away from Lyndon Johnson. Connally became Richardson's chief lobbyist.

Given all the cash Texas oilmen were funneling into Washington, it wasn't long before they sought a return on their investments. Texas Oil's new profit engine, natural gas, remained regulated by the Federal Power Commission, and in the summer of 1949, a liberal Roosevelt appointee named Leland Olds was nominated for a third term as the FPC's chairman. To Texas oilmen, Olds was nothing short of the Antichrist, a hardworking anti-business economist who once wrote that only "the complete passing of the old order of capitalism" could liberate workers from the tyranny of corporate ownership. "Olds was the symbol of everything they hated," recalled Posh Oltorf, George Brown's lobbyist. "He was just anathema to them because of his philosophy."[13]

As the FPC's chairman, Olds had the power to set natural-gas prices. He had twice been confirmed, in 1940 and again in 1944, but the explosion in natural-gas demand—and profits—from the new pipelines changed everything. Oilmen bombarded Lyndon Johnson with letters demanding that his reappointment be blocked. Even Roy Cullen and others who had long been skeptical of Johnson—who voted against him in 1948, in fact—begged him to do something; Olds's writings "are conclusive proof that he does not believe in our form of government," Cullen wrote in a telegram he sent to Johnson and twenty-one other senators.

"This transcended philosophy," John Connally told Robert Caro. "This would put something in their pockets. This was the real bread-and-butter issue to these oilmen. So this would prove whether Lyndon was reliable, that he was no New Dealer. This was his chance to get in with dozens of oilmen, to bring very powerful rich men into his fold who had never been for him, and were still suspicious of him. So for Lyndon this was the way to turn it around: Take care of this guy."

Johnson, after wangling the chairmanship of the Senate subcommittee that was to review the nomination, ambushed Olds in a hearing on September 28, 1949, in which a Corpus Christi congressman named John Lyle, a Johnson crony, using files borrowed from the House Un-American

Activities Committee, broadly attacked Olds as a Communist. Attorneys for the Texas natural-gas industry piled on, and Olds, despite angry denials, was doomed. "Leland Olds Labeled Crackpot and Traitor," read the *Houston Post* headline. Other papers followed suit. A week later the subcommittee voted 7–0 against the nomination.

It was a blatant smear, but by the time liberals realized what Johnson had done, it was too late. Eleanor Roosevelt and other allies rallied behind Olds, saying he was never a Communist. *The Nation* decried Johnson's surprise attack as "a flagrant attempt by vested interests to exclude from office a man who proved too consumer-minded." When the full Senate voted 53–15 against the nomination, the Texas newspapers, including those who previously attacked Johnson as too liberal, rained plaudits on him for saving the gas industry.

After the Senate vote Johnson returned to Houston on a Brown & Root airplane, where he took a Brown & Root limousine to the Brown & Root suite at the Lamar Hotel. There, in Suite 8-F, he accepted backslapping thanks from the Browns and a crowd of oilmen. Afterward he flew to St. Joseph's for a week with Richardson, Murchison, and others. It was a joyous victory lap for Texas Oil's man in Washington. Johnson was now on his way to real political power; the assembled oilmen could be forgiven for mistakenly thinking they were, too.

V.

The defeat of Leland Olds was the first indication of the political power Texas Oil could wield, and of the lengths it was willing to go to further its aims. The vast amounts of money the Big Four were plowing into national politics for the first time made Texas a regular destination for American politicians with higher aspirations. Some came holding their noses, especially those forced to sit through Roy Cullen's lectures. All had an eye on the 1952 election, when just about every Texas oilman was determined to defeat any reelection drive by the liberal Harry Truman. The first to arrive was the junior senator from Georgia, Richard Russell, an ally of Johnson's, whom Johnson brought to the weeklong fete at Sid Richardson's spread on St. Joe's in October 1949, a week after the Olds victory. There the men spent the days shooting ducks and strolling the sandy beaches, the nights sipping bourbon and talking about politics, the oilmen trying to measure whether Russell had what it took to defeat Truman. He didn't.

Dwight Eisenhower, the hero of D-day, was next to arrive, two months later, in December 1949, for the first of several visits. Roy Cullen, who was intrigued by Eisenhower, invited him after dining with the general in New York. In Houston the Cullens held a luncheon for Eisenhower at their home, and at a dinner the general surprised Cullen by awarding him the Freedom Foundation's medal of honor, an honor the oilman had been unable to accept in person. After Cullen showed him around the University of Houston the next morning, Eisenhower flew to St. Joe's for a weekend with Sid Richardson.

Eleven months later, in November 1950, Eisenhower accepted Cullen's invitation to return to Houston, to speak at the University of Houston. This time the general stayed overnight at the Cullen mansion, which gave Cullen time to sound out the general about his rumored run for the White House. Over cigars in the living room, Eisenhower remained cagey about his political ambitions. Cullen pushed hard. "General, the people of this country want you as their next president," he said. "I can assure you of this....But there is one thing I can tell you. You can be nominated and elected if you refuse to talk politics with anyone. Remain just what you are, a soldier."

After both his visits with Cullen, Eisenhower headed to St. Joe's to see Sid Richardson, who soon emerged as his wealthiest single backer. Where the general tolerated Cullen, he genuinely enjoyed Richardson. The Old Family Friend confirms rumors that Eisenhower invested with Richardson beginning sometime during World War II. "I know he did, Sid told me," says the Friend. "There's an old game in oil, you know, where your friends, they only invest in your good wells, not the bad wells? You understand? It was that way with Eisenhower. You could never prove it. But he did it."

While Eisenhower's visits to Texas did wonders for oilmen's egos, it did little to push the general toward the White House. All through 1951, in fact, Eisenhower wavered whether to seek the presidency. As he did, the Big Four joined Republicans around the country in reviewing suitable replacement candidates. That spring they thought they found one in another World War II hero, General Douglas MacArthur, who Truman "fired" that spring as head of American forces in the Korean War. H. L. Hunt had been championing a MacArthur candidacy for years. On the general's return from Korea in the spring of 1951, Hunt, Roy Cullen, and Glenn McCarthy all sent MacArthur invitations to visit Texas. A MacArthur aide responded to Cullen, saying the general could visit Houston in June. Cullen replied he would be attending his granddaughter's graduation that particular day. When the general's aide suggested a visit from the Hero of the Pacific might be more important, Cullen replied, "I admire General MacArthur very much, but I wouldn't break my word to my granddaughter for a dozen MacArthurs."[14]

MacArthur delayed his visit by a day. When the trip was announced, McCarthy mistakenly assumed it was a result of his own invitation and issued a press release that said so. Cullen was not pleased. When MacArthur arrived in Houston, checking into the Shamrock, the city threw an impromptu parade; Cullen, in a white summer suit, rode beside the general in an open car, McCarthy in the backseat. A photograph captured Cullen shooting a baleful glance at McCarthy sitting behind him. The real fireworks, however, began after the general left, when it was disclosed that twenty-three thousand dollars of his hotel and restaurant bills hadn't been paid. When the matter hit the newspapers, McCarthy claimed ignorance. Reached at his ranch, Cullen hit the roof. "This is embarrassing to Houston and to Texas," he barked. "Find out how much the bill is—I'll pay it." And he did. (The incident was later lampooned at the Houston press's gridiron dinner, with reporters portraying Cullen, McCarthy, and Houston's mayor squabbling over a dinner check; the skit ended with the mayor and McCarthy shouting "Roosevelt! Roosevelt! Roosevelt!" at Cullen until he fainted.)

All through late 1951 speculation ran rife as to whether Eisenhower, who had left his post as president of Columbia University to command NATO forces in Europe, would run; a Draft Eisenhower movement had sprung up in New York, and had begun holding rallies. Sid Richardson was among those determined to push him off the fence. That November Richardson invited George Allen, one of the general's closest friends, to St. Joe's to discuss ways they might push him into the race. They decided they would need to confront Eisenhower in person—in Paris. When they boarded the *Queen Mary* in early February 1952, Richardson carried with him two letters for Eisenhower, one from Clint Murchison, the other from Billy Graham.

Richardson and Allen arrived in France as speculation about Eisenhower's plans approached its zenith. They joined a growing crowd of well-wishers lingering at NATO headquarters. Among them was the famed aviatrix Jacqueline Cochran, who had brought a film of a massive Draft Eisenhower rally to show the general. When she aired the film for Eisenhower on the night of February 11, Cochran later claimed, Eisenhower became tearful and told her he would, in fact, seek the White House. A more plausible version of Eisenhower's decision comes from his longtime aide, General Lucius Clay, who told his biographer that Eisenhower had still not made up his mind when the two conferred after the funeral of England's King George, on February 16. Richardson and Allen, neither of whom left accounts of the trip, traveled to London with Eisenhower and were present at the home of a British general when General Clay took Eisenhower aside and urged him to make his decision. Glimpsing Richardson and Allen in the living room, Clay guided Eisenhower into an anteroom where, he said later, Eisenhower firmly stated his plan to run. Afterward Eisenhower emerged and told Richardson and Allen. While no account of the incident suggests Richardson's maneuvering played a role in Eisenhower's decision, he could claim to be the first civilian to learn of it.

Once Eisenhower declared his candidacy, every one of the Big Four boarded his bandwagon. Despite their dalliances with MacArthur and others, Texas oilmen smelled a winner in Eisenhower and, during his campaign against the Democrat Adlai Stevenson, pulled out all the stops to get him elected. By one estimate Richardson funneled about one million dollars into the campaign, not including two hundred thousand dollars to cover Eisenhower's various stays at the Commodore Hotel in New York or his expenses during the Republican convention in Chicago. Roy Cullen, meanwhile, implored Texas's governor, Allan Shivers, to lead Democrats into the Eisenhower camp. Murchison assembled a group of oilmen to hire a public-relations firm, Watson Associates, which created and distributed six

hundred thousand copies of an authentic-appearing anti-Stevenson newspaper called the *Native Texan*. It featured headlines such as "Adlai's Ideas
Aid Kremlin" and "Truman Stays Faithful to Stalin" along with a cartoon
of a maniacal-looking Stevenson sneering at a classroom of Texas school-
children. The *Native Texan* was mailed exclusively to rural Texas communities. "Whatever you might think about it, [these farmers] are anti-Negro on
the equality thing anyway," Murchison told a reporter. "My paper kind of
catered to these feelings."[15]

Eisenhower was elected to the White House that November, and while
Texas Oil money alone didn't put him there, it certainly helped. Washington
noticed. "When senators returned to Washington after the 1952 elections,"
Robert Caro has written, "there was a new awareness on the north side of the
Capitol. There was a vast source of campaign funds down in Texas, and the
conduit to it—the only conduit to it for most non-Texas senators, their only
access to this money they might need badly one day—was Lyndon Johnson."
This was only partly true. While Johnson's influence over the flow of Texas
Oil's donations was important, it was not total. Four years later a University
of North Carolina professor named Alexander Heard finished an exhaustive
study of political giving in 1952, and reported that by far the largest single
American donor that year not only wasn't one of Johnson's oilmen; he wasn't
even a Democrat. It was Roy Cullen.

The 1952 election was the dawning of a new age, one in which American
political power, heretofore centered in the Northeast, began to flow, along
with many Americans themselves, into the West and Southwest, especially
to Texas. "The first sense of the tilt in national leadership from Northeast to
Southwest can probably be dated sometime between 1952 and 1954," Theodore White wrote twenty years later, "when imperial New York sensed a serious financial trespass. The trespass came from Texas. There, a handful of
uncouth oilmen had begun to invest in congressional candidacies across the
nation.... The Texas intrusion, at least in New York, seemed outrageous."

As the oilman closest to Eisenhower, Sid Richardson found himself once
again a guest at White House dinners. Richardson, however, was less interested in selecting entrees than cabinet members. (When a White House
aide called to invite him to dine with the president, Richardson would quip,

"Well, what's for dinner?") The most important post, for Richardson and other oilmen, was secretary of the navy. The navy was the largest purchaser of oil in the world, while overseeing millions of acres of oil-bearing land, from Alaska to Colorado. Richardson prevailed upon Eisenhower to name as navy secretary an obscure Fort Worth attorney named Robert B. Anderson, who for the previous decade had run the estate of the rancher W. T. Waggoner, on whose lands both Murchison and Richardson had drilled wells. A rare glimpse of the benefits Richardson accrued during Anderson's tenure came in 1955, when a busboy at a California hotel found an envelope Richardson had left at poolside. Inside was a letter from Perry Bass. "Dear Sid," it read. "As you will see, we have a glowing report this month on our sales to the Navy"; it went on to list a monthly breakdown of Richardson's steeply rising sales of oil to the navy.[16]

No doubt it was with similar windfalls in mind that other Texas oilmen tended to Eisenhower's private interests. The president owned a farm in Pennsylvania, and after his election, to avoid potential conflicts, he leased it to George Allen. Allen, in turn, allowed two oilmen, including Murchison's pal Billy Byers of Tyler, to pay the farm's bills. During Eisenhower's years in the Oval Office, Allen and Byers transformed the farm into a presidential retreat, erecting a thirty-thousand-dollar show barn, three other barns valued at twenty-two thousand dollars and six thousand dollars in landscaping—all of which was later made public by the columnist Drew Pearson. Meanwhile Murchison, along with Byers and Wofford Cain, paid the upkeep on a 550-acre Virginia horse farm for Mamie Eisenhower's brother-in-law, Gordon Moore, who to that point had lived on a salary of eighty-five hundred dollars a year. Murchison maintained a stable of prize show horses at the Moore estate and paid Moore a large commission for his role as a middleman in Murchison's purchase of a West Virginia racetrack.[17]

While accepting their money and their favors, Eisenhower in private could be scathing about his new friends in Texas Oil. "Should any political party attempt to abolish social security, unemployment insurance, and eliminate labor laws and farm programs, you would not hear of that party again," he wrote his brother in 1956. "There is a tiny splinter group, of course, that believes you can do these things. Among them are H. L. Hunt [and] a few other Texas oil millionaires.... Their number is negligible and they are stupid."[18]

VI.

Much to their eventual chagrin, the national politician who became most closely identified with Texas Oil during the 1950s was not the statesmanlike Eisenhower but the blustery junior senator from Wisconsin, Joseph McCarthy. After an unremarkable four years in the Senate, McCarthy streaked into the national consciousness in February 1950 in the wake of a speech he made in Wheeling, West Virginia, in which he alleged that the State Department was riddled with Communists. It was a purely political ploy—McCarthy had no record of fighting communism, and was searching for an issue to buoy his reelection—but in the wake of the spectacular trial of Alger Hiss, his charges caused a national uproar. McCarthy's ensuing crusade against Communist "infiltrators" transformed the senator into a polarizing figure across the country.

The Big Four immediately embraced McCarthy—except for Sid Richardson. "I don't see how I could be friendly with Sam Rayburn and Lyndon Johnson," Richardson told a reporter, "and be friendly with Joe McCarthy too." Cullen, who had met McCarthy in 1948, was the first to bring him to Texas, introducing him at a speech at Sam Houston Colosseum in September 1950. "Senator McCarthy has done more than anyone [else] to throw the pinks and Reds out of the country," Cullen declared. "I hope [he] keeps all the Communist spies running until they get to Moscow." Many Texans agreed. Dallas held a hundred-dollar-a-plate fund-raiser in his honor, only the second such gathering in the state's history. In Houston a group of citizens raised money to give McCarthy a Cadillac when he married his former research aide, Jean Kerr, who by then had gone to work for Facts Forum. McCarthy was so popular in Texas, the media began referring to him as the state's "third senator."

Of the Big Four, it was Murchison who initially drew closest to McCarthy.

The senator had telephoned him in 1950 at the suggestion of a mutual friend and asked him to donate ten thousand dollars to help defeat a McCarthy rival, Senator Millard Tydings of Maryland.* Murchison agreed, ponied up another ten thousand dollars in 1952 to defeat another McCarthy opponent in Connecticut, then began placing his personal planes at McCarthy's disposal. In time he took to coaching the senator on his finances, slipping him stock tips. "I'm for anybody who'll root out the people who are trying to destroy the American system," Murchison told a writer in 1954. "Then along came this Marine, a man with a tough hide, I sized him up as the best tool in sight to fight Communism."[19]

McCarthy's most tangible impact, though, was on H. L. Hunt. When McCarthy came to Dallas in April 1952 to deliver a speech sponsored by the American Legion, Facts Forum aggressively promoted his appearance. Hunt telephoned McCarthy at the Dallas Athletic Club, where the senator was staying, then dropped by. The two men hit it off, taking off their coats and starting a game of gin rummy; McCarthy accepted Hunt's offer to have Dan Smoot introduce him that night. Reporters soon arrived—apparently alerted by a Hunt aide—and as Hunt pinned a MACARTHUR FOR PRESIDENT button on McCarthy's lapel, the two posed for a photo that appeared on page 1 of the Dallas Morning News the next day.

The two men met again that fall, and this time discussion centered on the common ground Facts Forum and McCarthy were plowing. Within weeks Hunt hired three onetime McCarthy assistants: Victor Johnson, an administrative aide; Robert E. Lee, a former FBI agent and one of McCarthy's most persistent investigators; and Jean Kerr, the researcher who the senator would soon marry. In short order the McCarthyites transformed Facts Forum into the very nationwide multimedia enterprise Hunt had dreamed of. They oversaw an eruption in new Facts Forum programming, including two syndicated radio shows and three television shows produced

---

*A Senate subcommittee later found that Murchison's money, along with five thousand dollars contributed by Roy Cullen's partner Jack Porter, were part of a sum not reported to the "appropriate authorities." Instead the money had been used to help pay for a tabloid newspaper distributed throughout Maryland that carried a fake photograph of Senator Tydings posing with the American Communist leader, Earl Browder. Tydings was defeated.

by a television veteran named Hardy Burt in a studio Hunt acquired on East Fortieth Street in New York City.

All the shows were essentially the same: moderated discussion programs purportedly devoted to airing "both sides" of a public issue. In fact, all copied Dan Smoot's format, skewing their "debate" heavily toward the right, strongly backing McCarthy even as they broadcast views that amounted to thinly veiled appeals to racism and anti-Semitism. In one program, for instance, the commentator argued against fair-employment legislation by stating: "Remember that the Negroes when first brought to America by Yankee and English merchants were not free people reduced to slavery. They were merely transferred from a barbaric enslavement by their own people in Africa to a relatively benign enslavement in the Western Hemisphere."

The broadcasts were augmented by a group of anti-Communist books and pamphlets Hunt mailed to hundreds of Facts Forum "participants." Typical was *We Must Abolish the United Nations*, by a man named Joseph Kamp, whose previous works included a book titled *Hitler Was a Liberal*. Wrote Kamp: "I pull no punches in exposing the Jewish Gestapo or any Jew who happens to be a communist." Hunt eventually pulled the Kamp book under pressure from the Anti-Defamation League, but replaced it with McCarthy tracts and books written by other well-known anti-Semites, including one by a retired general named Bonner Fellers, who once observed that "Hitler did Germany a world of good."

Facts Forum News, meanwhile, was a Hunt-funded newsletter that urged its readers to buy Facts Forum materials and join other right-wing groups. With a claimed circulation of sixty thousand, most of its articles were reprints, though a few were original, including one called "The Liberal Mind" by the up-and-coming William F. Buckley. Facts Forum also circulated a monthly poll of its readers, which McCarthy and other conservative politicians took to brandishing, sometimes without explaining who the poll's participants actually were. Hunt draped it all in patriotic rhetoric that, in the days before its real opinions became widely known, succeeded in drawing the support of not only leading right-wingers but a handful of nationally known moderates; its board included not only Norman Vincent Peale but the actor John Wayne.[20]

During the early months of 1953, few media critics took notice of Facts Forum's new direction. One who did was a fellow conservative, E. M. Dealey, president of the *Dallas News*. The *News* had been favorably disposed toward Hunt's activities until the day an aide brought an unexplained fifty-dollar check from Hunt to Dealey's attention. Upon investigation Dealey discovered that Hunt had been sending checks to *News* reporters whose articles pleased him, and to letters-to-the-editor writers who mentioned Facts Forum. Outraged, Dealey not only returned Hunt's check, he announced to the newsroom that, unless absolutely unavoidable, Hunt was never to be mentioned in the paper again.

Hunt, however, was not so easily deterred. He turned up one day at Dealey's office with a bizarre proposition that the paper split its reporters into two camps, "pinks" and conservatives, who could write side-by-side articles on major news events. Dealey all but threw him out. For months afterward, though, Hunt bombarded Dealey with tear sheets of *News* pages with overlays showing how his "pinks vs. conservatives" approach would look. When a magazine writer learned of the incident a year later, Dealey let Hunt have it with both barrels, terming him a "latent fascist." Asked about Facts Forum, Dealey growled, "That kind of stuff reminds me of the Ku Klux Klan."[21]

VII.

All the gaudiest elements of Texas Oil's new prominence—the Hollywood stars, the political maneuvering, and especially the developing albatross that was Joe McCarthy—came together in a single place, a new hotel Clint Murchison opened as a personal retreat in La Jolla, California, north of San Diego. Clint and his wife, Ginny, often accompanied by Sid Richardson, had been summering in La Jolla for several years to bet the thoroughbreds racing at the Del Mar Turf Club, whose season ran from July until Labor Day. They usually arrived at the track at midmorning with friends and stayed all day, downing mint juleps and ham sandwiches between races. Murchison loved La Jolla so much, he tried to buy the landmark Casa Manana hotel; when its owners wouldn't sell, he decided to build his own. He bought a riding academy on the edge of town, tranformed it into a fifty-room hotel, and turned the stables into eight two-bedroom cottages around a pool.

When the new Hotel Del Charro—*charro* is Spanish for a costumed horseman—opened for the 1951 season, Murchison's oilmen friends and their wives descended upon it en masse: Effie and Wofford Cain, Emily and Billy Byars, Jodie and Pug Miller, Sid Richardson. A Texas flag flew overhead, a Dow Jones ticker clattered in the lobby, and after the day's races a gin game was always under way by the pool. Tuxedoed waiters drifted through the rarified crowd, carrying trays of mint juleps and bourbon; riffraff was kept away by room rates set at eight hundred dollars a night at today's prices. (House rules excluded pets and, unofficially, Jews.) Within weeks the neverending party was joined by a rushing tide of movie people, including John Wayne, Elizabeth Taylor, William Powell, Jimmy Durante, Betty Grable, and Joan Crawford. Crawford, taking swigs from a flask of vodka, caused a stir on several evenings by focusing her attentions upon Richardson.

"Everyone around the country knew that Sid was a billionaire, and there had been a lot of press about him right at that time when we introduced Joan to him," Ginny Murchison remembered.[22] "She followed him around so much that he finally came and sat on the couch between me and Effie Cain, so that Joan couldn't get near him. He was very shy around women, and he didn't like it at all when they flirted with him."* Irksome actresses aside, Richardson relaxed at Del Charro as nowhere else, drinking bourbon, playing cards around the pool, and cursing so loudly that when Eisenhower's secretary of health, education and welfare, the Houston heiress Oveta Culp Hobby, was put in an adjacent cottage, she asked to be moved. Most mornings Richardson put in a call to Sam Rayburn before driving with Murchison to the track.[23]

Hobby and Rayburn were only the first in a stream of politicians to stay in touch with goings-on at Del Charro. Soon others, including Eisenhower and his vice presidential candidate, Richard M. Nixon, came to pay their

---

*Richardson's Old Friend says Crawford chased Richardson so fervently that she actually flew to Fort Worth unannounced. Richardson refused to see her, and the actress ended up spending an awkward evening with Perry Bass's family. "Sid so hated snobs," the Old Friend said. "She was the kind of person he hated most."

respects to Murchison and Richardson. "They spoke to Nixon like he was an office boy," the Del Charro's manager, Allan Witwer, recalled years later.[24] The Washington figure who most enjoyed the hotel, and who stayed the longest, was J. Edgar Hoover, who accepted Murchison's invitation to Del Charro in 1952 and returned every summer until his death in 1972. The two men, who met at a California fund-raiser in July 1951, became fast friends. Hoover and his longtime aide Clyde Tolson stayed in Bungalow A, one of the cabins reserved for Murchison's friends. The oilman looked after Hoover's every need. When the director mentioned one night that he loved Florida because he could step from his hotel room and pluck fruit from a tree, Murchison had a grove of plum, peach, and orange trees planted on Hoover's patio by the next morning. A sickly grapevine was hung, complete with healthy grapes the staff spent hours wiring to the vine.

Hoover was a famously buttoned-up man, but the oilmen around the Del Charro pool did their best to loosen him up. One evening, while dining on "caviar of chili" a Dallas millionaire had flown in from Ike's Chili Parlor in Tulsa, Sid Richardson spied Hoover sitting quietly beside the buffet. Suddenly Richardson's booming voice rang out across the poolside crowd: "Goddamnit, Hoover, get your ass out of that chair and get me another bowl of chili!"[25] (Richardson, too, flew in his own food, including steaks, melons, fish, and mesquite charcoal.)

Hoover, like Murchison, loved the races, but what he appreciated most about Del Charro, one suspects, was that he stayed for free. At the end of Hoover's initial monthlong visit, Allan Witwer recalled, "Hoover had made no attempt to pay his bill. So I went to Murchison and said, 'What do you want me to do?' 'Put it on my bill,' he told me. And that's what I did." In today's dollars, Hoover's bill came to almost twenty thousand dollars; Murchison, and later his successors, paid every single one for twenty years.[26] The two men even collaborated in business, when Murchison's publishing house, Henry Holt, distributed Hoover's best-selling book about fighting communism, *Masters of Deceit*.

In fact, as later became apparent, Murchison did more for the director than pay his bills and print his book. The extent of his favors remained secret until investigations into FBI corruption following Hoover's death.

"Hoover did have oil ventures with Clint Murchison," a Justice Department attorney confirmed in 1988. "If the drilling company hit a dry hole he'd get his money back. Everything was a sure thing. It had to be a sure thing. If not, he'd get his money back, be it stocks, bonds or oil ventures. It was extraordinary." Hoover, the former FBI assistant director William Sullivan later told an interviewer, "had a deal with Murchison where he invested in oil wells, and if they hit oil, he got his share of the profits, but if they didn't hit oil, he didn't share in the costs." The Justice Department, it later turned out, had actually investigated the Murchison-Hoover dealings during the Kennedy administration, but "there wasn't enough to make a criminal case," remembered William Hundley, head of the department's Organized Crime section at the time. "But it was wrong. He shouldn't have done it."[27]

In August 1952, the same month Hoover and Eisenhower visited Del Charro, Joe McCarthy joined the party. Reporters lurked in the bushes outside, forcing Murchison's security men to shoo them away. At night, sipping drinks around the pool, the oilmen listened in rapt silence as McCarthy railed against Communist influences in the government. Murchison ate up every word, and soon began welcoming the senator to his home in Dallas. The following May McCarthy joined Murchison at his Mexican ranch for a weekend of hunting, then accompanied the oilman to a speech at the Dallas Petroleum Club. Murchison would later insist he was already developing doubts about McCarthy: "After Joe came out with that figure of 205 Communists in the government, and then wasn't able to produce the names of more than thirty-five or so, I thought he ought to admit publicly that he'd been wrong. He could then have hammered hard on the right figures. I tried to get him to do this, but he wouldn't—I guess he figured it would be bad tactics."[28]

Whether Murchison knew it or not, he was playing with fire.

ELEVEN

"Troglodyte, Genus Texana"

I.

For five years, since the national spotlight first shone upon them in 1948, the Big Four had enjoyed unfailingly upbeat handling in the national press, one of the keys to their growing power in Washington. But the political winds shifted in the summer of 1953, as opposition to the tactics of Texas's "third senator," Joe McCarthy, began to mount. A turning point in McCarthy's career came that July, when one of his aides published a scathing article charging that communism ran rife through the ranks of Protestant ministers. This was too much, even for many McCarthy supporters. Reporters who had remained mostly neutral toward McCarthy finally began questioning his methods. From the outset the Big Four were drawn into the controversy.

The first salvo came on July 6, when the *New York Post* began a series of four articles detailing Texas Oil's support for McCarthy. For the first time Murchison and Hunt were identified among the senator's principal financial backers; the *Post* suggested they were grooming McCarthy for the presidency—a notion Murchison was at pains to deny.

"I like Joe McCarthy," Murchison told the *Post*. "I tell you, I think he's done the greatest possible service to his country. He fears nobody and he's certainly got those Communists feared to death of him. I hear complaints about the people [he] has hurt, not necessarily Communists, just people. Well, it's a war we have with communism. In Korea, that war has cost us 135,000 casualties. In McCarthy's war there are bound to be a few casualties too. They can't be helped."

The *Post* series, fueled by Murchison's uncharacteristically boneheaded quotes, was the first negative publicity any of the Big Four had sustained. Chastened, Murchison's ardor for McCarthy cooled in the following weeks, a period that saw the senator come in for increasing press criticism. When McCarthy returned to Del Charro that August, behavior that Murchison had once dismissed as playful he now found boorish. The senator was a heavy drinker; at breakfast he had a shot of whiskey with his orange juice. He drank all day at the races, then downed glasses of whiskey around the pool until three or four in the morning. When drunk, McCarthy told off-color jokes, sometimes in the presence of women, which offended Murchison. One evening, McCarthy challenged another guest to balance a marble on his forehead, then drop it through a funnel McCarthy inserted into the man's pants. When the guest tried it, McCarthy poured bourbon into his pants.[1]

For Murchison the final straw came one night at the pool when McCarthy, deeply drunk, began insulting his new wife, Jean. Suddenly McCarthy rose and pushed her, fully clothed, into the pool. As friends told it, Murchison got up and stalked to his bungalow without a word. The next morning he had an attendant deliver a written message to the senator, ordering him to leave. "I finally had it," a friend quoted Murchison saying, "when he pushed Jean into the pool."[2]

While making Murchison's life uncomfortable, the *New York Post* series was also among the first to take note of Hunt's burgeoning media empire. Soon other reporters began sniffing around. The most dogged was Ben Bagdikian, Washington bureau chief of the *Providence Journal,* who stumbled onto the story when one of Hunt's men offered another *Journal* reporter $125 to conduct a Facts Forum interview, explaining that Hunt was willing to pay cash to elect "our kind of guy." The reporter not only refused the offer, he wrote an article about it, which was picked up by the wire services. Bagdikian, who would go on to a lengthy career as an author and media commentator, spent weeks studying Facts Forum. The eight-part series he produced that November was a fiery exposé that laid out Facts Forum's right-wing agenda and questioned whether it deserved its tax-exempt status as an "educational" foundation. The series was widely noted by, among others,

Time magazine, which characterized Facts Forum's worldview as "isolationism, ultraconservatism and McCarthyism."

The *Post* and Bagdikian series drew a new wave of eastern writers to Texas, but where the first contingent in 1948 and 1949 had celebrated oilmen as cracker-barrel champions of free enterprise, this one came looking for bellicose right-wing nut jobs. The first of the ensuing broadsides erupted in the *Washington Post* on Sunday, February 14, 1954, when the newspaper's White House correspondent, Edward T. Folliard, began a six-part series on what a front-page editor's note termed "The Big Dealers, the fabulous moneymen of Texas who have been pouring part of their millions into American politics. . . . They are also in the Texas tradition—more money, more issues, and more noise. The unique thing about them is public ignorance of their motives, purposes and ideas. [The public] knows scarcely anything of the Texans." In his first article, "Lone Star Wealth Pours into Politics from Coast to Coast," Folliard focused squarely on the Big Four; he then profiled Hunt on the front page Monday, Facts Forum on Tuesday, Murchison on Wednesday, Richardson on Thursday, and Roy Cullen on Friday.

The *Post* articles were kind to the Big Four as individuals—Folliard termed Murchison a "genius" and lauded Cullen for having the "courage" to admit being conservative—even as they portrayed their political views, especially those of Hunt and Cullen, as slightly left of Hitler. "Hunt," Folliard wrote, "is convinced that the world situation today, with Russia having sway over 600 million people, was plotted in Washington, D.C., in the days of Roosevelt and Truman. . . . [I]f you didn't agree with him—well, you had something missing upstairs."

The only one of the Big Four to react to the *Post* was Hunt, who called a press conference that Wednesday—his sixty-fifth birthday—at the Waldorf-Astoria in New York. Before it began, as if to establish the oilman's credentials, a spokesman told reporters Hunt was worth two billion dollars and enjoyed an annual after-tax income of fifty-four million dollars. When Hunt ambled to the podium, he began telling stories of his childhood. The reporters countered with questions about Facts Forum. Hunt handled himself with aplomb, denying that it—or he—was pro-fascist, anti-Semitic, anti-Catholic, or anti-Negro. "There is a regular pattern to these charges," Hunt said. "In the *Communist Daily Worker* the motive is obviously to call us fascists. In the responsible press, I do not think it is due to a deliberate smear, but rather a failure to listen to our programs and understand what we are doing."

The *Post* series set Washington abuzz on the eve of what was destined to become Joe McCarthy's most controversial inquiry, the Army-McCarthy hearings. Two weeks later, on March 9, came the television broadcast that heralded the end of the senator's career, Edward R. Murrow's primetime denunciation on CBS. After that the floodgates opened, as reporters, politicians, and ordinary citizens not only piled on the anti-McCarthy bandwagon but looked to smoke out those who had been pushing it all along. They found their villains in the Big Four.

In the following months everyone from the *St. Louis Post-Dispatch* to *Fortune* carried articles probing Texas Oil's support for McCarthy's American inquisition. The most thoughtful, and most influential, critique was written by Theodore White and published that May in a now-defunct magazine called *The Reporter*. White, who was on his way to becoming the dean of presidential-campaign historians, had toured Texas that winter, and while newspapers such as the *Washington Post* remained on mostly neutral ground, he pulled few punches. In his two-part series, "Texas: Land of Wealth and Fear," White labeled Cullen "a genuine primitive," Murchison "a successful neurotic," and Hunt "mysterious" while gleaning "an almost monastic purity in Richardson's single-minded devotion to the pursuit of wealth."

Alone among the writers who visited the state that season, White attempted to take stock of the changes the Big Four and their brethren had wrought upon Texas since oilmen had thrust their first ultraconservative governor, Pappy O'Daniel, into the state house in 1938. White judged the state a paranoid right-wing fortress, run by an oil-supported governor, Allan Shivers, whose signature piece of legislation had been a proposal to make membership in the Communist Party punishable by death. Everywhere White looked, Texas appeared to be taking McCarthyism to new levels, enforcing loyalty oaths and promoting a campaign by housewives organized into chapters of "Minute Women" who hunted "pinks" in the public schools with the same zealotry McCarthy hunted Communists in the federal government. As White wrote:

This emotional climate would be no more than a matter of morbid or humorous interest to other Americans as they watch a growing community fumble its way to maturity were it not for another set of facts:

—That millions of Texans are convinced that their primary enemies are other Americans and that the American experience in this age and generation has been a total failure, their own prosperity notwithstanding.

—That within Texas the machinery of government, from the person of the Governor down through the structure of the major parties, has been captured by a nameless Third Party, obsessed with hate, fear and suspicion—one of whose central tenets is that "If America is ever destroyed, it will be from within."

—That a handful of prodigiously wealthy men, whose new riches give them a clumsy and immeasurable power, seek to spread this climate and their control throughout the rest of the United States.

Ardent and devout states' righters at home, bellowing and snorting that the "sovereign" privileges of Texas must not be disturbed, these men see no contradiction in a Texas political imperialism that intervenes with its money in the domestic politics of thirty other "sovereign" states from Connecticut to Washington, from Wisconsin to New Mexico.

This new portrait of Texas, of secretive oil billionaires plotting an ultraconservative takeover of America, sank deeply into the country's emerging post-McCarthy mind-set. That it no doubt overstated the Big Four's reach was no matter; the nation suspected a powerful cabal behind McCarthy and the new conservatism, and Texas oilmen fit the bill. This shift in public opinion, apparent in press coverage, comments of congressmen and letters to the editor, was immediate and powerful. The first to encounter its strength was Hunt, who on April 22 staged a second press conference at the Waldorf-Astoria, this time to announce that he was negotiating with a national television network—NBC—for a daily fifteen-minute program that would once again air "both sides" of an issue. A Socialist politician, Norman Thomas, fired off a letter of protest, and one week later NBC's chairman,

David Sarnoff, announced that the network had declined to run Hunt's program.[3] Hunt now found Facts Forum's every move closely monitored.

Roy Cullen confronted a more visceral reaction when, near the height of press interest that spring, he announced he was inviting McCarthy to deliver a speech on San Jacinto Day, April 22, at the San Jacinto Monument on Houston's Ship Channel. The news provoked outrage among liberal students at the University of Texas. Two, Bob Kenney, editor of the campus newspaper, and Ronnie Dugger, who went on to found the state's only progressive newspaper, the *Texas Observer*, circulated a petition calling on Cullen to retract the invitation. By the time the two young men headed to Houston to deliver it, the petition was forty feet long and contained fifteen hundred signatures. When Cullen agreed to see them, the two students faced him across his desk. There was a long pause. Finally, Cullen turned to Kenney.

"Are you a communist?" he asked.

"No," Kenney said.

Cullen turned to Dugger. "Are you a communist?" he asked.

"No," Dugger said.

"All right," Cullen said, "let's talk about this thing."

"So we talked about it," Dugger recalls in an interview, "and he said he couldn't withdraw the invitation, that it was too late, but he saw our point, and we left on pretty good terms. After he got his blood question out of the way, he was a southern gentleman." A gentleman, however, who still adored McCarthy. Afterward Cullen told a reporter he considered the senator "one of the greatest men in America."

For all the controversy, McCarthy's San Jacinto speech proved anticlimactic. It was a sad scene. A beautiful day, children off from school, the papers had predicted forty thousand people. Up on the dais, there was a nervous glance or two; the *Post* counted forty-two hundred in the crowd. Cullen, in a bow tie sitting beside McCarthy, looked tired. He had been hospitalized for a kidney stone two years earlier, during a vacation in Vancouver. It had been front-page news in the Houston papers—"H.R. Cullen Is Seriously Ill in Canada," read the *Press*'s banner headline—and photographers were at the airport when he was carried off the plane in a stretcher.

Later, when he entered Hermann Hospital for a checkup, reporters wrote that up, too. He was seventy-three now, still feisty as ever, but that day at the monument, his gray hair tousled by the breeze, was a sign that winds of change were blowing.

II.

The central question running through media coverage of the Big Four's political activities was simple: "What do they want?" The nation's business community was wondering the same thing. Only Murchison had aggressively diversified outside oil, and while his investments in real estate, insurance, and publishing had been quiet affairs, any of the Big Four could threaten General Motors or U.S. Steel if so inclined. These vague concerns turned to reality on February 26, 1954, just a week after the *Washington Post* series, when Murchison stunned Wall Street by confirming that he and Richardson were plunging into the nation's largest takeover battle, the fight to control the country's second-largest railroad, the $2.7 billion New York Central.

It was a tangled affair, at the time the largest proxy fight in American history, initiated by a suave Texas-born Wall Street financier named Robert Young, who had traded in his cowboy boots for a Park Avenue apartment and mansions in Palm Beach and Newport. That January Young, who with a partner controlled a competing railroad, the Chesapeake & Ohio, known as the C&O, had suddenly severed all ties with the C&O in order to satisfy federal dual-ownership laws and mount an attack against New York Central. Young left behind the C&O's 12 percent stake in New York Central—the largest single block of the railroad's shares. If he was to have any chance to win the shareholder vote scheduled at New York Central's annual meeting that May, Young had to get that block of shares into friendly hands. He called Murchison.

Murchison was happy to help, especially once he and Young structured a loan package that allowed Murchison to buy the shares with money borrowed from Young and his partners. To spread the risk, Murchison telephoned Richardson in Palm Springs, reaching him as he was heading out to play cards. It wasn't until he read the newspapers that Richardson realized that in his haste he had agreed to a twenty-million-dollar deal instead of one worth five million dollars. "What the hell did you say was the name of that railroad?" he called and asked Murchison. Kidded about it later, Richardson said, "Well, Clint mumbles so."

The investment drew Murchison into Bob Young's rarified East Coast world, a milieu in which he was never entirely comfortable. While visiting Young's mansion in White Sulphur Springs, West Virginia, which was decorated with the paintings of Young's sister-in-law Georgia O'Keeffe, a valet shadowed Murchison everywhere, even drawing his bathwater and hovering beside the tub. "That's okay, you can wait outside the door," Murchison quipped. "I've been bathing myself since I was three." Chatting with socialites at Young's Palm Beach mansion, Montserrel, he squirmed uncomfortably in an antique French chair as he tried to balance finger sandwiches and a cup of tea on his knee. Sensing his discomfort, Young asked if he would prefer something else to drink. "Yeah," Murchison snapped. "A double martini."[4]

All through the media tumult over the Big Four's backing of McCarthy that spring, the New York Central fight grew nastier, captivating the business press. The publicity catapulted Murchison onto the cover of *Time* in May, making him the second Texas oilman to be so honored in four years; the magazine's lead story—"Those Texas Millionaires"—focused on the Big Four's business rather than political ambitions. Murchison, it noted, "is the first of a brand-new breed of Texas oilmen. Having made his millions in oil, he is now using them to further the popular Texas ambition of buying up the rest of the U.S."[5] The liberal *Nation* got angry: the New York Central fight, it said, was pursued "much the same way that Murchison and others have backed McCarthy's challenge to the top leadership of the Republican Party." The Texans, it concluded, "are determined to muscle in on all the great roulette games where power is won and lost in our society. [They] will not take backseats; they demand front-row center."[6]

The New York Central fight raged through May. The railroad sued to block the transfer of its shares to Murchison and Richardson; it lost. It appealed to the Interstate Commerce Commission; it lost. It refused to hand over the shares until a judge forced it to do so on the eve of the annual meeting, at which point the writing was on the wall. With the backing of Murchison and Richardson, Young was victorious. The Texans sold their shares back to him just two months later, prompting a Senate investigation into the whole affair. Litigation dragged on for months, eventually forcing Murchison into a New York courtroom, where he politely told the judge he did not represent the views of Richardson, who he referred to with a smile as a "rich, fat old man."

III.

The scathing coverage the Big Four suffered in the wake of McCarthy's reversals led to a backlash against Texas Oil whose sting would be felt for years to come; its image, in fact, never really recovered. One can draw a straight line from that spring of 1954, when reporters who once painted the Texans as fun-loving and adventurous began portraying them as racist, ravenous, and conspiratorial, to the alternately kooky and villainous portrayals of Texans in *Dallas* and movies such as *Doctor Strangelove* and Oliver Stone's *JFK*.

At the beginning, at least, this new image owed as much to Elmer Fudd as to Simon Legree. An early signal came the same week in May that Murchison appeared on the cover of *Time*, when the Woman's National Press Club lampooned the Big Four at its gridiron show in Washington. That evening, as a band played Noel Coward's "The Stately Homes of England," four female reporters danced onto the stage at the National Press Club. Dressed as cowboys, flashing wads of money, they sang:

> *I'm Murchison, I'm Richardson, I'm Cullen, and I'm Hunt*
> *Our millions multiply like bees*
> *In Texas money grows on trees*
> *Here we are, the four of us*
> *And there are so many more of us*
> *Texas sons who did succeed,*

We know how oil wells come in
And how to parlay cash to win.
Apart from this, our education lacks coordination,
Though we're shrewd, observative,
Politically conservative . . .

We like to hunt for Commies,
And pinks of every hue,
But if we can't find Commies,
Plain liberals will do.

So if the liberals have to go,
We'll supply the cash . . .
For we're the millionaires of Texas![7]

Texas has always been the butt of jokes, mostly focused on its penchant for braggadocio, but in 1954 the media's focus on the Big Four unleashed a wave of unprecedented ridicule aimed at Texas millionaires who were seen as unlettered, uncouth know-it-alls. A *New York Times* reviewer, for instance, suggested a book of oil field photography "would make a wonderful Christmas present for a Texas oil millionaire who has not yet learned how to read."[8] Texas, a *Times* writer quipped in 1955, "is the place where oil millionaires overnight become experts on subversive books, atomic warfare, and how to get into heaven."[9]

Jokes about Texas millionaires began receiving wide currency in the television skit shows of the day. In one, *Mr. Peepers*, a group of tourists ogled Buckingham Palace as a Texan among them said, "I wonder if they'd sell. I'm looking for a country place." One of America's favorite comedians, Milton Berle, made Texas a regular target on his variety show. In one skit, set at a racetrack, Berle queried a boy singer from Texas named Charles Applewhite. "Well, Charlie," Berle asked, "how much money you gonna bet?"

Charlie: "Oh, Mr. Berle, where I come from in Texas everybody has so much money there's no fun betting money. We bet people."

Berle: "You bet people?"

Charlie: "Dad did real good last year. I had four mothers."

Berle then encouraged Charlie to sing.

Berle: "Don't be nervous, Charlie. Don't think of it as singing in New York. Think of it as singing in your own backyard in Texas."

Charlie: "That's easy. In Texas, New York IS our backyard."

Berle wasn't alone; during the mid-1950s every comic seemed to have a store of Texas-millionaire jokes. The radio personality Fred Allen had several favorites. One concerned the Houston millionaire who took away his son's pogo stick; the boy was jabbing oil wells in the backyard. "The Texas joke material is of enormous variety, ranging from the low-down to the exceedingly high-toned, and growing all the time," a Texas-born New York writer named Stanley Walker wrote in a *Times* feature on the phenomenon. "For the gag miners, it may turn out to be the Mother Lode, a Golconda of infinite resources." Walker advised Texans to grin and bear it: "The citizens of Texas . . . must, in this crisis, keep their shirts on. Things could be worse, though it's hard to see how. . . . If Texans have really forgotten how to laugh at themselves . . . then Heaven help them. The going will be even rougher."

During the mid-1950s the only Americans who weren't laughing at Texas oil millionaires seemed to be those who found their right-wing politics ominous, a view that soon brought a new caricature into Broadway shows and Hollywood movies. "Have you noticed the way Texas millionaires are being used as villainous types lately in fiction stories and television dramas?" the *Dallas News* asked in 1955. It was hard not to. A popular 1957 novel, *The Promoters*, told the story of a conniving Texas oilman's efforts to take over a railroad—an allusion to Murchison's bid for the New York Central. A 1959 Broadway musical, *Happy Town*, featured four evil Texas oilmen attempting to swindle the inhabitants of a small town. The 1957 movie *Written on the Wind* starred Robert Stack as a drunken Texas oilman who fights Rock Hudson over a girl and ends up shot—the *Times* reviewer termed it "another harsh inspection . . . of those Texas millionaires whose sad psychoses are subject for frequent literary concern [these days]."

By the end of the decade Texas editorialists had given up trying to rebut these stereotypes. "The rest of the nation demands a whipping boy," the

Dallas Morning News grumped, "and the distorted image of Texas as ten million blowhards wallowing in wealth from oil or cattle will not go [away] easily. It means too much to night-club comics, hack columnists, novelists and
Eastern politicians who delight in lumping Texas in with the big rich to
push their own liberal schemes."[10]

As its image in American popular culture began to darken in the spring
of 1954, few Texans were laughing. In Washington the backlash against
"Texas oil money" was very real. Democrats, in fact, began testing attacks
on Texas oilmen for use in the midterm elections that November. In May
the *Houston Post's* Washington correspondent, Elizabeth "Liz" Carpenter—
later a White House press secretary and author—reported that rumors in
the capital had "Texas oil money" funding campaigns against Democratic
senators in Alabama, Michigan, and other states. Noting the "epidemic" of
bad publicity afflicting the Big Four, Carpenter wrote: "The smart talk in
Washington political salons is to attribute every political trend most anywhere in the country to 'Texas oil money.' If you say it mysteriously enough,
it sounds sinister and as though you have the inside dope." The publicity
was so bad, she speculated, that unless reversed it might endanger Texas
Oil's Holy Grail, the depletion allowance. "The link between the 'Texas oil
money' and Sen. McCarthy," Carpenter wrote, "makes a vote for the depletion allowance [appear] a vote for maintaining McCarthyism."[11]

That summer each of the Big Four felt the sting of the new skepticism. For
Hunt, it meant mounting attacks on Facts Forum, which became snarled in
a congressional investigation into whether certain foundations deserved tax-
exempt status. A House committee chaired by the conservative B. Carroll
Reece of Tennessee was using the probe to "investigate" Communist influence in the Rockefeller and Ford Foundations. Each time Reece requested
documents from those foundations, however, his Democratic opponent,
Wayne L. Hays of Ohio, requested data from Facts Forum. During hearings
that May, and again in a report he issued in September, Hays demanded
an IRS probe of Hunt's activities. Hunt's producer, Hardy Burt, denounced
Hays for "the violent campaign of vilification against Facts Forum [that] was
triggered by the Communist press." The squabbling dragged on for months,
until an IRS review cleared Facts Forum of wrongdoing.

Roy Cullen, meanwhile, found himself the target of senators who didn't appreciate the checks he was mailing their opponents. As in 1952, Cullen was again the largest donor to American politicians in 1954. One of his most vocal critics was a senator from Maine, Margaret Chase Smith, a McCarthy opponent who was facing an unusual primary challenge from a political unknown named Bob Jones. That spring Cullen's son-in-law, Douglas Marshall, received a letter from an editor in Maine. The editor wrote that while interviewing Bob Jones,

> one of the questions I asked Bob was where he was getting his financing. He said from two sources in Maine, but would not identify them. It has come to my attention that you wrote Jones a few days before his announcement, informed him you had talked with Senator McCarthy, and that there would be adequate financing for him, plus the services of two public relations men, if he would oppose Mrs. Smith. . . . What I would like to know is, are you, your father-in-law, Mr. Cullen or any other friends of McCarthy's in Texas backing Jones financially in the current campaign? . . . All of us believe that he hasn't got money of his own."[12]

Similar scenes were being repeated across the country, in Montana and Alabama and Maryland. After the primary, which she won, Senator Smith emerged as a focal point of anti-Texas sentiment in Washington. "H. R. Cullen and his Texas oil and gas associates sent money into Maine in 1954 to try to destroy me politically," she wrote an oil lobbyist. To a man in Dallas, she added: "The Texas oilmen have a perfect right to contribute to the national Republican party and to the Eisenhower campaign. But they are reaching just a little too far when they send their money into Maine to attempt to buy the Maine Senate election and to dictate to the people of Maine who their Senator should be. You can be sure that the people of Maine have never attempted to tell the people of Texas who should be the Texas senators."

It was into this tempest that Cullen had the bad timing to drop his authorized biography, *Hugh Roy Cullen: A Story of American Opportunity.* He had been noodling with the idea of a book ever since Glenn McCarthy had published his own authorized biography, a windy mishmash called *Corduroy*

Road, in 1951. Two writers, including one who authored fiery right-wing editorials for the *Houston Post,* interviewed Cullen at length; their book, while steering clear of its subject's more infamous views, was a straightforward telling of his life story. A few reviews were kind. But many favored the tone of Stanley Walker's withering take in *The Nation.*

It was headlined "Troglodyte, Genus Texana":

There is a dangerous ailment in Texas which has been named Cullen's Syndrome, after the subject of this book. Its concurrent symptoms are these:

The patient is almost always an oil man, not a cotton man or a banker or a cowman or a merchant.

He believes his riches were in no way the result of luck but of his own foresight, courage, and initiative—all made possible by the American Way of Life.

He thinks one way of showing his appreciation for America is to chip in with like-minded patriots and buy Joe McCarthy an automobile.

Although he may never have got as far as high school, he is an authority on textbooks, the tariff and winning football formations, the Constitution, geophysics, currency inflation, and how to get rid of warts.

He is fond of writing letters to office-holders and potential office-holders advising and/or threatening them about the course they should follow. Given half a chance, he will, out of his accumulated wisdom, drop homilies, maxims, aphorisms, texts, proverbs, and parables for the benefit of his fellowman, whom he professes to love dearly.

Thanks to articles like that, by mid-1954 the new stereotype of the ignorant, ultraconservative Texas oilman had taken firm hold in the American imagination. It was Clint Murchison, the oilman most attuned to public opinion, who tried to turn the tide. On July 12 Murchison announced that he and Richardson were buying their beloved Del Mar racetrack and transforming it into a nonprofit foundation, Boys Inc., that would channel 90 percent of its profits to Boys Clubs in California and elsewhere; it was a measure of the Big Four's notoriety that the *New York Times* carried the story on

Oil derricks at Spindletop, 1930.

Hugh Roy Cullen, for years the richest oilman in Houston and perhaps the United States, was a fifth-grade dropout who used his millions to joust with politicians from Wendell Willkie to Dwight Eisenhower. Above, the "Big White House" he dreamed of as a child and built during the Depression.

Clint Murchison, the brainiest of Lone Star oilmen, introduced the toys of wealth to mid-century Texas—swimming pools, private airplanes, private islands, California resorts, and sprawling Mexican ranches, not to mention politicians from J. Edgar Hoover to Richard Nixon. Above right, Murchison clowning with friends at his compound on Matagorda Island. Below, Franklin Roosevelt, Sid Richardson (at right), and friends during the president's 1936 visit to Matagorda.

Sid Richardson, the most secretive of Texas oilmen, created the fortune that his great nephews, the Bass brothers of Fort Worth, turned into the state's largest. Top right, Richardson's modernistic home on St. Joseph's Island. Bottom right, Richardson with his close friend Dwight Eisenhower.

H. L. Hunt (above in 1963, and left as a young man), the greatest of all Texas oilmen and for years the wealthiest man in America, sired three separate families, two in secret. The foundation of his fortune was the mammoth East Texas oil field, the heart of which he purchased from the itinerant wildcatter Dad Joiner, top right, with Hunt and his drilling crew on the day they met at the Daisy Bradford well in 1930. Bottom right, Hunt's Dallas mansion, Mount Vernon.

Glenn McCarthy, the model for
the stereotypical Texas millionaire,
a bourbon-swilling, fistfighting,
damn-the-torpedoes Houston oilman
who rocketed into the national
imagination in the late 1940s. Above,
fireworks above McCarthy's dream
project, a symbol of the new postwar
Texas, the legendary Shamrock
Hotel, built, opened, and lost in a
scant five years.

Nelson Bunker Hunt, H. L. Hunt's best-known son (top left, with his family at dinner), a man who squandered what was probably the world's greatest fortune attempting to corner the silver market in the late 1970s. At left below, another son, the popular Ray Hunt, who rebuilt Hunt Oil into one of the nation's largest independents. Above, Bunker and his brothers Lamar and Herbert at the 1975 federal wiretapping trial in Lubbock. At right, Frania Tye, H. L. Hunt's secret second "wife," at the civil trial in Shreveport, Louisiana, where she attempted to reclaim a share of the Hunt family fortune.

John Murchison and Clint Murchison Jr., symbols of modern Texas, fought their way onto the national business stage in 1961 but quickly withdrew, ultimately squandering much of their father's vast fortune. They became best known as principal owners of the Dallas Cowboys football team. Left, the Murchison brothers on the cover of *Time*, 1960. Top right, the brothers in conference with their father, 1952. Bottom right, Clint Jr. and his stepmother Ginnie, at his private island in the Bahamas, Spanish Cay.

Above, Herbert and Bunker Hunt testifying before Congress, 1980. Below, Sid Bass, who with Richard Rainwater pulled off one of the greatest investment runs of the twentieth century, building a $50 million fortune into more than $5 billion.

page 1. The purchase of five more tracks around the country was under way, and Murchison hoped to lure J. Edgar Hoover to head the foundation if and when he retired.

The Texans' philanthropy, however, was met by a gale of protest, most of it from a group of fundamentalist Christians in California, who told reporters that the nation's youth shouldn't be sullied with gambling profits. The real problem, though, was the IRS, which launched a multiyear investigation into whether profits earmarked for Boys Clubs should in fact be taxed. The IRS review killed Murchison's plans to buy additional tracks in Michigan and Illinois, but the Del Mar foundation eventually channeled more than one million dollars to Boys Clubs—that is, after the four years it took to repay Murchison and Richardson for purchasing the track. The relationship lasted until 1968.

Murchison's bid for media approbation, however, did little to halt the flow of vitriol aimed at Texas Oil. And if the Big Four could brush it off, a number of Texas politicians worried what it meant for the state's future. "National politics is where the Texas reputation has suffered most severely," a former aide to Lyndon Johnson wrote in the *Houston Post* in 1955. "Some Washington advisors regard it as 'unsafe' for political figures to let it be known that they number Texans among their friends. . . . It may be unpleasant to face, but a long list of positions of importance in national affairs could be drawn which, under prevailing conditions, no Texan is likely to fill simply because he is a Texan." Congressman Brady Gentry told the East Texas Chamber of Commerce the backlash was hamstringing all efforts to help the state. "One of the contributing causes to the deplorable state in which we find ourselves," Gentry said, "is caused by the tremendous publicity that has been given the Big Rich of Texas. . . . If it still continues, I feel certain that the time will come when the tax depletion will be greatly lessened, if not entirely eliminated."

One of the few journalists to analyze the anti-Texas backlash was George Fuermann of the *Houston Post*. In his third book in six years on the "new" Texas, 1957's *Reluctant Empire,* Fuermann defended most oilmen, blaming the rise of anti-Texanism squarely on the political activities of Cullen, Murchison, and Hunt. "The increasing suspicion of Texas oil has made it seem to others that oilmen are dishonest and crafty, yet the customs and [honesty] of independent oilmen are among their chief merits," Fuermann wrote. Of Cullen, Murchison, and Hunt, he noted, "[these] three men, with little help from others, have made Texas oil, and thus Texas, a national antipathy."[13]

IV.

The year that followed 1954's surge of anti-Texan sentiment was not an election year, so the Big Four's withdrawal from public view in 1955 could be viewed as either a return to normalcy or a respite to lick their wounds. Their retreat from the headlines, however, didn't mean they had scaled back their political ambitions. It was in 1955, in fact, that Sid Richardson—a man who just twenty years earlier could not afford to buy groceries—conceived of a coup that would represent the apex of the Big Four's influence in Washington: the replacement of Dwight Eisenhower's vice president, Richard M. Nixon.

Carping about Texas oil power meant little to Richardson. Unlike Hunt and Cullen, he was on the *inside*—inside the Oval Office, *inside* the inner circle, *inside* Eisenhower's head. No one could touch him there. In the autumn of 1955, Richardson fell in with a group of Eisenhower friends, including George Allen, who felt that Nixon would hamper chances for the president's reelection in 1956. Dark and dour, Nixon had long been unpopular with many of Eisenhower's advisers, and Eisenhower himself had real doubts whether he was presidential material. "The fact is," Eisenhower told his speechwriter Emmet Hughes, "I've watched Dick a long time, and he just hasn't grown. So I just haven't been able to believe that he *is* presidential timber." Among Eisenhower loyalists, both inside and outside the White House, there was already mumbling about replacing Nixon on the Republican ticket when, in September 1955, Eisenhower suffered a heart attack. Overnight the notion of a Nixon presidency became much more concrete, a possibility any number of Eisenhower's friends wanted to squash.

Among them was Richardson, who had a man he thought perfectly suited to replace Nixon: his old friend, the Texas Oil attorney Robert B. Anderson.

After serving as secretary of the navy for two years, Anderson had retired as assistant secretary of defense just months before, complaining that it was difficult to live on a government salary. In a conversation in late 1955, Richardson mentioned to Eisenhower the possibility of Anderson replacing Nixon on the 1956 ticket. Eisenhower agreed Anderson would make a fine vice president. There was just one problem: Anderson wasn't willing to return to Washington, even as vice president, if it meant a government salary.

At that point, Richardson put in motion a complex scheme whose sole purpose was to make Bob Anderson a very rich man. A group of four oil companies, including one owned by Clint Murchison, was drilling wells on land owned by Richardson in Texas and Louisiana. Richardson asked each of these companies to assign royalty interests to one of his oldest friends, the onetime Gulf Oil executive Jay Adams, now an independent oilman based in Fort Worth. Acting purely as a go-between, Adams then assigned his interests to Anderson. Anderson then sold them to a company owned by Murchison's friend Wofford Cain, who later returned them to Richardson's hands by selling them to his nephew Perry Bass. Though little more than an exercise in paper shuffling, Richardson's scheme made Anderson $970,000.

A few days before Christmas 1955, Richardson flew to Washington in one of his DC-3s laden with steaks, quail, and ducks for Eisenhower. During the visit he made clear that Anderson was now willing to accept a spot on the ticket; Eisenhower agreed to consider it. In a face-to-face meeting that spring, in fact, Eisenhower suggested to Nixon that he step down and become a cabinet member, the better, he argued, to run for president in 1960. According to both men's biographers Nixon declined, and Eisenhower didn't have the heart to push him out. In the end, in June 1956, Eisenhower named Anderson secretary of the Treasury, a position where he remained a great friend to the men of Texas Oil.[14]

V.

All through the 1950s the fundamentals of Texas Oil continued to deteriorate: rising production costs and competition from lower-priced Middle

Eastern oil were slowly strangling the industry. Oil's decline, however, was more than offset by the astronomic profits oilmen were making in natural gas. Thanks in large part to a malleable bureaucrat named Mon Wallgren, who had replaced Leland Olds as chairman of the Federal Power Commission, gas prices had risen steadily, from six cents per one thousand cubic feet of gas in 1948 to ten cents in 1955. Texas gas producers, chief among them Sid Richardson and H. L. Hunt, saw their profits skyrocket, while the publicly held pipelines, such as the Brown Brothers' Texas Eastern, watched as their stock prices doubled.

The impetus for the rise in prices was an FPC rule that excluded independents from controls on natural gas, which allowed Hunt, Richardson, and others to request higher prices simply by telling the commission their costs were rising. Their customers, however, were far from happy. By 1954 midwestern states, who were paying 40 percent more for gas than in 1949, could take it no more; Michigan and Wisconsin sued the FPC, demanding protection from further price increases. The matter landed before the Supreme Court, which held for the states, ordering the FPC to regulate the independents. When the FPC dragged its heels, Congress intervened. A bill to nullify the Court's ruling was introduced in the House in early 1955 and was passed by a small margin that July. A companion bill—which held the future of Texas Oil's gas bonanza—was introduced in the Senate, where a vote was scheduled for February 1956. Once again it appeared Lyndon Johnson, now Senate Majority Leader, held the fate of Texas Oil in his hands.

The Harris-Fulbright Natural Gas Bill, as it was called, heralded a gathering of Texas Oil power such as Washington had never before seen. The stakes had never been higher: if the bill went through, gas prices were expected to rise between two hundred million and four hundred million dollars a year, an increase that would boost the value of southwestern gas reserves between twelve billion and thirty billion dollars. Lobbyists from every gas producer in Texas and Oklahoma—Humble, Phillips Petroleum, William "Kill 'em" Keck's Superior Oil, and many more—could be seen prowling the halls of the Capitol in search of arms to twist. Sid Richardson's man, John Connally, orchestrated the campaign from a war room at the Mayflower Hotel, where he worked with one of Johnson's oldest pals, the lobbyist Ed Clark. Even the oilmen themselves made appearances. "I saw Hunt here today," Texans whispered in the Mayflower lobby. "Sid's here, too. I saw him. And Old Man Keck in his wheelchair."[15]

From Connally and Clark on down, everyone in the Texas camp realized this would be a difficult battle to win. The media's attacks on the Big Four remained fresh, and northern newspapers were up in arms over what they viewed as price gouging by gas producers; the New York Times termed the bill wrong "socially, economically and politically." Keenly aware of public opinion, Johnson did everything possible to avoid anything remotely controversial that might provoke further headlines, ordering the oilmen to keep out of sight and urging his senatorial supporters to avoid any contentious debate. By the end of January, his strategy appeared to be working. With a vote scheduled for Monday, February 6, Johnson was sure he had enough votes to win.

Then, on Friday, February 3, as Senate debate was winding down for the weekend, one of the bill's supporters, Francis Case of South Dakota, rose at his desk and made a stunning announcement: a lobbyist for the natural-gas industry, he said, had paid a visit to his office in Sioux Falls and left an envelope containing twenty-five hundred dollars in cash.[16] Down at the Mayflower, Ed Clark immediately knew what this meant: At the very least, a Senate investigation. At worst, the bill's defeat. All that Friday afternoon and into the night he prowled the corridors of the Mayflower, pinholing lobbyists and urging them to leave town—that night—before subpoenas flew. Into the wee hours limousines ferried oilmen to the airport, where private planes waited to fly them back to Texas, Oklahoma, and, in Sid Richardson's case, Palm Springs.

Somehow, despite the charge of bribery, Lyndon Johnson managed to get the bill passed the next week. But the taint lingered. Editorials across the country condemned oilmen for brazen corruption. Then on February 17, disaster struck: Eisenhower announced that while he favored the bill, he could not abide the "arrogant" behavior exemplified by the Case situation. In a show of courage that cost the Republican Party millions in oil-industry donations, he vetoed the bill. "There is a great stench around the passing of the bill," he wrote in his diary. This is "the kind of thing that makes

American politics a dreary and frustrating experience for anyone who has any regard for moral ethical standards." In a letter to Sid Richardson, the president wrote, "I could not possibly sign the bill in view of the questionable aura that surrounded its passing, which was, of course, created by an irresponsible and small segment of the industry."

Pleasant words aside, it was the worst reversal Texas Oil had sustained, and oilmen had no one to blame but themselves. In the wake of Eisenhower's veto there were calls for an investigation into the role Texas Oil money played throughout the affair. A Senate committee was formed to look into the matter, but Johnson emasculated it, and the committee soon dissolved. Still, lasting damage had been done. The days when Sid Richardson could plot a vice president's overthrow were over; for the rest of Eisenhower's term, Richardson's political involvement remained limited to funding a presidential library. Everywhere, oilmen made themselves scarce. All that remained of their power in Washington was Lyndon Johnson, but after 1956 their roles were to be reversed. It was Johnson, now with his sights on the presidency, who led the oilmen, who were thankful to have any friends in the capital at all.

The collapse of Texas Oil's power in Washington coincided with the death throes of Cullen's and Hunt's political careers. Facts Forum was already flagging in 1955 when Dan Smoot left to head his own small media company. Hunt shuttered what remained in November 1956; outside Dallas, no one seemed to notice. In Houston, Cullen, now seventy-five, was still delivering the occasional fire-breathing speech—in a June 1956 address he called for the impeachment of the entire Supreme Court—but the press, even the Houston newspapers, rarely treated his remarks as news anymore. Cullen's brand of ultraconservatism did not go away, nor did the political donations of wealthy Texans, but 1956 marked an end to the era in which Lone Star oilmen appeared poised to have a significant impact on the nation's political direction. Never again would there be a presidential election as in 1952 in which they would be involved at such high levels. For the Big Four, and for their friends around the state, all their money, and all the adoring coverage they had enjoyed during their media honeymoon, had been squandered. It happened in large part because Murchison, Richardson, Hunt, and Cullen were not only politically naive but badly out of step with a postwar America that, like Texas, was fast maturing, where a majority of Americans were beginning to accept civil rights and the federal powers born during the New Deal. As for Hunt and Cullen's desire to spread ultraconservatism, many of the same qualities that made them successful oilmen—self-reliance and a stubborn streak—hamstrung their political activities. Their need for personal control rendered them unable to build lasting political alliances, or to see the value in investing in any intellectual enterprises but their own. This shortsightedness was vividly illustrated by William F. Buckley's strenuous efforts to build a partnership with Hunt. In 1954, as Buckley was planning the journal that would become the influential National Review, he appealed to Hunt for backing. Hunt refused, and his son Bunker agreed to invest ten thousand dollars but never paid. It is a measure of Facts Forum's reach at the time that Buckley offered to excerpt its articles in National Review or vice versa. Already a rising star in the conservative firmament, Buckley even agreed to write for Facts Forum and appear on Dan Smoot's show if Hunt would help fund National Review. Nothing, however, could pry open the oilman's bankbook. Buckley, his biographer John B. Judis concluded, "was too Catholic, too eastern and too moderate for most of the Texas Right."

"I talked to Hunt about National Review and he just wouldn't do anything," remembered Karl Hess, one of Buckley's fund-raisers. "Dan Smoot was Hunt's ideologue.[17] People like Smoot didn't really want to change anything. They wanted to lay the curse on the benighted. Buckley really wanted to change things." As a result, while Buckley went on to become the champion of modern American conservatism, Cullen and Hunt, as political figures at least, were consigned to the dustbin of history. Had they built bridges rather than guarded them, they and other Texas oilmen might have eventually been embraced as prophets. Instead, historians have correctly dismissed them as fools.

TWELVE

The Golden Years

It's been a hard day all around. First, my wife's pet kangaroo has to go and get poisoned, and then somebody stole my midget butler's stepladder.
—VERBATIM QUOTE OF A TEXAS OILMAN AS HE STEPPED OFF A
SANTA FE TRAIN IN HOUSTON, 1957[1]

I.

The wave of publicity that inundated Texas Oil in the early 1950s surged late into the decade, feeding the nation's appetite for insights into the strange and flamboyant new world of the Texas Big Rich. Much of the curiosity was spurred by depictions such as Edna Ferber's *Giant*, offered to the nation a second time as the 1956 movie starring James Dean as Jett Rink. But the exploits of rich Texans encapsulated many of the changes the country was undergoing. There was a sense that America's population and power were shifting south and west, and Texas oilmen presented a colorful window through which to view these changes. "Texas is such an alluring subject for New York writers," George Fuermann wrote in the *Houston Post* in 1956. "First it was cowboys, six-shooters and longhorns. Now it's millionaires."

Any number of newspapermen, magazine writers, and authors wrestled with questions of what the new Texas wealth meant. For starters, all these new millionaires scrambled popular perceptions of the nation's power structure. If H. L. Hunt was really America's richest man, what did that make the Rockefellers? Or the Du Ponts? And did Hunt really have more money than Sid Richardson or, for that matter, Howard Hughes? Guessing who ranked where became a staple for authors touring Texas. Theodore White, writing in 1954, judged Richardson "far and away the richest American, with the possible exception of his Dallas neighbor H. L. Hunt, who may be his only rival in the billion-dollar bracket."

In an attempt to introduce structure to the confusing new world of America's ultrawealthy, a number of publications began compiling lists of the country's richest people—the genesis of a phenomenon that continues today with lists such as the Forbes 400. The first two of these lists, appearing in 1957, codified the notion that Texas Oil had reshuffled the status quo. *Ladies' Home Journal,* in an attempt to name the ten richest Americans, judged Richardson the country's wealthiest man, pegging his net worth at seven hundred million dollars. Murchison, who ranked sixth, downplayed the idea that the Big Rich kept score. "After the first hundred million," he quipped, "what the heck?"

The second list, in *Fortune* magazine, was far more comprehensive, naming the seventy-six Americans whose net worth topped seventy-five million dollars; more than a third, twenty-six, derived their fortune from oil, fourteen from Texas Oil. *Fortune* ranked H. L. Hunt the fourth-wealthiest American, behind J. Paul Getty and two others, relegating Richardson to fifteenth place, just below Howard Hughes and the Rockefellers. Richardson's "downgrading" was subjective, the magazine noted; if oil reserves were counted, Richardson's wealth topped seven hundred million dollars, rivaling Getty's fortune as the nation's largest. Murchison ranked just twenty-third, behind two Houston oilmen, John Mecom and Jim Abercrombie.

One message echoed through all the coverage. With the glaring exception of the all-but-forgotten Glenn McCarthy, the Big Rich seemed to be living proof that money could buy happiness. Because despite their political reversals, despite the fact that many in the nation didn't believe they were to be entirely trusted, the one thing that came ringing through all the newspaper profiles and lavish magazine articles was that the Big Rich were having one heckuva good time.

II.

The 1950s was the Golden Age of Texas Oil, an era when the Big Rich seemed to be swimming in money and the toys of wealth, when the headiest oilmen, intent on living up to every Ferberesque myth, collected airplanes and ranches and works of art as if they were candy. In Houston, where Jim West could be seen tossing silver dollars on the sidewalks, one oilman wore a hundred-dollar bill as a bow tie; when asked, he would take it off and throw it in the air, then tie another. Another took to riding a pet lion to meet the mailman; yet another tried in vain to keep penguins in a walk-in freezer. One wrote Pablo Picasso asking to buy ten paintings; he didn't specify color or type, just the size of his wall. A Houston oilman's wife wrote the Smithsonian to ask whether the Hope Diamond was for sale. Then there were the two oilmen who loved playing practical jokes on each other; the high point of their duel came when one took a European vacation and his rival erected a full-size roller coaster in his front yard.

There was the Houston heiress who always flew to Paris with two extra first-class tickets for her two toy poodles, each of whom traveled with jeweled collars and chinchilla furs—furs being something the ladies of Texas Oil knew lots about. In 1951, when a ranch home owned by the oilman L. M. Josey burned to the ground, the *Houston Press* reported that Mrs. Josey fought the fire while wearing her mink stole. Irked, Mrs. Josey had her secretary write the paper. "Your story says Mrs. Josey battled the blaze clad in nightgown, robe and mink stole," the secretary wrote. "We wish to correct this. Mrs. Josey was wearing her marten furs." When the ladies of a New England garden club toured Texas, one asked an oilman's wife how she kept her azaleas so radiant. "Mink manure," came the answer.

It was no accident many of these stories arose from Houston, the city that came to embody the new Texas myth; it seemed to prize vulgarity and ostentation even as Dallas and Fort Worth clung to what appeared to be the last rungs of taste in the state. In Houston oil was *everywhere* and *everything*. They hit a gusher at the city dump, at the city prison farm, and in backyards. Downtown, skyscrapers sprouted like crabgrass; between 1940 and 1960 the city grew from a 500,000 to 1.25 million people. In the newspapers columnists debated whether Texas now had more millionaires than New York or California (apparently not); in 1954 Mayor Oscar Holcombe actually threw a press conference to announce the fact that he, too, had belatedly become a millionaire. Unsurprisingly, northerners poured in, thinking it was easy to become rich. For a time it seemed true: one of Roy Cullen's gardeners went into oil and eventually became a millionaire. Looking back years later, a onetime Houston reporter sensibly asked: "Was *anybody* poor?"

On the whole, the Dallas Big Rich lived tamer lives than their Houston counterparts. The city's nearest answer to Shamrock-style revelry were the mammoth street parties the oilman D. H. Byrd threw every year when his beloved University of Texas football team squared off against the Oklahoma Sooners at the Cotton Bowl. Byrd, who made his fortune in East Texas—he was there the day Hunt first met Dad Joiner—was one of the biggest Longhorn boosters in the state. His signature gift to the university was Big Bertha, the world's largest bass drum, eight feet across, so big it took four men to pull its carriage.

Parties were one way the Normal Rich applied for Big Rich status. In fact, some of the most talked-about fetes of mid-1950s America were thrown by Lone Star oilmen yearning to be noticed. On New Year's Eve 1952 a little-known Dallas oilman named Tevis F. Morrow rented out the Sunset Boulevard nightspot Mocambo for a dusk-to-dawn affair where the guests—including Edith Piaf, the heiress Doris Duke, and Conrad Hilton— were given ten-gallon hats; *Time* carried a photo of the goggle-eyed Morrow accepting kisses from a trio of actresses led by Joan Crawford. For the next four years a string of oilmen took turns topping Morrow's fete, climaxing with a New Year's Eve 1956 party thrown by the Dallas oilman D. D. Feldman, a publicity lover who had once briefly changed his name to D. D. Fontaine. Feldman's party, for which he transformed a Hollywood restaurant into a replica of New York's famed turn-of-the-century Delmonico's, attracted Fred Astaire, Gary Cooper, and Bing Crosby. He threw it, Feldman told reporters, to prove Texas oilmen were nothing like their portrayals in *Giant*. "I wanted to show the world that Texans can compete with the best in gentility," Feldman said. To which a wire-service report replied: "The wealthy Texan objected to a scene in the movie which showed millionaire Texans as boisterous fun lovers. By 2 A.M. of Jan. 1, 1957, when the party started dwindling, the movie looked pretty authentic."

III.

When it came to the toys of Texas Oil, every conversation began with ranches and ended with airplanes. Then as now, a ranch is to the Texan what a house in the Hamptons is to the New Yorker, or a Malibu beachhouse to a denizen of Beverly Hills. Ranches could be luxurious or rustic, so long as they came with horses, cattle, a wet bar, animal heads—antlers were a must—and shotguns. Roy Cullen's spread north of Houston was a simple affair, a few hundred acres and a stone house; he and Lillie enjoyed watching the deer. One of the more unusual ranches belonged to John Mecom's son, John Jr., who populated his acreage near Laredo with a bustling menagery of African wildlife, including elephants, giraffes, wildebeests, gazelles, and hippopotamuses.

Many of these retreats were decorated by the Big Rich's unofficial arbiter of taste, Neiman Marcus of Dallas, a retailer that earned national notice during the 1950s catering to the whims of oil money. Neiman's was typically the first stop for any Texan who struck it rich, a notion enshrined in a 1956 *New Yorker* cartoon showing a Texan and his wife dancing in a gusher's black rain. "It's wonderful, Harry!" the wife cries. "How late does Neiman Marcus stay open?" The Neiman's Christmas catalog offered baubles so outrageous, store executives received annual calls from the likes of Walter Cronkite and Edward Murrow snooping for the latest tidbits of Big Rich gauchery. In 1955 Neiman's offered a jewel-encrusted stuffed tiger. Another year it was his-and-her Jaguars, then his-and-her mummy cases, then, for oilmen tired of their fishing boats, authentic Chinese junks. The peak came the year the store offered a full-scale replica of Noah's ark, complete with French chef, Swedish masseur, German hairstylist, Park Avenue doctor, and a Texas A&M veterinarian—to care for pair after pair of animals, including ninety-two mammals, ten reptiles, fourteen freshwater fish, twenty-six birds, and thirty-eight insects. Alas, priced at $588,247, the ark went unsold.

When it came to vacation retreats, no one in Texas could compete with Clint Murchison. The state, in fact, didn't have enough free land for the kind of open space Murchison craved, so in the 1940s he began acquiring land in northern Mexico. It started with a small hunting lease during the war, but the more Murchison visited, the more he came to love the anonymity and frontier-style living he found in "Old Mexico," as Texans of a certain age called it. His first purchase was a new island. After losing Matagorda to Toddie Lee Wynne in 1944, Murchison wasted little time snapping up a replacement, nine-hundred-acre Isla del Toro, off the Mexican coast near Tampico. He shipped in a herd of Brahman cattle, built a sprawling hacienda-style mansion, and transformed the island into a private fishing and duck-hunting resort.

Word of inland ranches for sale slowly filtered down to Murchison's new spread, and all through the 1940s he bought up Mexican land, eventually more than a half million acres, including a three-hundred-thousand-acre section it was said he never got around to actually seeing. The centerpiece of Murchison's holdings was Hacienda Acuna, a grouping of a half-dozen interconnected ranches covering seventy-five thousand acres in northern Mexico's rugged Sierra Madre. As Murchison told the story, he had set off to scout the Acuna site alone, renting a car in a hilltop town, then driving until the road stopped, at which point he trundled the last twelve miles on a farmer's oxcart. What he found on arrival was a striking mountain wilderness, a verdant land dripping with Spanish moss and orchids, awash in wildflowers, waterfalls crashing from the hillsides. The only building was a tumbledown bunkhouse. Afterward he told his wife, Ginny, "I have found the paradise of the world."

He bought it all for ten thousand dollars, then returned with Ginny on horseback, where they slept under the stars and the next morning selected the site for their dream house. They hired every peasant in the area— Murchison built them houses with running water, and a school—and began clearing land for the mansion and the airstrip. Once the strip was finished, Murchison flew in brick masons, carpenters, architects, and construction workers by the score. Everything, Murchison ordained, had to made locally. Peasants built the furniture from walnut and cedar trees. Stone for the house was quarried on site. As the walls rose, Murchison reveled in his primitive Eden, laughing as well-heeled guests from Dallas teetered to the outhouse at night and rode horses down to the waterfall to bathe. Finally, after more than a year of work, the house was nearly finished, a twenty-room red-roofed hacienda in the shape of a horseshoe, set around a courtyard where Murchison had an enormous barbecue pit dug. Rocking chairs lined the surrounding porches. Bougainvillea and flowering vines crept up the support beams.

Clint and Ginny flew to Spain to buy the antiques to fill the house. On the floors they tossed ocelot and mountain-lion skins, all shot on the ranch; the walls were dotted with the bleached-white skulls of longhorn cattle. After a road was finished, hundreds of beef cattle were trucked in, along with dozens of dairy cows for milk and butter. Acuna's staff, several dozen butlers and cooks and vaqueros, lived in cottages Murchison had built in the fields. Everything was nearing completion in late 1949 when Murchison's Wall Street friend, Bob Young, mentioned that Acuna would make a perfect stop for his friends the Duke and Duchess of Windsor, who had expressed an interest in seeing Texas.

The visit was quickly confirmed—so quickly that the house wasn't entirely finished. Clint and Ginny embarked on a whirlwind of last-minute preparations, finishing the last of the guest bedrooms just as word of the Windsors' visit broke in a New York gossip column. News that European royalty was coming to see the Big Rich for themselves was an oddity almost every newspaper in America reported. In the month leading up to the February 1950 visit, Texas newspapers carried items almost every day; the *Dallas Morning News* devoted a half page to photographs and copies of the Acuna blueprints.

It fell to Ginny Murchison to prepare the waitstaff for the man who had once been king of England. Every afternoon she rehearsed with the Mexican butlers and maids the proper way to serve tea. The hardest part, Ginny always said, was teaching them the English words. Duke and Duchess didn't easily translate; in the end, they agreed the duke would be called "El Rey," the king, the duchess, the former Wallis Simpson, "La Reina," the queen. At the last minute, Ginny had a group of friends flown down from Dallas for a "rehearsal party" in which Clint's pal T. B. Cochran played the part of the duke.[2]

Finally, on February 4, the duke and duchess, having traveled three days from Palm Springs to Tampico in one of Bob Young's private trains, arrived in the town of Gonzales, thirty miles from the ranch. Murchison staff members were there to greet them as the Windsors, exhausted, stepped onto the dusty platform. With them was Bob Young and his wife, along with Charles Cushing, of the Boston Cushings, and the New York socialite Edith Baker, along with 150 pieces of luggage, 104 of which belonged to the duchess, including twenty enormous steamer trunks. The bags were thrown in a cattle truck. The guests crammed into station wagons for the four-hour drive up the rutted roads to Acuna.

The ranch remained so primitive that Murchison had neither telephone nor radio, and thus no way of knowing his guests had arrived. Instead, he had one of his pilots fly ahead from Gonzales. As he flew low over the ranch, a single roll of toilet paper dropped from the plane, unfurling into a long ribbon that wafted to the ground. A servant retrieved the cardboard tube. Inside was a message: "They've arrived in Gonzales." Several hours later, Ginny and a girlfriend, their hair still in curlers and wearing no makeup, were working in the kitchen, preparing hors d'ouevres. Someone walked in and began to help. Not till the butler, Jose, shouted "El Rey!" did Ginny realize the Duke of Windsor had arrived and wandered into the kitchen unannounced.

An avid sportsman, the duke appeared to thoroughly enjoy his week at Murchison's side. The men spent the days hunting, firing at mountain lions, turkeys, pheasants, pigeons, and quails. The duke bagged two turkeys and stood for multiple snapshots with his prey. For doves, the group boarded one of Murchison's planes to another ranch, where Murchison used his trusty shotgun to bag scores of white birds.

The women's days were more challenging, thanks to the duchess. She was a tad needy, always sending her maid to pester Ginny for bobby pins or cosmetics. At one point she demanded to have her hair done, an impossibility, but when she insisted Ginny had a plane ferry her to Tampico, where a Mexican hairdresser did the best she could. The duchess, however, managed to redeem herself, at least in Ginny's eyes, the night the furnace went out, reaching deep beneath the boiler herself to reignite the pilot light. All in all, things went swimmingly, at least until the party reconvened to Murchison's island ranch of El Toro for a few final days of fishing. The duke and Bob Young, accompanied by Edith Baker, were out on a boat whose engine suddenly gave out. The captain was unable to fix it; there was no radio aboard to call for help. Night fell. Back at the ranch house, Murchison, fearing the worst, dispatched all his remaining boats in search of the duke's. Meanwhile, the duke, who tended to imbibe more freely when away from the duchess's unyielding eye, passed out in an inopportune spot.

"What do you say to the man who was once the king of England when you have to use the restroom and he's lying in front of the door passed out on the floor?" Edith Baker later asked Ginny. "I finally just stepped over him into the bathroom." After a long, anxious night Murchison spent pacing the beach, one of the search boats finally found the stricken vessel, returning the duke to land just before dawn.[3]

"I really rather enjoyed myself," the duke told Murchison afterward. "Very relaxing, a dead motor."

As much as he enjoyed the duke's visit to Acuna, Murchison harbored a special love for his other retreats, especially an exclusive fishing and hunting club outside his hometown of Athens called, of all things, the Koon Kreek Club. Koon Kreek represented the antithesis of material wealth, a chance for wealthy country boys to rediscover their childhoods. The club itself was nothing much, a collection of five man-made lakes arrayed around a beat-up wooden clubhouse where the members, almost all Dallas oilmen, drank and played gin and served themselves heaping dinners of fried chicken, ham hocks, and turnip greens. There were no butlers or valets or neckties at Koon Kreek, just fishing guides, one for every boat. This was a man's man's club, the members in straw hats, dungarees, and hip waders, spittoons on the card tables, up every morning at dawn to fish for bream. In duck-hunting season the men retreated into the club's eleven blinds. Murchison never tired of extolling Koon Kreek's virtues. "Brings a man back to his roots," he said. He kept another ranch nearby, Gladoaks.

Like Murchison, Sid Richardson never tired of hunting, and both men loved to emphasize their "Texan-ness" by taking eastern visitors along. When John McCloy, the chairman of Chase Manhattan Bank, came to St. Joe's to shoot quail, Richardson informed him they would be hunting from the comfort of a Range Rover. McCloy, wielding an antique British shotgun, insisted on walking. It took only a few minutes before McCloy let out a yelp upon spying a rattlesnake. He blasted the snake, then another and another, before finally joining Richardson in the car.[4]

Richardson had interests in five or six ranches in addition to St. Joe's, three in partnership with a brother-in-law. He had thousands of cattle, but the ones he was most proud of weren't his. Their story involves Richardson's friendship with the author Frank Dobie, a longtime University of Texas professor and folklorist. During the 1920s Dobie became alarmed that the number of Texas longhorn cattle—the unofficial symbol of the state—was dwindling. Ranchers had discovered longhorns produced better beef when impregnated by British bulls, leading to a shrinking of the longhorn population; by the late 1920s many of the remaining longhorns were being taken to slaughterhouses to eliminate an epidemic of tick fever. Without help, the national symbol of Texas faced extinction.

Neither Dobie nor Richardson ever spoke of how they teamed to save the longhorn; it's not even clear how they met. But by 1939 the two were close enough that Richardson allowed Dobie to closet himself at St. Joe's to finish one of his best books, *The Longhorns.* By then, Dobie was talking with a onetime Texas Ranger named Graves Peeler about the prospect of locating rare, pure-bred longhorns in Mexico and bringing them to Texas. For years this effort was cloaked in mystery, much of it the product of Peeler's failing memory during interviews in the 1970s. In fact, Dobie's papers indicate the idea to launch a concerted effort to save the longhorn came from Richardson in about 1939. In the next three years, Richardson paid for Peeler to travel into Mexico to buy the cattle, which were herded into state parks near Corpus Christi and Brownwood. These animals formed the nucleus of the so-called Texas Herd, the state's semiofficial group of cattle. Saved from extinction, longhorns eventually become a sought-after animal, their lean beef ideal for low-calorie diets.

Of the Big Four, only Hunt eschewed the ranching life; in matters of style, he was always more a southerner than a Texan. Not that he didn't have the land: over the years Hunt amassed a million acres of undeveloped acreage in at least ten states, from Florida to Montana. In the late 1950s he began an effort to harvest it, using everything from East Texas pecans to Florida oranges to launch a new food business he called HLH Products. HLH bought up a dozen or more processing plants and began selling canned vegetables, meats, and peanut butter in grocery stores around the country. Though it would take up increasing amounts of Hunt's focus in coming years, HLH Products was a consistent money-loser, and in time would cause serious problems for the Hunt family.

IV.

To reach his ranch, every oilman worth his salt had a plane, still a novel idea
to most Americans in the 1950s. Murchison, among the first to go airborne
in the 1930s, led the way after the war, purchasing an army-surplus DC-3 he
converted into a luxury plane and rechristened *The Flying Ginny*. The plane
originally sat thirty; once Murchison had a wet bar and swivel chairs put in,
it held sixteen. On its inaugural flight in 1946, Murchison took his son John
and two other couples and flew to Alaska, where they chartered a yacht to
cruise the inward passage, a trip considered so unusual at the time that it
received extensive coverage in Texas newspapers. *The Flying Ginny* almost
single-handedly kept Murchison's Mexican ranches supplied with ice, whiskey, and other essentials; the family thought nothing of using the plane to
pop down to Juarez for lunch. By the mid-1950s Murchison had augmented it
with another half-dozen planes he kept at Dallas's Love Field.

Sid Richardson followed Murchison's lead, acquiring matching DC-3s.
The Bass family joked that he bought the second one just so he wouldn't
be forced to fly with his sister Annie, whom he found needy; Annie's plane
did little more than ferry her to Colorado Springs every summer. In time
so many oilmen sought their own aircraft that Neiman Marcus offered his-
and-her private planes in its 1960 Christmas catalog. The largest single fleet
probably belonged to Houston's Big John Mecom, who owned a $1.3 million
Lockheed JetStar and nine other planes. The most storied gathering of Big
Rich aviation took place at a ranch near Austin in 1959, when an oilman
named Pat Rutherford threw a party for Lyndon Johnson; the eight hundred
guests arrived in fifty-three separate private planes.

By and large the Big Rich didn't go in for yachts, preferring to charter
when the need arose. One of the few exceptions was Mecom, the Houston
wildcatter whose fortune by the mid-1950s began to rival those of the Big
Four. Son of a Spindletop-area roughneck, Mecom had begun drilling in
Southeast Texas in the mid-1930s, hit it big near Galveston during the war,
then struck the massive Lake Washington field south of New Orleans. The
first of the Big Rich to buy a serious yacht, Mecom purchased the same 315-

foot ship in which Franklin Roosevelt had fished alongside Sid Richardson and Clint Murchison in 1937, renaming it the *Nourmahal*. He had owned it barely three months when it caught fire at its berth in Texas City and sank. "The whole thing rolled over and lay down like a big cow," the Texas City fire chief observed.

The more refined wildcatters—there were a few started art collections. Clint Murchison's son John began acquiring modern art in the late 1950s and in time amassed what was probably the state's largest collection. Sid Richardson began buying art during the war. Goaded by his friend Amon Carter, Richardson hired Newhouse Galleries of New York to help him quietly assemble what became one of the greatest American collections of Western art, much of it works by Frederic Remington and Charles Russell.* He eventually acquired more than fifty Remingtons and Russells, many of them displayed at St. Joe's.

For a time, Texas Oil's greatest art collection belonged to Alger Meadows of Dallas, a onetime wildcatter who began his career in the East Texas field and by the 1950s was chairman of General American Oil. Meadows began buying paintings during vacations in Europe, found it addictive, and began to entertain dreams of greatness. "I kept thinking, what if I could have, in Dallas, Texas, a collection of art that might be considered a tiny Prado?" he recalled. "I might be the only person in the country who could do this." After donating scores of paintings to Southern Methodist University, Meadows laid out a half million dollars for forty-one Spanish sculptures that formed the basis for the university's sculpture garden. Then, in 1964, he went on a tear, snapping up fifteen Dufys, seven Modiglianis, three Matisses, two Bonnards, a Chagall, eight Derains, a Gaugin, and a Picasso. Finally he had the makings of a Texas Prado—that is, until an assessor judged them all forgeries. When the story leaked, *Life* magazine named Meadows "the man who owns what may be the largest private collection of fake paintings in the world." Undaunted, Meadows spent years replacing every forgery with the genuine article, eventually amassing a sterling collection of Cezanne, Renoir, Goya, and Pollack. "Oh, we've got our hands on some of the prettiest things you ever saw," he beamed.

---

*Richardson's collection remains on display today at Fort Worth's Sid Richardson Museum.

V.

Just as dukes and duchesses and bank chairmen came to Texas to see what all the fuss was about, the Big Rich began venturing out into the world for much the same reason. The world's tourism industry welcomed the new millionaires with open arms. In 1954 the elegant French Line began offering "Texas cruises" out of Houston aboard the SS *Antilles*. Wealthy Texans in cowboy boots and ten-gallon hats tromped on board but, as the *New York Times* reported, were disappointed by the selections of wines, preferring bottles of bourbon brought to the dinner table, and especially the food, which they complained was altogether lacking in barbecue, greens, and black-eyed peas. In July 1952 the government of France held "Texas Week on the Riviera"; one casino adorned its menu with Brownsville Delight—some sort of melon—plus Abilene mutton, Greenville beans, and Jacksonville raspberries. A passel of oilmen flew in to throw parties, prompting a Communist newspaper to gripe that they had only come to hunt French peasants. One memorable shindig that week, thrown by oilman Jimmy Radford of Abilene, featured champagne spewing out of miniature oil derricks. For party favors, Radford handed out Texas horned frogs.

The women of the Big Rich tended to travel in packs. Mrs. James Abercrombie of Houston and four of her friends called themselves "The Flying Five" and periodically took a family plane for jaunts in the Caribbean or Europe. Mrs. Ralph Fair of San Antonio preferred taking girlfriends, along with the obligatory hairdresser and masseuse, to the Fair Ranch in Montana. For many of the Texas nouveau riche, these kinds of trips, still unusual in America, were somewhat jarring. "There I was working in the oil fields," Mrs. Bruno Graf of Dallas told *The New Yorker*'s John Bainbridge, "and the next thing I knew I was going all through Europe with all my diamonds and my personal maid."

For the Big Rich's most avid hunters, African safaris became a rite of passage. D. H. Byrd, who hunted alongside his close friend, the military aviator James Doolittle, added an entire wing to his Dallas mansion just to house his animal heads, including springbok, gazelles, ostriches, wildebeests, gemsbok, hartebeest, lechwe, and cape buffalo, plus a Sumatran tiger and the heads of two polar bears. Closer to home, Alaska and northern Canada became favorite locales for weekend hunting expeditions during the '50s; before his collapse, Glenn McCarthy maintained a massive hunting lease in North Dakota. Texans were just as serious about their fish. In 1951 the Houston oilman Alfred C. Glassell Jr. opened the Caso Blanco Fishing Club on the Pacific coast of Peru, adjacent to a game-fishing area so rich it was known as "Marlin Boulevard"; membership fees started at ten thousand dollars.

Two years later, in August 1953, Glassell reeled in a fifteen-hundred-pound black marlin, at fourteen feet, seven inches long still the largest fish ever caught on a rod and reel. *Sports Illustrated* put Glassell on its cover, while film of his titanic struggle with the fish—it leaped from the ocean forty-nine times during a two-hour struggle—was spliced into *The Old Man and the Sea*, a movie based on the Ernest Hemingway novel; Glassell later presented his trophy to the Smithsonian.[5] News of his catch transformed Caso Blanco for a time into the sport-fishing capital of the world, drawing such celebrity anglers as John Wayne, Jimmy Stewart, baseball great Ted Williams, and Hemingway himself.*

When it came to oilmen's appetite for big game, the more exotic the catch, the better. But for sheer mystery and adventure, no Texan could top the expeditions mounted by the San Antonio oil heir Thomas Slick Jr. Graduating Phi Beta Kappa from Yale in 1938, Slick spent the next twenty years in a whirlwind of oil prospecting, art collecting—he owned Picassos and O'Keeffes—and amateur science, founding two separate research institutes in San Antonio. He was drawn to many of the world's most isolated lands and stranger mysteries. During college he took a group of Yale pals to Scotland to investigate the Loch Ness Monster. In 1956, while on a diamond-hunting trip in British Guiana, his plane crashed, forcing him to live with the remote Waiwai tribe.

Slick became best known for the expeditions he launched in search of the world's most elusive primate, the Abominable Snowman. One of the first serious efforts to investigate reports of a strange creature high in the Himalayas,

*Glassell's club was forced to close after a Peruvian military coup in 1968. Its clubhouse still stands, abandoned and decaying, the oilman's first thousand-pound marlin still hanging over his crumbling fireplace.

Slick's two expeditions between 1956 and 1958 took dozens of porters, guides, and scientists on treks deep into Nepal to collect Yeti footprints, artifacts, and sightings. It was Slick's team that discovered what some believed to be the first physical evidence of a yeti, the so-called Pangboche Hand, a mummified hand discovered at a Nepalese monastery. The find, however, was never authenticated. Afterward Slick sponsored two of the first expeditions that sought to verify a rash of new sightings of a yeti-like creature in the American Northwest, what became known as Big Foot. His investigations came to an abrupt end when he died in a plane crash in Montana in 1962.

VI.

All these tales, from the original *Life* and *Fortune* articles in 1948 to chronicles of the Big Rich that populated periodicals as diverse as *Sports Illustrated* and *The New Yorker*, painted Texas as a new promised land, a place where a young man with energy and guts and a little luck could make himself a millionaire. Northerners had begun streaming into the state during World War II, but their numbers mushroomed after the media boomlet of 1948–49, rich and poor alike coursing into Dallas, Houston, and, increasingly, the new boomtowns out in West Texas, Midland, and Odessa. A new oil formation, the Spraberry Trend, had been discovered there in 1948, and within three years had become the single biggest oil play in America. In 1952 one of every fourteen rigs in the country was drilling around Midland and Odessa. By 1950 more than 215 different oil companies had opened offices. In the decade after the war Midland alone, a town of barely nine thousand people in 1940, more than sextupled in size, its population reaching sixty-two thousand.

Many of the new arrivals were the sons of eastern wealth, second- and third-generation heirs looking to make names of their own. Young Gettys, Rockefellers, and Mellons all came to Midland, which despite no university of its own and an atmosphere thick with the rotten-egg smell of natural gas, soon sported a Harvard Club, a Yale Club, and a Princeton Club. Among these ambitious young Ivy Leaguers was one young man who would come to symbolize a new breed of Texas oilman, and in time a new Texas:

a Connecticut senator's son named George Herbert Walker Bush. Bush was intoxicated by the *Life* and *Fortune* stories, and by tales of easy money Texas friends told him in the navy. With his wife, Barbara, he had driven to Midland after his graduation from Yale in 1948, rejecting a Wall Street career in favor of a job offered by a family friend, the chairman of the oil field–services company Dresser Industries. For most of the next three years, Bush hustled the sand-blown backroads of West Texas, hawking Dresser goods and services from Winkler County all the way to Muleshoe.

In time Bush tired of working for others and, after raising a half million dollars from an uncle, Herbie Walker, and family friends—one was the owner of the *Washington Post*—he and a partner opened an office in the Midland Petroleum Building to trade oil leases. A neighbor, Hugh Liedtke, also eastern-educated, persuaded him the real money was in finding the oil itself, so in 1953, with still more money from Uncle Herbie and friends in the Astor and Rockefeller families, Bush merged his company into Liedtke's and renamed it Zapata Petroleum, after a Marlon Brando film, *Viva Zapata*. On its very first well, in nearby Coke County, Zapata struck oil. Five more strikes followed, then sixty-five more in the next eighteen months. By 1955 Zapata was producing 1,250 barrels a day, making Bush and Liedtke, on paper at least, minor millionaires.

While Liedtke and his brother Bill searched for oil on land, Bush started a new subsidiary, Zapata Offshore, and with bonds sold on Wall Street by his Uncle Herbie, bought two experimental drilling rigs whose designers promised they could find oil in a new frontier: underwater. Bush's offshore rigs, leased to large oil companies, were among the first to begin drilling the shallow floor of the Gulf of Mexico. Over the next several years Zapata acquired more rigs and sent them farther and farther afield, off the shore of Cuba, the Persian Gulf, even the coast of Borneo. Bush's success, however, led to an amicable parting with Hugh Liedtke, who wanted to focus on land drilling. Liedkte took Zapata Petroleum, Bush Zapata Offshore. It made sense, friends said. Liedtke was all about money. Bush saw other mountains to climb.

In the summer of 1959 Bush, now CEO of his own company, moved Barbara and their sons to Houston, where George joined the Houston Country Club and the exclusive Bayou Club. There he began amassing a clutch of friends who would remain at his side for decades. He met Jim Baker, a

Princeton man whose grandfather founded Houston's giant Baker & Botts law firm, on the tennis court at the country club. Bob Mosbacher was a young wildcatter who had struck oil in South Texas. In time Bush befriended two prominent attorneys, Leon Jaworski and Bob Strauss. He had always known he would try politics someday, and with the help of friends he began testing the waters in Houston.

Bush did so at a turning point in Texas political history. After almost twenty years of ultraconservative rule, the last of the true hard-liners had been swept from state office just two years earlier, in 1957, the same year Lyndon Johnson finally secured control of the state Democratic Party, the same year Texans actually voted into office a liberal senator, Ralph Yarborough. The era of Pappy O'Daniel and Roy Cullen and Joe McCarthy was ending, people said, swept away by new ideas, new people, new times. These were the years of the Organization Man, and George Bush was an Organization Man for Texas, an eastern-bred country-club Republican who didn't want to change the status quo so much as run it. Not for Bush hillbilly bands and bags of cash; he looked silly in a ten-gallon hat and knew it. The cities of Texas were filling with gray-suited executives who thought the way Bush did, and they would boost him to his first political post, chairman of the Harris County Republican Party, in 1964. The rest was history.

If George Bush represented the future of Texas politics, Hugh Liedtke was the future of Texas Oil. Liedtke became an oilman in the Murchison model, a Wall Street–savvy CEO who over the next twenty years used Zapata Petroleum as a battering ram to engineer a series of mergers and hostile acquisitions that produced Pennzoil, one of the largest independents in Texas, housed in a glittering glass tower in downtown Houston. Liedtke was the consummate example of what *The New Yorker*'s John Bainbridge in 1959 called the Texas "wheeler-dealer," a man who found oil on the telephone and the stock market. (The tag eventually gained wide usage, inspiring a 1964 movie of the same name, a lark starring James Garner as a high-rolling Texas oilman.) Like Bush, Liedtke could adopt the good ol' boy style when needed. But unlike the Roy Cullens and Sid Richardsons who made their fortunes tromping through marshes and deserts looking for oil under the ground, Liedtke was simply a businessmen—whose business happened to be oil.

VII.

And then, suddenly, it was over.

It's difficult to select a date or any one event when it happened—the selling of the Shamrock in 1954, the uproar over the 1956 natural gas bill, the first decline in Texas oil production in 1957—but by the late 1950s the golden age of Texas Oil had come to an end. Stylistically, the backlash against the excesses and political extremism of the Big Four had taken its toll. In 1957 the *Houston Post* published an entire series of front-page articles exploring why so many Americans hated Texas. In 1958 Big John Mecom and a group of oilmen took out a full-page advertisement in the *Post* defending Texas Oil and declaring that attacks on it were un-American—an early sign of an industry-wide persecution complex that would endure for decades.

It wasn't just the rest of America that was tiring of the Big Rich's gaucheries. The George Bushes and Hugh Liedtkes of the new Texas were frankly embarrassed by the Silver Dollar Jim Wests and Glenn McCarthys and D. D. Feldmans. By the late 1950s, while ranches and airplanes and exotic wildlife were still prized, the days when they were brandished with unabashed pride were fast falling away. "Evidence of individual Texas wealth was being secluded by the middle 1950s, and there are fewer anecdotes to feed the 'Big Rich' legend," George Fuermann wrote in 1957. "The legend has become obnoxious; being from Texas has become a distinction for some to worry about."

But for Texas Oil, the real problem wasn't image. It was economics. The days of Texas fortunes "hatching like mayflies," as *Time* memorably observed in 1950, were quickly fading. Never again would one man working alone find a major oil field in Texas; they had all been found. By 1960 costs to drill in the continental United States had risen so high, a solo operator could barely afford to sink a wildcat well. A number of deep-pocketed wildcatters had begun looking overseas, John Mecom in Honduras, Colombia, Jordan, and Yemen, H. L. Hunt's son Bunker in Pakistan. Worse, competition from Middle Eastern oil was slowly strangling the industry. The majors saw little reason to buy more Texas crude when they could buy Saudi oil for half the price. European demand, long a foundation of Texas sales, peaked during the Suez Crisis of 1956 and fell sharply thereafter. In a bid to buoy prices, the Railroad Commission ordained that no Texas field could pump oil more than twenty-one days a month. It reduced this allowance again and again, until in 1962, Texas wells could operate only seven days a month. Needless to say, no one was looking too hard to find new oil in Texas.

The titans of the golden age, meanwhile, had begun passing from the stage. Sid Richardson's close friend, Amon Carter, the Fort Worth newspaper publisher who struck the Ellenburger Lime, died in 1955; some said Richardson was never the same after that. Patillo Higgins, the man who started it all, died in 1956. Silver Dollar Jim West died of diabetes in 1957. They found $290,000 in a vault beneath his River Oaks home, much of it in silver dollars; it took seven armored cars to haul it all out. The old West family ranch, sold to Humble years earlier, was deeded to the federal government as the new headquarters of NASA's manned space program. Big Jim West's Italianate mansion on Galveston Bay became the new Lunar Science Institute. Outer space, not oil, appeared to represent the future of Texas.

VIII.

Nothing marked the end of the golden age so much as the dimming of the men who had created it, the original Big Four oilmen. Clint Murchison was the first to retreat. In February 1956, while visiting Bob Young's Palm Beach mansion and spending his days with Ginny at Hialeah, Murchison began complaining he didn't feel well. He had trouble maintaining his balance, and his speech was slurred. He and Ginny flew to New Orleans, where he checked into the Ochsner Clinic. Doctors there deduced he had suffered a slight stroke, the result of a partial blockage of an artery in his heart. Afterward Murchison went to Baylor University Hospital in Dallas to have the artery cleared.

The stroke signaled the end of Murchison's active business career. In the months to come, while he remained engaged in the day-to-day supervision of his investments, his memory began to fail. He asked his secretary to listen in on his phone calls and take notes, in case he missed something. It was embarrassing, and by 1957, when he turned sixty-two, Murchison was spending more and more time away from the office, all but moving to his East Texas ranch, Gladoaks. There, splayed before a card table in his pajama bottoms, a cigarette in one hand and a cup of coffee in the other, he began each morning around five with a call to Sid Richardson. "What's the dope?" were usually Murchison's first words, and the two would spend an hour discussing everything from investments to the Gladoaks peach crop.

There was a second stroke in 1960. At that point, Murchison passed his last management duties to his sons, John Dabney and Clint Jr., and moved into Gladoaks full-time, where he busied himself buying cattle, a dozen more ranches nearby, and the odd company, a chain of supermarkets one year, a group of feedlots the next. He and Ginny still traveled, spending the summers in La Jolla and betting the horses at Del Mar. But after another series of strokes in 1965 Murchison was relegated to a wheelchair. He stopped flying down to Acuna then; it was too cold. Instead they bought a cliffside mansion in Acapulco, but in time even that became a burden. Murchison spent his last years mired at Gladoaks, many days sitting in a battered old station wagon as a chauffeur drove him through the fields and orchards of his many ranches. He finally died, of pneumonia, in June 1969.

They buried him in Athens, next to his parents. There was a massive funeral at the Methodist church, a thousand mourners or more. Lyndon Johnson and Richard Nixon called with condolescences. The New York Times ran his obituary on page 1, terming Murchison a one-man conglomerate. "His entire life was devoted to making money," the Times wrote, but it wasn't really true. Clint Murchison had torn through life with gusto, and those he left behind were uniformly thankful to have known him. At Acuna the peasants built a fifteen-foot cross of solid ebony and hauled it to the ranch's tallest peak, where they anchored it to the rock and bowed their heads. Then they buried a pair of Big Clint's mangy old boots, and returned down the mountain.

IX.

Sid Richardson spent his twilight years alone, shuttling between a new bungalow at the Thunderbird Club in Palm Springs, his retreat at St. Joe's, and his rooms at the Fort Worth Club. Wherever he was, his routine rarely varied. He arose around five and spoke to Murchison, a talk that inevitably began with one saying he had been awake for hours waiting for the other to get up. Mornings he lolled around, typically attired in boxer shorts, pajama tops, a ratty gray sweater, and a bathrobe, sipping coffee between calls. In Palm Springs he would retire after lunch to a lounge dubbed the Snake Pit, where he spent the afternoon playing poker with a group whose members included the actor Phil Harris and Ray Ryan, the gambler who had played with H. L. Hunt.

There was scant new business to be done. The glut of Middle Eastern oil meant there was little need to drill, but he did some anyway, at one point partnering with John Mecom on what was then the deepest well ever drilled, a dry hole bored nineteen thousand feet beneath a Louisiana swamp. The growing aversion to American oil, in fact, almost cost Richardson his Louisiana fields. He couldn't sell all the oil, and his leases, even in the giant Cox Bay and Port-la-Hache fields below New Orleans, remained alive only so long as he pumped: if the fields stopped producing, the leases could be voided. In the late 1950s, his Old Friend says, Richardson avoided disaster only by cutting a deal to sell all his Louisiana production to Ashland Oil at cost. In 1957 *Time* magazine reported speculation that he was poised to cash out, selling his empire to one of the majors. "A damn lie," Richardson snapped.

His health was failing. He had high blood pressure. The doctors at the Ochsner Clinic told him to quit drinking and cut the salt from his diet. They made him stop smoking. President Eisenhower sent a condolence letter or two. Richardson seemed happiest when he could fly back to Fort Worth, where his first call would inevitably go to Perry Bass's wife. "What's for dinner?" he would ask. She always fixed him his favorite East Texas meal, sliced tomatoes, black-eyed peas, collared greens, and a single ham hock.

Still, he was lonely. A longtime Fort Worth newspaper reporter, Carl Freund, remembers seeing Richardson many days on the sidewalk outside the Texas Hotel, chatting with bookies about a horse race. On Saturdays he would summon Bass and his teenaged grandson Sid down to his office just to shoot the breeze; later, on the way home, Bass would tell Sid the old man just didn't want to be alone. When the Basses were busy, Richardson had trouble finding people to join him for dinner. "He'd get lonesome, and he'd call me up; he'd want me to come get him," remembers the Texas Hotel's manager at the time, Andy Anderson. "He loved to go to a Mexican café or some . . . rib joint. And he never would have any money in his pocket. He always expected me to pay for it. He knew, of course, that I would charge it to the company."[6]

A few days before Christmas 1958, Richardson sent his DC-3 to Washington to ferry his lawyer, John Connally, to Palm Springs. Connally sensed trouble. He had taken a call from Perry Bass, who had seen Richardson's will. To his dismay, it left almost nothing to the Bass family. Richardson's share of their oil fields was to go to a new Sid Richardson Foundation. In Palm Springs, Connally fell into Richardson's easy rhythms of coffee and cards, and after a few days brought up the matter. "Mr. Richardson," he said one morning, "you need to leave some substantial money to your family and Perry's children."

Richardson appeared startled. "Why should I?" he snapped. "Bass is rich. He could leave them plenty." But Connally pressed, and when Richardson returned to Fort Worth he agreed to change his will. In the new will he left two million dollars to Perry Bass and each of Perry's four sons, along with St. Joe's Island and stock in several corporations. His share of the oil fields, however, still went to the foundation.

On September 30, 1959, Richardson flew down to St. Joe's. He planned to fly on to tour a pair of ranches he and Connally had purchased near San Antonio. The next morning a servant found him dead in his upstairs bed beneath one of his beloved Remingtons. He had suffered a massive heart attack and died in his sleep. He was sixty-eight.

Sid Richardson received far more headlines in death than he ever had in life. His body was flown to Athens for burial. President Eisenhower sent a cross adorned with white carnations, along with regrets that he was unable to attend. His sister Annie was there. She cried. His chauffeur was there, and the cattle wrangler from St. Joe's, and the houseman and his family, and his pilots. Perry Bass stood by with his four boys as Sam Rayburn, Lyndon Johnson, and John Connally watched Billy Graham deliver the eulogy. "He was a loyal American and passionately loved his country and maintaining the American way of life," Graham said, because he couldn't say much about oil or money or cattle or playing cards, the things Sid Richardson loved most.

X.

Roy Cullen spent his last years paying more attention to his growing crop of grandchildren than politics or oil. His time had passed. He knew it. In their seventies now, he and Lillie spent more time at their ranch north of Houston, playing cards and dominoes on the back porch and driving out at dusk to watch the deer. They built homes there for their daughters and their husbands, and in time even acceded to their son-in-law Corbin Robertson's desire to purchase a family plane, an old converted DC-3 the grandchildren nicknamed "Big Red."

Most of their brood had taken the plane to a Gulf Coast beach vacation in February 1957 when they got the call. "Gampa" had suffered a stroke in his sleep at his River Oaks mansion. Rushed to Hermann Hospital, Cullen lingered for four months but never regained consciousness. He died in June 1957, Lillie at his side. She died two years later, never adapting to her beloved husband's absence.

In Houston the coverage of Cullen's death dwarfed that of any Texas oil-man before or since; it was as if a president had died. The stories took up every inch of every front page, and entire sections inside. Much of it dwelled on his philanthropy; no one wanted to talk too much about his political views. Indeed, for all the sorrow and warm eulogies, there was a sense, however slight, that Houston was in some way relieved the days of Hugh Roy Cullen were over. A new Houston was blooming, and a new Texas, and crusty old oilmen who griped about liberals and New York Jews were irksome anachronisms, reminders of a time men like George Bush and Hugh Liedtke would just as soon forget.

THIRTEEN

Rising Sons

I.

By the autumn of 1959 H. L. Hunt, who turned seventy that year, was the only one of the Big Four still active, and even he was passing off his corporate responsibilities to aides and his sons Bunker and Herbert, who by then had reached their thirties. It had been a long, tumultuous decade for Hunt, lived now in the spotlight; whether it was accurate or not, he was widely regarded as the world's richest man. But he was an oilman now only on paper. His men sensed he had lost all interest in oil. "He enjoyed looking at the old wells and remembering the primitive methods employed to bring them in," his nephew Tom Hunt recalled. "But he never cared to look at anything that Hunt Oil had done without him. . . . We had new wells that were producing ten or twenty thousand dollars a month but to him they weren't his."[1]

The year 1955 was a turning point, when Hunt not only closed his beloved Facts Forum but suffered the wrenching loss of his wife, Lyda. She had suffered a stroke. Not trusting Texas hospitals, he had her flown to the Mayo Clinic, but she died within days. Despite the other women in his life, Lyda had been Hunt's rock. He wept on the plane back to Dallas. "I don't know how I'll ever get by without her," he said. Afterward Hunt disappeared on a six-month tour of Latin America, where he later said he had studied local governments. In all likelihood he simply needed time to deal with his grief. "Daddy went downhill from the day Mother died," his daughter Margaret remembered.[2] Bunker tried to get him to play cards, but he wouldn't. Nothing interested him.

To the dismay of his first family, what saved Hunt was the love of his mistress, Ruth Ray Wright, and his secret "third" family.* He and Ruth had been together almost fifteen years by then. A simple, sweet, religious woman, Ruth still lived in her house on Meadow Lake Avenue, as did the couple's four children. Their son Ray, who bore a striking resemblance to Hunt, turned thirteen in 1956; they also had three girls, aged twelve to seven. With Lyda gone, Ruth began demanding to be recognized as Hunt's wife; Hunt procrastinated, knowing it would mean rumors about their secret life. By the mid-1950s a handful of Hunt's sons, including Bunker and Herbert, appeared to have learned of Ruth; others, notably the oldest sibling, Margaret, did not. When she first encountered Ruth at Mount Vernon after Lyda's death, Margaret thought she was a secretary. Margaret's husband, Al Hill, who was close to Hunt, wasted no time pulling her out of the house into their car.

As Margaret recalled the moment years later, she demanded to know why they were leaving. "Sweetheart," Hill began, "now, I know this is going to make you unhappy—"

"Al," Margaret demanded, "what is going on?"

"Ruth Ray," he said, turning toward her, "is your father's, well, she's his wife without being married to him. They have four children together." As Margaret recalled her thoughts: "I was astounded, disbelieving, horrified, brokenhearted. As I was about to demand, 'Al, why didn't you tell me?,' I heard Mother asking me the same thing (twenty years before) and my reply: *Why would I?*"

After Lyda's death, Ruth and her children all but moved in to Mount Vernon, fixing Hunt's meals, packing his brown-bag lunches, and singing to him when he became depressed. Neither branch of the family was happy with the new arrangement; Margaret, for one, stopped speaking to her father. The first family, led by Margaret and Bunker, considered Ruth's family interlopers and wanted nothing to do with them. Ruth's family, especially

---

*Hunt maintained minimal contact with his "second family" by Frania Tye, who remained in Atlanta.

teenaged Ray, thought it high time their parents were finally married. As Hunt later told the story, Ray came to his office one day to lay down the law. "You *will* marry my mother," he told Hunt. "She is a good, religious person, and you *will* marry her."

In time he did, sneaking off one Sunday afternoon in November 1957 to the home of Ruth's minister, who wed them. The six children of Hunt's first family only learned of their father's marriage when a short item appeared in the *Dallas Times Herald*. All across Dallas, tongues clucked. Nobody knew of Hunt's bigamist past, but reading between the lines, it was clear the world's richest man had been up to *something*. When Hunt adopted Ruth's four children, changing their last names to Hunt, everyone in Dallas knew the truth. Hunt didn't care. He was happy with Ruth, who now joined him full-time at Mount Vernon.

His new life with Ruth changed Hunt. She considered gambling a sin, and tended to cry when Hunt spoke of his wagers. Soon after their marriage he gave up gambling altogether—to placate her, Hunt said. Another motivation, however, might have been a federal investigation linked to the gambler Ray Ryan. Hunt was among dozens of gamblers subpoenaed before an Indiana grand jury, though he avoided testimony when his doctor claimed he had a throat ailment. A rumored probe into whether Hunt had paid taxes on his winnings never reached a court.

The biggest change, though, was Hunt's embrace of religion. Ruth had joined the Dallas Baptist Church, the South's largest Baptist congregation, and had taken to holding prayer meetings at Mount Vernon. In time Hunt began joining her at services, though it was a right-wing minister named Wayne Poucher who later took credit for Hunt's turn toward Christ. As Poucher told the story, Hunt was dining at the minister's home in suburban Washington one evening in 1959 when the family invited him to join their nightly prayer service. Hunt said he'd rather observe, but as the prayers wore on he slid from his chair and kneeled on the floor beside Poucher. By the time the prayers came to an end, Poucher said, tears were sliding down Hunt's face.

Afterward, "I took him to [his] hotel and for two hours we talked about him and his soul," Poucher recalled. "I finished by telling him that I wanted to take him to the church building and baptize him." Hunt was torn. "Wayne,

I want to," he said, "but I have been an evil person and I don't feel I can ask God to forgive me until I have lived better for a little longer time." Soon after, Hunt was baptized at Dallas Baptist. Finding God, Hunt said later, "was the greatest trade I ever made. I traded the Here for the Hereafter."

Though Facts Forum had been defunct for four years by then, Hunt had never given up the idea that he was destined to educate America about the dangers of communism and liberalism. His religious conversion obliged him to incorporate Christianity into his philosophies, and the more he studied the Bible, the more Hunt felt it held the key to solving America's many ills. And so, in the summer of 1958, Hunt announced he was resurrecting his old Facts Forum apparatus, infusing it with religion and renaming the new organization LIFE LINE. While LIFE LINE's pamphlets and radio broadcasts would essentially cloak Hunt's right-wing propaganda in Christian robes, LIFE LINE was in many ways ahead of its time. Its wedding of fundamentalist Protestantism and right-wing politics, not to mention its crusades against "big government" and Wall Street greed, came twenty years before the Christian Right's emergence in the late 1970s.

LIFE LINE's offices, staffed by two dozen clean-cut young conservatives, each throughly vetted, were tucked away on two stories of a downtown Washington building. Hunt's staff quickly got to work selling right-wing books, pamphlets, and a three-times-weekly newspaper, Life Lines, now augmented with religious writings, but the centerpiece of LIFE LINE's efforts was the fifteen-minute daily commentary it offered radio stations for a minimal fee; the Sunday show was free. Typically, half a LIFE LINE broadcast was composed of traditional hymns and sermons, but the remaining commentary, delivered by Wayne Poucher and other right-wing ministers and onetime FBI agents, consisted of straightforward ultraconservative, John Birch–style rhetoric—visceral attacks on Socialists, Democrats, liberals, the United Nations, Wall Street, and anyone who criticized the oil industry. Hunt, like the Birchers, believed the secret hand of Communist Russia and China was everywhere, in American universities, pulpits, government offices, even hospital wards. In one memorable memo, Hunt told Poucher to use one broadcoast to expose a conspiracy in which wealthy Americans were being subverted by Socialist nurses and mistresses.

Hunt's ideas were classic paranoid right-wing fantasies, but the fact his ideas were stupid didn't mean he was. Hunt kept his name out of all of LIFE LINE's published materials; he realized he had become a lightning rod for criticism. Moreover, by wrapping himself in Christianity, he was able to attract to LIFE LINE's advisory board a number of leading ministers and others who had backed Facts Forum, including John Wayne. Concerned that too many right-wing organizations undercut their credibility on communism with attacks on blacks and Jews, Hunt warned Poucher and his other commentators to avoid criticizing both groups. In at least one case, he told them to go on the air with kind remarks about a well-known Jew, so that "LIFE LINE would be given the credit of extolling and memorializing a Jew."[3]

However silly LIFE LINE's message appears today, it struck a chord in the late 1950s, especially in the rural South, where dozens of small radio stations were happy to accept its cut-rate commentary. *Life Lines* debuted on twenty outlets in 1958 but grew steadily; by the early 1960s its broadcasts could be heard on 354 stations in forty-seven states. Fifty stations ran them twice a day. Hunt was always LIFE LINE's biggest backer, but its tenuous status as a tax-exempt "educational" foundation—the same status Facts Forum had so assiduously defended—secured donations from others as well, chiefly oil companies, wealthy right-wingers, and Hunt's bank, First National of Dallas, all of whom were able to deduct their gifts from their federal income taxes.

Under IRS guidelines, an "educational" foundation was only allowed tax-exempt status so long as it avoided partisan political commentary, a staple of *Life Lines* broadcasts. A 1962 review by the IRS's Baltimore office found LIFE LINE in clear violation, but the case went nowhere. The Federal Communications Commission launched a similar review in 1963, but its case languished as well, frustrating Democratic congressmen bewildered by Hunt's ability to deduct the money he was spending on ultraconservative causes. "There is probably no one," declared Senator Maurine Meuberger of Oregon, "who gets more right-wing propaganda for his tax dollar than Haroldson Lafayette Hunt."

LIFE LINE's rise coincided with Hunt's reemergence as a public figure. He had been firing off letters to newspapers for years, but in the late 1950s his output began to soar. He dictated them to secretaries, sometimes five and six a day, then dispensed them to newspapers from the *Duluth Herald News*

Tribune to the *Alabama Baptist*. He began giving speeches, often to small religious and right-wing groups, and in time developed a certain following among what is today called the religious right. But even his most ardent supporters, one suspects, were left scratching their heads at the project he unveiled in January 1960, a self-published novel.

It was called *Alpaca*, and it told the story of Juan Achala, citizen of a mythical Latin American country called Alpaca. A slim, volume, just 158 pages, printed on cheap paper and cheaper binding, *Alpaca* chronicled Juan Achala's long trip through Europe in search of love and a new constitution. Its centerpiece was Hunt's idea of a utopian constitution, one in which each citizen received a number of votes in line with the amount of taxes he paid. The book got a smattering of reviews, mostly scathing, and though Hunt claimed it had somehow been responsible for democratic reforms in, of all places, Iraq, the general reaction was eye-rolling ridicule. Newspaper readers across the nation chuckled at a book signing he held at Dallas's Cokesbury Book Store. While Hunt looked on, smiling, his youngest daughters by Ruth, eleven-year-old Helen and ten-year-old Swanee, bowed and, as newsmen flashed their cameras, sang a ditty Hunt had written:

> *How much is that book in the window?*
> *The one that says the smart things . . .*
> *How much is that book in the window?*
> *The one that my daddy wrote. . . .*

After *Alpaca*, while Hunt continued his barrage of letters and mailings, it was difficult to find anyone outside the Far Right who took him seriously. Still, he had enough money to cause trouble when he wanted to. One of his most renowned adventures came during the 1960 Democratic National Convention in Los Angeles, when Hunt arrived determined to defeat John F. Kennedy's nomination. Hunt loathed Kennedy, who had vowed to "review" the depletion allowance. Worse, Hunt deeply believed in a sermon delivered by his Dallas minister, Reverend William Criswell, who thundered that the election of a Catholic president would mean "the end of religious freedom in America." Hunt believed a Kennedy White House would be run, in essence, by the pope.

During the convention, Hunt was seen wandering in and out of Lyndon Johnson's and other hospitality suites, a sad, vaguely pathetic figure looking for someone to listen to him. But he had a plan. As the convention climaxed, more than two hundred thousand pamphlets carrying copies of Reverend Criswell's anti-Catholic sermon arrived at newspapers and Protestant churches around the country. Hunt expected it to trigger an anti-Kennedy outcry. The only thing it triggered was a Senate investigation. The pamphlets carried no hint of who had paid for or printed them, an apparent violation of federal campaign laws. When word leaked in Washington that a certain unnamed Dallas oil millionaire was behind the mailing, Hunt and his top security man, a onetime FBI agent named Paul Rothermel, disappeared, actually slipping out of Dallas and shuttling among a series of West Texas hotels for several weeks. When Hunt's name finally surfaced publicly, the ridicule was withering. "Come out, Big Daddy, wherever you are," one Texas editorial writer chided.

In time the whole thing blew over—the pamphlets turned out to be technically legal, after all—but the damage had been done. The stereotype of the dim-witted right-wing Texas oil millionaire lived on, and H. L. Hunt was viewed around the world as its embodiment. He was an embarrassment, to Dallas and to Texas at large. In time even some of the religious commentators at LIFE LINE began to veer away from him. By 1963 Reverend Criswell felt Hunt's far-right-wing views had come to dominate LIFE LINE's offerings at the expense of religion and said so, publicly. The same year, one of LIFE LINE's most loyal commentators, Wayne Poucher, refused Hunt's order that LIFE LINE publicly criticize a piece of oil-industry legislation. Hunt promptly fired him.

"I thought I knew Mr. Hunt, but I didn't," a frustrated Poucher told a reporter afterward. "No one does."

II.

From 1959 on, the most notable exploits of the Big Four families would belong to their second generations, ambitious young men struggling to escape their fathers' shadows. In the short run, only two sets of brothers were to lead public lives. The Cullen daughters, their fortune diminished by their father's philanthropy, withdrew from view, as did Perry Bass, a worn man in his forties by then, left with a fraction of Sid Richardson's riches. For the next decade Bass would settle into a caretaker's life in Fort Worth, raising his boys and dreaming of ways to return to his first love, sailing. He and his wife, Nancy Lee, were close friends with Margaret Hunt and her husband, Al.

For the moment, the mantle of the Big Rich fell to five men, all in or approaching their thirties: Clint Murchison Jr., who turned thirty-six in 1959, his older brother John, thirty-eight, and the three sons of Hunt's first family, Bunker, thirty-three; Herbert, thirty-one; and Lamar, twenty-six. At least initially, it was the Murchisons who attracted the wider notice, especially the gifted Clint Jr., a man who by all rights should have become a titan of commerce but never would. Clint Jr., people said, was almost too much like his father. He inherited Big Clint's gift for numbers, scoring genius level on IQ tests, but also his social awkwardness. Like so many sons of successful families, Clint Jr. spent years searching for something to make his own, but when he found it something went out of him, as if he no longer had to work at life, and his later years would devolve into the most sordid of any of the second-generation Big Rich.

Father and son certainly looked alike. Clint Jr., like Big Clint, grew up to become a squat, roundish man, five feet six, given to flattop haircuts, and the same horn-rimmed glasses his father wore. As a teenager he had been sent to the Lawrenceville School in New Jersey, where he maintained an A-plus average and impressed the faculty; after his freshman year, the headmaster wrote Big Clint that his son might be the brightest in the school's history. Clint Jr. spent much of World War II in a Marine Corps officer-training program at Duke University, graduating from college there, and afterward enrolled in the master's program in mathematics at the Massachusetts Institute of Technology, where he talked of pursuing a career in academics. He was married in 1945, to a vivacious girl named Jane Coleman whose mother grew up in Athens. The two had dated since they were teenagers, and with Clint at MIT they settled outside Boston, where in 1946 Jane gave birth to their first son, Clint III. They later added three more children.

During those postwar years Clint Jr. wrapped himself in a rarified world of scientists and their equations, poring over technical journals the way his father studied stock-market tables. He was a gifted student who remained uncomfortable with strangers. Jane watched her husband many evenings at dinner or a party at a professor's house, talking only of math while other students joked and argued about Truman or Alger Hiss or any headline of the day. Clint Jr.'s insecurity was more than a Middle American swimming in eastern waters. It was, like his father's intellectual insecurities—the big words, the dictionary hunts—something deep within him, always there, always nagging, a sense that no matter what he achieved he might never measure up to expectations.

Clint Jr.'s dreams of mathematics greatness ended in late 1949 when he received a note from his father. "Dear Clint W." it read. "Come on home. Dad." It was time to join the business. A similar message went out to Clint Jr.'s older brother, John Dabney. The two siblings were nothing alike, then or ever. Tall, dark, and tightly controlled, John looked nothing like the other Murchison men. At sixteen he had gone to Hotchkiss and then Yale, leaving school the day after Pearl Harbor to enlist. He became a fighter pilot, flying more than fifty missions over the skies of China, North Africa, and Italy. The war took its toll on John. The stress, and a poor diet, caused him to lose weight, and by late 1945 he was released from service on the edge of nervous collapse. Big Clint met him with a present: his own fighter plane. One weekend John flew it to a party on Matagorda, now owned by the Wynne family, where he was smitten with a pretty Hockaday girl, Louise Gannon, known as Lupe. They were married in 1947 when John returned to Yale. Like all the Murchison wives, Lupe was everything her husband was not, outgoing, outspoken, and bursting with life.

When John graduated the following year, the couple returned to Dallas, but Big Clint thought his eldest still seemed somehow nervous; he had developed asthma as well. Clint had a Mayo Clinic doctor come down to El Toro one weekend to discretely observe John, and the doctor said that while his nerves would heal, the asthma wouldn't. John needed a dry climate. Back in Dallas, Big Clint put John and Lupe on the first flight to Santa Fe, New Mexico, telling them to take it easy for a while. The newlyweds bought a small house outside town and John, as his father had done forty years earlier, took his first real job as a bank teller. While in Santa Fe, Lupe gave birth to the couple's first son, John Dabney Murchison Jr., known as Dabney. Like Jane and Clint Jr., they later added three more children.

When the Murchison brothers returned to Dallas in 1949, they began their apprenticeships alongside their father at 1201 Main. John turned twenty-eight that year, Clint Jr. twenty-six. Big Clint hadn't been around his sons much since they were teenagers. John had grown to be kind, careful, and intellectual, given to studying a business opportunity so long that he sometimes lost it. Clint, happy to be home and freed from the constraints of academia, emerged as just the opposite, freewheeling and cocky, given to jumping into deals he barely understood. "We've got vice and versa," Big Clint took to saying. "One of my boys makes up his mind too fast, the other one won't make it up at all."[4]

Clint Jr.'s impetuous nature, his father saw, masked a complex personality. By his mid-twenties his son had developed two distinct personalities, in fact, one for friends, another for everyone else. He could be warm and witty in the bosom of family, though it would take years to develop the appetite for jokes and pranks he displayed in later life. At work he could be insufferably full of himself; years before he had anything to boast of, Clint Jr., no doubt mindful of his MIT education, seemed convinced he was the smartest man in Dallas. Around strangers he was typically uncommunicative; if he said anything at all, it was often rude. Asked why he hadn't acknowledged an employee he passed in the hallway, he said, "Why should I? I said something to him yesterday." He cared nothing for niceties, especially on the phone, and he had his father's gift for the put-down. To an investor he disliked, Clint Jr. remarked, "You have all the characteristics of a dog but loyalty." His wife, Jane, knew his brusqueness was a mechanism to cope with his terrible shyness, but over the years more than one outsider walked away from a chat with Clint Jr. muttering "asshole."

Big Clint felt a job in sales might draw his youngest out of his shell, so he put him to work selling modest-priced homes in a subdivision one of his subsidiaries was building eight miles north of downtown. The tiny cement-block homes had been an idea Big Clint had during the war; they were intended for returning soldiers and their families. To guide his son, Big Clint had him work alongside a party-hearty character named Robert Thompson, a onetime Washington lobbyist who had known the family since the 1930s. Thompson was the life of every party he entered, the type of man who tap-danced on tabletops.

Under Thompson's tutelage Clint Jr. didn't sell many houses—the exact number appeared to be zero—but he did begin to loosen up. Between the war, an early marriage, and his years buried in books at MIT, the young heir hadn't taken much time to explore life's hedonistic side. Thompson took care of that, introducing him to taverns high and low, from Mexican dives on the east side of Dallas to New York's '21' Club, which in later years became Clint Jr.'s home away from home. Some suspected it was Thompson who introduced Clint Jr. to the secret thrills of extramarital sex, a pastime with which the youngest Murchison would grow increasingly preoccupied as the years passed.

Big Clint intended his sons to use 1201 Main as an incubator to start their own businesses, and in 1952 Clint Jr., at the age of twenty-nine, took the plunge, purchasing the City Construction Company of Dallas for thirty-two thousand dollars and a promissory note. He renamed it the Texas Construction Company, known as Tecon, and brought in Thompson as his partner. Tecon started small, repaving Dallas-area streets, but Clint Jr. had dreams of building it into an international construction conglomerate. He bought a string of competing companies, and in time Tecon grew to ten million dollars in assets. Clint began bidding on projects all around and outside the country, and in early 1954 outbid seven of the world's largest construction companies to win the prize job he needed to advertise Tecon's services: a contract to remove two million tons of dirt and rock from a hillside slowly crumbling into the Panama Canal. It was a risky undertaking. The job required one million tons of dynamite, and any debris that fell into the canal would be removed at the contractor's expense. More than one of Big Clint's men warned him Clint Jr. was risking everything he had. Big Clint stayed out of it. "If he makes a mistake," he said, "he'll learn from it."[5]

Tecon survived the Panama Canal job, and under Clint's leadership busied itself building subdivisions and highways across the country. His brother

John, meanwhile, more comfortable with banking and finance, spent the early 1950s under the guidance of one of Big Clint's men buying insurance companies. To Atlantic Life, a seventy-one-million-dollar insuror Big Clint had bought in 1941, John added Lamar Life of Mississippi, then Life and Casualty Insurance of Tennessee. Bit by bit, Big Clint began turning over assets to the boys, and it was often John who watched over them, taking a seat on Henry Holt's board and supervising the half dozen rural Texas banks Big Clint had picked up over the years.

Though opposites in every way, John and Clint Jr. operated seamlessly. They owned their assets in a partnership, Murchison Brothers, which Big Clint had set up for them in 1942. Though they plied separate spheres, and though their partnership was never put into writing, every investment was shared equally. After seven years of apprenticeship, Big Clint's first 1956 stroke precipitated an organizational break between father and sons. The elder Murchison moved his interests into a new building on Mockingbird Lane in north Dallas, leaving his sons to swim on their own. There were a few early missteps—John's investment in a uranium mine was a total loss—but for the most part the two brothers proved able and active investors.

Unlike the parsimonious Hunts across town, but very much like their father, John and Clint Jr. enjoyed the finer things. John drove a Porsche to work and piloted his own Beechcraft Twin Bonanza; when he tired of slogging back and forth to Love Field, he built an airstrip two miles from his North Dallas home, then discovered so many friends asking to use it that he turned it into a private airport. As Big Clint's health deteriorated, he began complaining about getting around the Big House. John's growing family was bursting out of its three-bedroom house, so father and son switched homes—not that John's wife, Lupe, was all that excited about a bar with tarpon scales on the wall. She launched an extensive remodeling, peeling off the scales, removing the screened-in porch out front, and turning the mansion into a vast showcase for entertaining. In time John and Lupe's parties would become legendary in Dallas, occasions other families discussed for years afterward. One of the most famous was John's fortieth birthday party, for which Lupe brought in a 120-piece orchestra from Houston, an all-black gospel choir, and a Dixieland band from New Orleans. At midnight fireworks burst overhead and the choir and the musicians launched into a full-throated rendition of "When the Saints Go Marching In," as hundreds of men in tuxedos and women in long gowns and diamonds, champagne flutes held aloft, danced late into the night.

In terms of Texas-style hedonism, though, John had nothing on his brother. Early on, Clint Jr. had declined his father's invitation to build a home on a hundred-acre parcel Big Clint purchased across the road from the Big House. It wounded his father, but as with everything else, Clint Jr. wanted his *own* land, his *own* mansion. In 1954, after two years of searching, he purchased twenty-five acres of land three miles from his father's estate. He wanted to design every detail of the home and the landscaping, and Jane, having just delivered the couple's fourth child, relented, though she made clear she was eager to move out of their cramped three-bedroom Tudor in Highland Park.

Clint Jr., promising to move quickly, began work on the blueprints. The home he envisioned was to be even larger than his father's, a 43,500-square-foot structure in the shape of a horseshoe. Hewn from hundreds of tons of stone borne to Dallas on railcars from Big Clint's Acuna ranch, the house alone covered an entire acre. The main swimming pool, complete with an underwater viewing room, was so big it "could float the *Queen Mary*," as one wag put it; a second, smaller pool was dug alongside the master suite, just for Clint and Jane. Then the problems began. Clint insisted on overseeing the tiniest details, from the width of molding in the bathrooms to the size and shape of each of hundreds of live oaks he was having planted. The electronics alone held up work for years. Decades before it became practical, Clint Jr. wanted a fully automated home packed with the latest gadgetry. Appliances would slide in and out of walls, curtains slid at the push of a button, even the bar was outfitted with an automated dispenser. "I don't care for tending bar," he explained. Stereos, lights, lamps, a security system—Clint Jr. insisted on doing it all himself. According to Jane Wolfe, he spent years perfecting a single big-screen television, long before such items reached stores.

By 1959, five years after planning had begun, the home was still long from finished. Yet even before Clint Jr. had his own mansion, he needed his own private island. Having grown up on Matagorda, he knew what he

wanted, and after another two years of searching he settled on a tiny, undeveloped Spanish cay, a two-mile patch of white sand surrounded by turquoise
water northeast of Abacos Island in the Bahamas. It was Matagorda Act
Two, sans rattlesnakes. Clint oversaw construction of a five-thousand-foot
runway, a radio tower, generators, and roads, then a luxurious main house
and six guesthouses. Big Clint, who felt his son spent far too much time on
such things, was unimpressed. "Well," he grumped, "I guess the next thing
the little s.o.b. will want is a string of racehorses and a mistress." When the
comment was relayed to Clint Jr., he quickly telephoned his father. "Hey
Dad," he asked, "do you know where I can get a string of racehorses?"[6]

Not that the pursuit of luxury kept John and Clint Jr. from the pursuit
of profits. By 1957 the Murchison Brothers offices at 1201 Main thrummed
with all the energy of Big Clint's heyday. In the late 1950s the brothers
expanded aggressively into real estate development, following a crowd of
savvy young Dallas developers like Trammell Crow who were busy throwing up subdivisions all across the Sun Belt. ("You know who built Atlanta?"
went the joke. "Dallas.") The Murchison subdivisions, dotting the suburbs
of San Diego, Los Angeles, New Orleans, Texas, and Florida, were built by
Murchison-owned companies using Murchison-owned building materials
trucked in on Murchison-built roads. "We like to make our money coming
and going," Clint Jr. quipped. Within two years the brothers had consolidated control over almost all of the family's investments—Big Clint's and
their own—giving them stakes in more than one hundred separate companies, from insurance, oil, and construction to publishing, real estate, bus
lines, banks, and BB guns. Their assets topped five hundred million dollars.

For the most part, no one knew it. Big Clint's failing health remained
a family secret, and the press continued to portray their empire as solely
their father's. When they were mentioned at all, John and Clint Jr. were
dismissed as "Clint's boys." That image was to change dramatically, however,
after September 1960, when John and Clint Jr. were shoved into combat
on the national stage. Their opponent was a dour sixty-nine-year-old East
Coast industrialist named Allan P. Kirby, and the fight was a convoluted
sequel to Big Clint's cameo in the struggle for control of the New York Central Railroad six years earlier. The railroad's chairman, Big Clint's friend Bob

Young, had rewarded him with a large stake in another company, the three-billion-dollar Investors Diversified Services of Minneapolis, known as IDS. It had been a sweetheart deal, the shares acquired at below-market prices, and afterward it became the focus of a series of irksome shareholder lawsuits filed against the Murchisons, Young, and his partner, Kirby.

Big Clint didn't get along with Allan Kirby—not many people did—but as long as Bob Young was around, it didn't really matter. Unfortunately, the mercurial Young had been plagued with depression for years. In January 1958, while staying at his winter mansion in Palm Beach, he slid a shotgun into his mouth and pulled the trigger; the next morning the Murchisons awoke to find themselves the uneasy partners of Allan Kirby. In the wake of Young's death, Kirby moved to rid himself of the shareholder lawsuits. The cost was steep, ten million dollars, but the hardest part was satisfying the plaintiffs' demands to retrieve Big Clint's IDS stock. The shares had risen in value, but somehow Kirby prevailed upon the senior Murchison to turn them over in exchange for an equal number of nonvoting shares. It was the only way to end the lawsuits.

The new arrangement, finalized in December 1959, made the Murchisons minority shareholders in an IDS now controlled by Allan Kirby; in effect, Kirby became a feudal lord while the Murchisons found themselves serfs nervously protecting a fifty-million-dollar investment in his fields. Things went bad from the start. Within weeks, and without so much as a phone call, Kirby abruptly threw out IDS executives the Murchisons had installed. It was a slap in the Texans' face, but with little but seats on the IDS board, they were at Kirby's mercy. At that point, the brothers had to decide whether to fight or sell their shares. Reluctantly, they girded for war. Grasping for leverage, the Murchisons swung a deal with Bob Young's widow to buy the stock she held in Kirby's holding company, the Alleghany Corporation; the package of common and preferred stock, if fully converted, would give the Murchison brothers about 20 percent of Alleghany's voting shares, roughly the same as Kirby's. It was a daring challenge. The serfs had now barged into the manor and thrown up tents in the living room.

Because Big Clint's strokes remained a secret, the press initially believed it was the elder Murchison who had orchestrated the move. "At 64," an ill-informed *New York Times* reporter wrote in a major profile of Big Clint in February 1960, "he has yet to slow down and consolidate his empire." In fact, this was John and Clint Jr.'s fight from the outset. The brothers first sent word to Kirby demanding a say in how IDS was run. The imperious Kirby, however, refused to listen. The Murchisons brought in John Connally, now an attorney for hire, to mediate some sort of solution, but Connally got nowhere.

For six months things simmered until, that September, Kirby ousted the Murchisons from IDS's board. John and Clint Jr. immediately declared war, announcing that they intended to seek a vote of Alleghany shareholders to remove Kirby as chief executive officer of his own holding company. It was to be not only the largest, most expensive proxy fight in Wall Street history to date—Alleghany controlled five billion dollars in assets, including IDS and the New York Central—but precisely the Texas Oil versus Eastern Establishment battle many had been expecting since the Big Four burst onto the national scene. It had taken a decade to arrive, and a new generation to fight it, and in the balance hung not just the fate of one investment but a judgment as to whether John and Clint Jr. would be viewed as legitimate players in the corporate world or merely pretenders to their father's throne.

That this was to be a battle between Old America and New was obvious to all. Allan Kirby was a quintessential symbol of old money, the son of the founder of Woolworth's, a prickly, withdrawn man who during a four-decade career had leveraged his vast inheritance into the largest single-ownership stakes in Woolworth, Alleghany, the New York Central, and the giant New York bank, Manufacturer's Trust Company. His family, then as now, lived in the horse country around Morristown, New Jersey, in a twenty-seven-room mansion hung with Rembrandts and Gainsboroughs. Kirby was strong and mean and very capable. He was worth three hundred million dollars.

Kirby was the perfect foil for a Texan, and the press ate it up. *Life* enthused about a battle shaping up between "two great American economic baronies—the rough-hewn Texas beef, oil and money combine and the elegant Eastern financial syndicate." The accompanying photos said it all, Kirby's Rolls-Royce adorned with APK 1 license plate, Kirby glowering on a New York ferry, the Murchisons symbolized by a Brangus bull bearing the family's 7L brand. One *Life* photo, of John Murchison standing with his family beside a bronze sculpture in the foyer of the Big House, suggested that for all the hoopla about a clash between East and West, the cultural chasm between the two sides—and the two Americas—was no longer as vast as it once had been. John, though a native Texan, was a Yale man. His state was growing up.

It was John, the brother more attuned to banking and finance, who took the lead against Kirby. A proxy fight is a bit like a political campaign. Alleghany shareholders were the voters, and all would need to be romanced if the Murchisons were to have any hope of winning. The odds were against them; Kirby had long, strong ties on Wall Street, and already controlled more than 20 percent of the shareholder votes. John readied for a siege. In the fall of 1960 he rented offices at Forty-eighth and Madison Avenue and shifted much of the Murchison Brothers corporate staff from Dallas to New York. He augmented his forces with dozens of New York lawyers, two separate proxy-solicitation firms, and, as his aide-de-camp, the noted investment banker Gustave "Gus" Levy of Goldman Sachs. Anticipating a long, nasty fight, John and his wife, Lupe, moved to Manhattan, settling into a seventeen-thousand-dollar-a-month suite at the Carlyle Hotel, leaving their four children behind with a governess in Texas. Clint Jr. flew in once or twice a week to help out.

John fired the war's first shot, slamming Kirby with a lawsuit charging him with fraud and conspiracy in connection with Big Clint's turnover of IDS stock. Kirby yawned; that he considered the Murchisons unlettered yokels went without saying. One of his men sniffed to a reporter that the Murchison empire was a "vehicle put together of glue, string, turpentine and wind." With a quarter of Alleghany's stock in his pocket, Kirby could easily have warded off the Murchisons by buying a majority of its shares. But he didn't. From all appearances, he never thought he had to. In time Kirby got around to his own lawsuit, charging Big Clint with insider dealings at IDS. That winter John and Clint Jr. countered with a public relations offensive, sitting for long interviews with eastern reporters who appeared surprised they weren't wearing Stetsons and string ties. The brothers played up their status as southwestern underdogs, telling the *New York Times* man with a straight face how surprised they were to find "rustlers and brand-changers" in the corridors of Wall Street.

For all their Sturm und Drang, many Wall Street fights, like courtroom trials, are designed to force both parties toward settlements. In early March 1961, with the Alleghany shareholder vote just two months away, Kirby consented to peace talks in his Park Avenue offices. John demanded two seats on the Alleghany board of directors, and Kirby appeared to agree. The next morning's headlines declared the war over before it had really started, but the settlement fell apart later that day when Kirby refused John's demands to replace IDS's CEO with a man more to his liking. At that point both sides filed papers with the Securities and Exchange Commission to begin active solicitation of proxy votes for the shareholder meeting.

Now the real fighting began. In his Madison Avenue war room, John and his advisers divided the country into eighty blocks and assigned men to telephone Alleghany shareholders in each. He and Clint Jr. spent hours on the phones, patiently explaining to investors large and small how they could run Alleghany and IDS more efficiently than Kirby. Kirby, meanwhile, hired private detectives and sent them snooping through years of Big Clint's complex business deals. They compiled thick dossiers charging Big Clint with a potpourri of financial chicanery, from sweetheart real-estate deals with Richard Nixon (true) to cheating the federal government on housing contracts (probably not) to allegations that he did business with the Mafia (doubtful). The dossiers were then forwarded to journalists including the columnist Drew Pearson, who by and large ignored them, telling Kirby's men to come up with something concrete or shut up.* But if few of Kirby's charges surfaced publicly, they circulated widely on Wall Street, and would pop up regularly in tales about the family for years to come.

Through it all, John worked sixty-hour weeks, and the stress and long hours began to take their toll. Never possessed of the strongest constitution—John's wartime stresses and asthma were never far from his mind—

---

* The work product of Kirby's detectives can still be found among Drew Pearson's personal papers at the Johnson Presidential Library in Austin, Texas.

his energy flagged. His eyes grew puffy. He began smoking. Lupe tried to divert him with weekend tours of Manhattan art galleries, which kindled in John a lifelong love for contemporary and avant-garde art. He began buying dozens of canvasses and shipping them back to Dallas. But art-collecting alone couldn't revive him. He felt increasingly estranged from his children and finally, fearing a nervous collapse, John returned to Dallas for three full weeks to recover. In his absence, Clint Jr. did his best to keep the fight going.

By early April, with the vote a month away, the brothers could point to some encouraging signs. Several major shareholders, including two onetime Alleghany directors, had announced their support. Both sides were buying stock, and thus votes, as fast as they could—trading in Alleghany shares was the heaviest in any stock since the crash of 1929—but the betting on Wall Street was that Kirby would find a way to beat his upstart challengers. Then, just eleven days before the shareholder vote, Kirby blundered. Alleghany hadn't declared a stock dividend in its thirty-two-year history, but Kirby now announced it would, five cents on every share. It was a transparent attempt to curry favor with stockholders, and a sign of Kirby's desperation. John wasted no time capitalizing, taking out full-page advertisements in the *Wall Street Journal* and *New York Times* featuring a cartoonish Kirby tossing handfuls of "five-cent chicken feed" to Alleghany shareholders.

The shareholder meeting, held at the Lord Baltimore Hotel in Baltimore, proved anticlimactic. Kirby made polite remarks, refusing to mention the Murchisons by name. John stood in the audience and answered a question or two. Afterward, both sides told reporters they were confident of victory. It was supposed to take a full week to count the votes. It ended up taking three, but by then the word had leaked:

Texas won.

The final tally gave the Murchisons 54 percent of shareholder votes, a very tight race by Wall Street standards. For John and Clint Jr., the ensuing media acclaim matched anything lavished on the original Big Four. The coverage peaked when they followed their father onto the cover of *Time*. Much like NASA's arrival in Houston two years earlier, the *Time* cover marked a turning point in the image of Texas and its oilmen. Where seven years earlier Big Clint had been portrayed in a rancher's straw hat, John and

Clint posed for *Time* in business suits; the only nod to Texas was the lasso-draped skyline behind them. According to the article:

> The Murchisons' victory on Allan Kirby's home grounds was dramatic notice of the changing role of Texas in the U.S. economy. Easterners still like to think of Texans as illiterate oil millionaires who wear ten-gallon hats—and, when they are in Wall Street looking for money, some Texans shrewdly play the expected part. . . . But at home in Dallas or Houston, today's Texas tycoon is more apt to wear a Brooks Brothers suit than Texas boots; though his poke may have started in oil (and gained by the 27½% depletion allowance), much of it now comes from electronics, real estate, insurance or shipping. And for the new Texan, Texas is no longer big enough. Ranging across the nation like eager bird dogs, Texas businessmen are supplying capital, entrepreneurial vigor and acumen in nearly every area of the U.S. economy.

After years in the shadows, the Murchison brothers now emerged as not only symbols of the "new" Texas, but its most visible young magnates, their next move a source of boundless speculation in the national business press. John was named Alleghany's interim chief executive. The Murchison brothers now controlled five billion dollars of new assets, including IDS and the country's largest railroad. John told Lupe he wanted to move to New York for good; he even had his eye on a Sutton Place town house. He loved everything about city life, the museums, the restaurants, the art. Lupe resisted. She loved Dallas. Most of all, she loved being a *Murchison* in Dallas. She was royalty in Texas, a nouveau riche housewife in Manhattan. And that, she saw, was the point. John didn't want to be royalty. He wanted the anonymity only Manhattan could bring. In the end they put off a decision, shuttling between Dallas and New York while John searched for a permanent CEO.[7]

If John foresaw a happy new life meandering between art galleries and his triumphant perch inside Alleghany's boardroom, he was badly mistaken. It was almost as if he hadn't thought things through. The Murchisons had won glory and acclaim and a sprawling new financial empire, but it was an empire in which the vengeful Allan P. Kirby still owned a full third of the stock. From the moment his defeat was announced, in fact, Kirby began a guerrilla campaign designed to thwart John's wishes at every turn. During the proxy fight John had promised shareholders he would recapitalize IDS via a ten-for-one stock split, but when he proposed it Kirby sued and won. John wanted to sell off New York Central shares. Kirby opposed him. Kirby opened his doors to the press, sniping at John's inability to boost the price of Alleghany's stock, which was sagging in the face of the internecine warfare.

"The Murchison boys want me to sell out," Kirby told one writer, "but I won't. That's one thing I won't do." With a smile, he added, "I'd say the Murchison boys are way, way out on a limb."

John was trapped. He tried in vain to hire a top-flight CEO to run Alleghany—the man who Sid Richardson had helped become Nixon's treasury secretary, Robert B. Anderson, was one rumored candidate—but Kirby's scorched-earth insurgency turned the company into a battlefield no one was eager to navigate. A stronger man might have perservered, but John was simply overwhelmed. There was no way to rid himself of Kirby, and in the end he saw but one option. In September 1963, barely two years after taking control of Alleghany, John confirmed to reporters he was in talks to sell out. A few weeks later he followed through, selling most of Murchison Brothers' Alleghany stock to a Minneapolis investor who promptly resold it to Kirby. Kirby basked in the glory of taking back the company he had lost. He did it for "pride," he said. "Family pride. As nearly as I can remember, until the proxy fight with the Murchisons in 1961, I never got licked."

And like that it was over, not just John and Clint Jr.'s dealings with Alleghany but their brief tenure as actors on the nation's business stage. Though they would continue investing in dozens of American companies, never again would the Murchison brothers venture into the corporate spotlight in any serious way. John limped back to Dallas amid rumors that he and Clint Jr. were liquidating their non-Alleghany holdings. John wasn't suited to be CEO of a major company; he loathed the attention, and for the rest of his life he would do his best to avoid it. Not so his brother. Clint Jr. still burned with the need to be noticed, and by the time John returned to Dallas, Clint was on his way to becoming a Texas icon. But it wasn't the boardroom or the oil fields where he would make his mark. By 1963 Clint Murchison Jr. had only one thing on his mind: Football. Football, football, football.

III.

Unlike the high-living Murchisons, or for that matter their own idiosyncratic father, the children of H. L. Hunt's first family lived unremarkable upper-middle-class lives in the late 1950s. Margaret Hunt Hill was a homemaker who lived in Highland Park with her husband, Al, dined at the Brook Hollow Country Club, and was accepted into Terpsichorean, an exclusive dance club. Her sister Caroline lived just as quietly. Drab, stolid Herbert had earned a geology degree at Washington & Lee and joined Hunt Oil. Reliable, with a good head for numbers, he was leading the family's first real diversification, buying up thousands of acres of real estate across North Dallas and its burgeoning suburbs. Herbert lived with his wife and family in University Park, drove to work in a Chevrolet, and mowed his own lawn.[8]

Only the second son, Bunker, followed his father's path into the oil fields. In later years everyone around Dallas would say it: of all the many, many Hunt children, only Nelson Bunker Hunt seemed to be his father's second coming, a man of vision and dreams, ambitious, outspoken, combative, self-possessed, and, like his father, a bit odd. An enormous twelve-pound infant when born in Arkansas in 1926, Bunker had never outgrown his baby fat and, thanks to a steady diet of junk food over the years, never would. By adolescence he had already grown the big moon face that, with his thick glasses, would in time lead more than one writer to describe him with the word *porcine*. His favorite meal was a Texas delicacy: a can of chili poured over Fritos.

His father, who had invested so much in Hassie, showed little interest in raising Bunker and his brothers. When he was twelve, his mother handed Bunker a train ticket and told him he would be sent to Culver Military Academy, as Hassie had been. But Bunker, who was at best ambivalent about his appearance, never took to shining his shoes and creasing his bedsheets. When school officials sent disapproving letters to his mother, Lyda packed him off to the Hill School outside Philadelphia. There the rumpled teenager made a good living as the school bookmaker, running craps games and taking bets on college football games, a hobby that got him briefly expelled when he watched a game one Saturday instead of working off demerits. He dropped out of the University of Texas after one semester to join the navy, spending the last months of World War II washing decks on an aircraft carrier.

After the war Bunker married an SMU girl and took a job at Hunt Oil, where opinions of him seemed evenly split: some of H.L.'s people felt he showed real promise as an oilman; others thought Bunker, who was dyslexic, wasn't the sharpest pencil in the pack. When Bunker, then only twenty-two, sniffed out a seven-million-dollar oil field in Scurry County, H.L. placed him in charge of Penrod, the family's drilling contractor. But in truth, Bunker never had a chance. Maybe it was his weight, maybe his casual attitude toward work—he was known to take naps, and liked to stop by Las Vegas during business trips—but there was something about him that irritated his father no end. Bunker's smallest infractions could drive H.L. to apoplexy, as when H.L. screamed at him the time someone at a ranch Bunker was overseeing stole a load of seed. According to family lore, H.L. fired Bunker from Hunt Oil, but he may have left of his own accord. Bunker knew, as only the sons of disapproving fathers can, that the only way to gain H.L.'s love, or at least his respect, was to make it on his own. That meant finding oil, and lots of it. But this was the 1950s. If he was to unearth a field even half the size of East Texas, Bunker realized it would mean looking overseas. He first hired a University of Illinois geology professor and asked him to prepare a report on the most promising places an "elephant" might be found. The professor's report listed the leading candidates as the Middle East, where the most promising concessions had long been snapped up, North Africa, and Pakistan. Bunker, then twenty-seven, flew to Karachi in 1953, and two years later he reached an agreement with the government for a forty-two-million-dollar drilling program along the Makran Coast. He spent the next several years drilling dry holes. His losses were estimated at eleven million dollars. His father was not impressed.

In 1955, as Bunker began drilling in Pakistan, the North African country of Libya announced a drilling-rights auction. Bunker jumped into the thick of it, competing directly against the majors. He emerged with two tracts, a promising coastal stretch near the Egyptian border known as Concession #2 and a patch of remote desert three hundred miles inland, Concession #65, in an area known as the Calansho Sand Sea. Bunker drilled the coastal tract first. It was a nightmare. All his geologists had to guide them were out-of-date Italian maps. Worse, Bunker's concession lay amid the World War II battlefields east of Tobruk, which remained pocked with land mines; one of his men was badly hurt when his Jeep hit one. By 1959, after three years of searching, Bunker had yet to find commercial amounts of oil. Worse, he had run through the money in his trusts. To keep going, he sold a 15 percent stake in his concessions to his brothers Herbert and Lamar.

Then, in 1961, with his money almost gone, British Petroleum offered to sink a wildcat in the distant Concession #65 in return for half of any oil it might find. Bunker agreed; he no longer had the resources to do it himself. It was the smartest decision of his career. In November 1961 the exploratory well blew in at a strong 3,910 barrels a day; it was the first gusher in what came to be known as the Sarir Field, whose reserves would eventually be estimated at eight to eleven billion barrels, almost three times as large as East Texas. It was one of the ten largest oil fields ever found to that point, and Bunker's half interest, valued at six to eight billion dollars, made him, on paper at least, the richest man in the world—richer than his father.

And, much as happened with his father three decades before, almost no one knew it. Bunker's windfall made no headlines in Dallas or anywhere else. It was, after all, the most tenuous of fortunes. To get the oil to market would require building a three-hundred-mile pipeline across open desert to the coast, where an oil port would need to be built. British Petroleum seemed to be in no rush; Bunker suspected it feared the sudden torrent of oil onto world markets would drive down prices. He reluctantly coaxed a five-million-dollar loan out of his father, but it would be years before he would see steady income from his Saharan bonanza.

When in Dallas, Bunker lived with his wife, Caroline, in a simple colonial in Highland Park, driving to work in a colorless American sedan adorned with a GO SMU MUSTANGS bumper sticker. Like his father, he was notoriously cheap, taking the subways when in New York and complaining about the cost; he favored hotel rooms the size of walk-in closets. Just as he was the only Hunt to follow his father into the oil fields, he was the only child to forge his way into right-wing politics. Bunker, in fact, embraced the very worst elements of his father's harsh anticommunism, down to and including anti-Semitism; as he told more than one friend over the years, he actually believed there was a Jewish-Communist conspiracy to take over the world.

In 1958 an Indiana businessman named Robert Welch formed an organization called the John Birch Society, whose central tenets included a belief that Jewish industrialists led by the Rothschild banking family were conspiring with the Russians and Chinese for world domination. Almost overnight, the Birchers, as they were called, emerged as the largest and loudest group of American ultranationalists, with their strongest chapters in Southern California and Texas. Though fellow travelers, H.L. never joined the Birchers; when a reporter asked why, he said, "I always thought they should have joined me." Bunker, however, embraced the Bircher doctrine as his own, adopting Robert Welch as his political mentor, hosting Birch Society meetings at his home and becoming one of its most important financial backers. A Baptist, Bunker also began donating to religious causes.

As famous as Bunker would become in later life, it was the youngest Hunt, the shy, well-mannered Lamar, whose exploits first made national headlines. Tall, thin, and soft-spoken, initially overshadowed by his older brothers, Lamar was a boyhood box-score enthusiast who had ridden the bench as a third-string end at Southern Methodist University. He had soft edges; people liked him, and would his entire life. After a starter marriage and divorce, he had settled in Dallas, and many evenings he could be seen shooting baskets in his driveway.

Like Bunker and many Texas oilmen, Lamar's favorite sport was football, whose professional teams had emerged as a national preoccupation following the dramatic 1958 championship game between the John Unitas–led Baltimore Colts and the New York Giants. Like Glenn McCarthy a decade earlier, Lamar thought football represented the future of American sports, and he grew determined to bring a team to Dallas; the city's previous team, the Dallas Texans, had folded after one season, in 1953. In early 1959 Lamar, then just twenty-six, approached the owner of the Chicago Cardinals, who refused to sell. So did the league's other owners, many of whom were just as upbeat about the National Football League's future. What Lamar didn't realize was that he had run smack into an identical effort by Clint Murchison Jr.

In the annals of the Big Rich, the Big Four families seldom crossed paths in any serious way, and where they did, as in the lifelong friendship between Sid Richardson and Clint Murchison, relations tended to be amicable. Texas, after all, was big enough for all of them. It was the rare occasion when a Hunt squared off against a Murchison, but when it finally happened, in Lamar and Clint Jr.'s pursuit of an NFL team for Dallas, all Texas paid attention. Murchison had a head start. He had been a season ticket holder for the Texans—he had twenty seats, in fact—and had tried without success to buy the team before it left Dallas. In 1955 he tried to buy the San Francisco 49ers but couldn't. Finally, in 1958, he reached an agreement to buy the Washington Redskins, but the talks fell through when the team's owner, George Preston Marshall, sought a change in terms. When the NFL announced it would expand to fourteen teams in 1961, Clint Jr. changed tack and set his sights on starting a new team. In early 1959, just as Lamar was sending out his first feelers about buying a team, he began meeting with the owner of the Chicago Bears, George Halas, chairman of the league's expansion committee.

Once he realized Murchison was in the picture, Lamar realized it was unlikely he would win a franchise of his own. Like his father, though, Lamar was a creative thinker. After months of mulling over the issue, the answer came to him one night on an airplane. When he reached Dallas, he telephoned one of his father's friends, a Houston oilman named K. S. "Bud" Adams, who had also attempted to buy the Chicago Cardinals; Adams wanted them for Houston. The two men had a long dinner at a steak house Adams owned in Houston, complaining about the NFL's stodgy ways, but it wasn't until Adams drove Lamar to Hobby Airport that H. L. Hunt's youngest son turned to him and revealed his cards. "Bud," he said, "I'm thinking about starting a new league. Would you be interested in joining me?" Adams's reply: "Hell yeah."

And that was all it took. On August 3, 1959, Lamar Hunt and Bud Adams, representing two Texas oil fortunes with one famous father between them, held a press conference in Adams's Houston office and announced formation of the American Football League. Lamar had just turned twenty-seven the day before. The league, they announced, had exactly two teams, the Dallas Texans and the Houston Oilers. Twelve days later, after fielding calls from dozens of interested owners, they would unveil four additional franchises, in Los Angeles, Denver, New York, and Minneapolis. For the most part, NFL owners snickered. Sportswriters quickly dubbed the two Texans and their pals "the Foolish Club."

At 1201 Main, however, Clint Jr. wasn't laughing. If Lamar went forward, his new team would divert attention from the NFL franchise he hoped to win; it would divide the market for season tickets and, worst of all, Lamar was making noises about leasing Dallas's only major football venue, the Cotton Bowl. Clint Jr. hustled to Chicago and pressed George Halas to grant him a franchise immediately. Halas understood and prevailed upon his committee to give it to him. The news was flashed across television screens all across Texas, including one in a ranch house sixty miles southeast of Dallas, where an old man frowned at his set. It was the first Big Clint had heard of his son's plans. He thought professional football was a silly, money-losing proposition and had said as much, many times, and loudly. In Big Clint's mind, it was just another example of his son's inability to focus on the business that really mattered. But Clint Jr. was intent on building something of his own, something his father hadn't given him.

"That's gonna break that boy," Big Clint murmured.

Dallas, it appeared, pending a final vote of NFL owners, would now have not one but two new professional football teams. But Clint Jr was determined to prevent that from happening. He had met and liked Lamar at a dinner party when Lamar was still in college. When he got the Halas committee's approval, Clint Jr. arranged a meeting at Lamar's office. There Murchison offered him 50 percent of the NFL franchise if Lamar would agree to kill the Texans. Lamar thanked him, but said he simply couldn't abandon his partners in the aborning AFL.

The game was on. Both Lamar and Clint Jr. forged ahead with their plans, Clint hiring as general manager a young CBS executive Halas had suggested, Texas E. "Tex" Schramm. Schramm tutored Clint Jr. on the ins and outs of sports ownership, emphasizing the chain of command, a polite way of telling Clint Jr. to stay out of football decisions. For the moment, Clint Jr. was more focused on securing final approval for the new franchise from the NFL owners. It required a unanimous vote, and the Redskins' owner, George Marshall, had spread the word that he might blackball Murchison, whom he reportedly found "personally obnoxious." The problem, it turned out, was that one of Murchison's men, Tom Webb, had quietly purchased the rights to the Redskins' fight song, "Hail to the Redskins," from the song's writer, embittered after Marshall fired him. Webb thought of it as a bargaining chip. Marshall didn't care; he just wanted his song back, and badly.

The NFL owners gathered at Miami's Kenilworth Hotel for the vote in January 1960. At a gathering the night before, Marshall sent a wandering accordion player to George Halas's table to serenade him with "Hail to the Redskins," a pointed reminder of his intentions. Halas promptly sent the musician to Marshall's table, where he played "The Eyes of Texas." The next morning Clint Jr. went to Marshall's room, introduced himself, and phoned Tom Webb. The two then performed an elaborate charade for Marshall's benefit, Murchison begging and wheedling the silent Webb to turn over the song. Murchison hung up. Marshall implored him to try harder. Murchison called a second time, again begging Webb to hand over the song, until finally Murchison put down the phone and told Marshall the rights were his. Marshall promptly pledged to back the birth of the new Dallas team.*

Back in Dallas, Clint Jr. and Tex Schramm hired their coach, a stern New York Giants assistant named Tom Landry, then set about selecting the team's name. Murchison insisted on the Dallas Rangers. Schramm resisted, pointing out Dallas already had a minor-league baseball team called the Rangers. Murchison prevailed. A press release went out, announcing the name. Schramm, however, wouldn't give up, and finally persuaded Murchison to rename the team the Dallas Cowboys. It took years for Clint Jr. to warm to the name.

---

*Clint Jr. appears to have gotten help arranged by one of Lyndon Johnson's aides, Bobby Baker. In 1973 Baker told Playboy that he arranged for Clint to pay the powerful Tennessee senator Estes Kefauver a twenty-five-thousand-dollar bribe, ferried to him by Clint's pal Robert Thompson. Kefauver, Baker charged, had then put pressure on George Marshall. Clint Jr. denied the story. Thompson didn't. For years he delighted in telling the story to friends.

Five years later he issued a press release announcing the team might change its name back to the Dallas Rangers. The reaction was immediate. Murchison counted 1,148 phone calls to the Cowboys office. As he wrote in a note to a Dallas sportswriter, the tally came in at "Keep the name Cowboys, 1,138. Change the name to Rangers, 2. Murchison is stupid, 8."

In an effort to compete with Lamar Hunt's Texans, the NFL announced the new Cowboys would begin play a season earlier than planned, in the fall of 1960. Lamar and the other AFL owners promptly slapped the league with a ten-million-dollar antitrust suit, calling the move "sabotage." In fact, despite Lamar's barbed quotes in the press, he and Clint Jr. remained friendly rivals. At a luncheon just before Christmas 1960, Clint Jr. surprised Lamar by wearing a bright red Dallas Texans blazer. A week later Clint was hosting a gathering at his home when two friends dragged in a massive six-foot-high gift-wrapped box. Clint stepped over, unwrapped the bow—and was startled when who should emerge but a smiling Lamar. The two men bore a passing resemblance, and more than once Clint Jr. found himself being introduced to people as Lamar. At one point, when a tall bespectacled man appeared at Cowboys offices asking for Tex Schramm, a new receptionist asked, "Are you Mr. Murchison?" The man smiled and said, "Lamar Hunt." Afterward Clint Jr. presented the woman with pictures of both of them captioned, "This is Lamar" and "This is Clint."[9]

During that first season in 1960, both the Texans and the Cowboys played at the seventy-five-thousand-seat Cotton Bowl. Neither drew the crowds Lamar and Clint Jr. needed. The Cowboys mailed two hundred thousand letters to prospective season ticket holders; 2,165 signed up. Barely twenty thousand people appeared for their first game, and attendance plummeted after that, falling as low as two thousand one Sunday. It didn't help that, forced to field a team of NFL castoffs—one a rodeo cowboy, another an art teacher—Tom Landry's Cowboys failed to win a single game, finishing with eleven losses and a tie. The tie, in a December game against the Giants in New York, got Clint Jr. so excited he scurried around the El Morocco night club trying in vain to find a Texan to tell. He arrived back in Dallas to find a Love Field welcoming throng of exactly two fans.

At one Cowboy home game barely eight thousand people appeared, and when it began to rain, all sought shelter beneath the press box. From his perch inside, it appeared to Clint Jr. that the entire stadium was empty. It stung, though Clint kept his spirits up. When the New York restauranteur Toots Shor wrote seeking box seats for a Giants game in Dallas, Clint sent him the tickets along with a note. "In case you want to bring any of your friends with you," it read, "I am also sending you Sections 1, 2, 3 and 4." Shor opened an accompanying box to find every ticket in those sections, ten thousand in all.[10]

Clint Jr. lost seven hundred thousand dollars that first year, but Lamar had it worse. Though his Texans won eight games and lost six against the other new AFL teams, Texan crowds were even smaller than the Cowboys' and their tickets cost less. At year-end, a Dallas sportswriter guessed Lamar had lost about one million dollars. "At that rate," he concluded, "he can only afford to lose for the next one hundred years."* Clint Jr. once said the only game he truly enjoyed in those early years was between the teams from Philadelphia and Pittsburgh: "It was the only game I'd seen since joining the NFL that hadn't cost me $50,000."

The Cowboys and Texans fought for the hearts of Dallas football fans for three long years, but nothing they did, not even the AFL Championship Lamar's Texans won in 1962, could fill the Cotton Bowl. Finally, in 1963, Lamar ran up the white flag. He wanted to relocate the Texans to a city within easy commuting distance, and was poised to move them to New Orleans when, at the eleventh hour, its mayor refused to let the team play at Tulane Stadium, fearing the loss of Tulane fans. Instead, Lamar negotiated a one-dollar-a-year stadium lease (for two years) with the mayor of Kansas City and in May 1963 announced that the Texans were moving to Missouri to become the Kansas City Chiefs.

The AFL, plagued with dwindling crowds, remained shaky until later that year, when Lamar ensured the league's survival by negotiating a thirty-five-million-dollar television package with NBC. Three years later, in June 1966, he would spearhead the AFL's merger with the NFL and the creation of a title game between the two league champions. The NFL commissioner,

---

*In later years the quote would be widely, and mistakenly, attributed to Lamar's father.

Pete Rozelle, wanted to call the game "the Big One." But it was Lamar, after seeing his children bouncing a Super Ball, who came up with the name that stuck: the Super Bowl. His Chiefs would lose Super Bowl I to the Green Bay Packers in January 1967, but Lamar's place in American sports history was secure. He was thirty-five.

IV.

With the notable exception of Bunker Hunt, the second generation of the Big Rich paid scant attention to politics, though the day was fast approaching when they all would grapple with what their fathers had wrought. Gauging what if any political legacy the original Big Four oilmen left their children is a chicken-and-egg dilemma. Was it their efforts—Facts Forum and LIFE LINE, Murchison's anti-Adlai newspapers, Cullen's fiery telegrams—that explained why so many Texans became ultraconservative? Or were oilmen just a product of the state's innate attitudes? The reality is probably a bit of both.

Whatever the case, there is no denying that by the early 1960s, while Texas ultraconservatives no longer held any major state office, their numbers were growing. Far below the media's radar, Houston had emerged as a stronghold of the paranoid Birchers, while Dallas bristled with racists and right-wing demagogues, many of them avid LIFE LINE listeners. (George Rockwell of the American Nazi Party once said Dallas had "the most patriotic, pro-American people of any city in the country.") As the nation, and the state's own politics, moved away from their hard-core values, Texas ultraconservatives grew restive. When George Bush ran for a Houston congressional seat as a moderate Republican, he found angry Birchers shouting him down at public meetings. When Lyndon Johnson accepted John F. Kennedy's invitation to run as vice president in 1960, Texas ultraconservatives denounced him as a traitor. During the campaign that November, Johnson and his wife, Lady Bird, came to Dallas for a speech, only to be met by an unruly crowd outside the Adolphus. After Johnson had words with a demonstrator, he waded into the crowd and was spat upon. Afterward Johnson called it "a mob scene that looked like some other country. It was hard to believe that this was happening in Dallas, and in Texas."

Not all Texas conservatives were right-wing nuts, of course; as George Bush's ascendancy attested, the state's population of mainstream conservatives was growing as well. By 1961, thanks in no small part to the organization-building Roy Cullen had initiated with Jack Porter in the 1940s, many Texas conservatives had defected to the Republican Party. That year, in fact, the state elected its first Republican senator, the conservative John Tower. The growth of the Texas GOP paralleled the emergence of a national conservative movement whose arrival, driven by the writings of William F. Buckley and his peers, had been but a twinkling in the eyes of Texas oilmen a decade before. At least initially, many liberals couldn't tell the difference between the Buckleys and the Birchers, both of whom vaulted into public view to fill the vacuum left by Dwight Eisenhower's retirement and the defeat of Richard Nixon in 1960. The status quo's reaction was swift.

On November 18, 1961, during a fund-raising speech at the Hollywood Paladium, President Kennedy lashed out at what he termed "the discordant voices of extremism" emanating from the Far Right. The media, dominated by liberal-leaning journalists, smelled something new; the ensuing weeks brought a torrent of exposés into the new "Radical Right"—in *Time*, *Newsweek*, and the *New York Times Magazine*—that drew little distinction between moderate conservatives and the wild-eyed Birchers. None of this had any obvious connection to the "new" Texas—until December 4, when *Newsweek* adorned its cover with a retired major general named Edwin Walker, who had resigned from the army after being accused of teaching troops John Birch dogma.

Walker had settled in Dallas, where in speeches around the city he had drawn a small but vocal following. The *Newsweek* cover—"Thunder on the Right: The Conservatives, the Radicals, the Fanatical Fringe"—portrayed him as a would-be Mussolini preaching to throngs of ultraconservative Texans. In the blink of an eye, as only the American media's herd mentality can make it appear, the country, and especially Texas, seemed awash in right-wing nut jobs. With the news that H. L. Hunt had been among Walker's backers—Hunt, in fact, felt Walker should run for president in 1964—many in the press, recalling the McCarthy-era backlash against the Big Four eight years earlier, blamed the Radical Right's "emergence" squarely on Texas Oil money.

In a special edition devoted to the "rise of the radical right," *The Nation* came right out and said so: "Virtually every Radical Right movement of the postwar era," it argued, "has been propped up by Texas oil millionaires." In *The Nation's* hands, all the old skeletons came tumbling out of the closet: John Henry Kirby, Vance Muse, Martin Dies, Roy Cullen, Facts Forum, Murchison, and Sid Richardson hobnobbing with Joe McCarthy around the Del Charro pool. According to the article:

> Favored by exemptions granted no other segment of American society [*The Nation* wrote] Texas oilmen have amassed incredible millions—in some cases, actual billions—and have become an arrogant economic oligarchy immune to the ordinary influences, superior to the ordinary needs and desires of America. The influence of Texas oil today is all-pervasive. Its millions, piled up thanks to the bounteous depletion allowance, spread through every section of American industry; one can hardly turn around in the publishing field in New York, without bumping into a Texas oil millionaire who bought himself a share—often a controlling share of an established periodical or book-publishing firm. The influence of overpowering wealth is a supremely potent force indeed, and this influence today is working its benefices on behalf of the Radical Right.

The Nation overstated its case against Texas Oil, and especially against Hunt, to whom it assigned much of the blame for the growth of the Radical Right. But this and similar reviews insinuated themselves into mainstream press coverage, and by 1963, almost a decade after the anti-McCarthy backlash, the entire state of Texas had been tarred with an image of right-wing extremism once reserved exclusively for the Big Four. Texans have a strange tendency to embrace their myths, even the dark ones, and so it was with the "new" ultraconservatism. That ultraconservatives remained a quarrelsome minority in the state didn't matter. They were loud, and they wanted to be noticed, as Adlai Stevenson discovered when he visited Dallas in October 1963.

Stevenson was to speak on "United Nations Day," in favor of an institution ultraconservatives loved to hate. General Walker mounted a counter-demonstration dubbed "United States Day," and to the dismay of many, the state's new governor, Texas Oil's old friend John Connally, issued a proclamation saying so, too. Walker's supporters infiltrated the Memorial Theater audience for Stevenson's speech and attempted to hoot him down, while others picketed outside and waved American flags. When Stevenson tried to placate some of the demonstrators afterward, a woman struck him. A young man spit on him. As he wiped the spittle from his cheek, Stevenson muttered, "Are these human beings or are they animals?"

The incident was splashed across front pages around the country. The subtext was clear: Forget the astronauts. Forget those freshly scrubbed Murchison boys on the cover of *Time*. *This* was the new Texas, same as the old Texas. For the moment, no one gave much thought to the fact that it was exactly one month until President Kennedy was scheduled to tour the streets of downtown Dallas in an open limousine.

V.

One month earlier, on the night of September 12, 1963, Clint Murchison Jr. finally opened his new home for guests. The party, almost ten years in the making, marked his fortieth birthday. Hundreds turned out to marvel at the great stone mansion, the largest in Dallas, probably the greatest in the state. The big-screen television, the robot bartender, the miles of walnut paneling that hid every electronic gadget out of sight—it was all worth it to Clint. Their elementary-school-aged children were now teenagers, but for one night at least, the long delay didn't matter to his wife, Jane. The guests oohed at the swimming pool and its underwater viewing chamber, aahed at the intricate moldings and gorgeous live oaks, and tried not to dwell on the fact that Clint's Cowboys were still mired in last place. It was a wondrous evening, the kind people would be talking about for years, and it was the last party anyone in Dallas would enjoy for a long time.

VI.

On the morning of November 22, 1963, John F. Kennedy woke in a suite at the Texas Hotel in Fort Worth, where Sid Richardson had lazed away many hours talking with his bookies. Some of the president's advisers had warned him about coming to Texas, but the crowds in San Antonio and Houston the last two days had been welcoming. Kennedy and Governor Connally were set to drive through the streets of downtown Dallas. They would pass near Clint Murchison's headquarters at 1201 Main, then slide directly beneath H. L. Hunt's office window, before rounding the Texas School Book Depository, recently purchased by D. H. Byrd.

As he left for the brief flight to Dallas's Love Field, where demonstrators were waiting with signs that read YANKEE GO HOME and YOU'RE A TRAITOR—an aide handed Kennedy a copy of the *Dallas Morning News* and pointed to a full-page advertisement draped in a funereal black border. "Welcome Mr. Kennedy to Dallas," it read. The text was an anti-Communist screed that broadly painted the president as a Communist dupe. "WHY" it asked, "has Gus Hall, head of the U.S. Communist Party praised almost every one of your policies and announced the party will endorse and support your re-election in 1964?" One of the men who paid for the ad, it turned out, was Bunker Hunt.

The president read the ad, then turned to his wife, Jackie, grinned, and said, "Oh, we're heading into nut country today."

FOURTEEN

Sun, Sex, Spaghetti—and Murder

I.

A few minutes past noon, H. L. Hunt stood in his office window watching the presidential motorcade pass beneath him on Main Street. He lost sight of it as the cars rounded the Texas School Book Depository Building. As they eased down into Dealey Plaza, shots rang out. By nightfall John F. Kennedy was dead, a native Texan, Lyndon Johnson, had been sworn in as president, and Texas Oil had changed forever.

Hunt was the first oilman to be swept up in the ensuing pandemonium, but by no means the last. Within minutes his son Herbert and his security chief, Paul Rothermel, hustled into his office. As they did, the phone rang. It was an FBI agent advising Hunt that, as a vocal critic of Kennedy's, his life might be in danger. He advised Hunt to leave Dallas. Hunt objected. Herbert and Rothermel urged him to go. Hunt relented, saying he planned to travel east. "I believe I can do better going to Washington to help Lyndon," he said, as if Johnson would have anything to do with him. "He's gonna need some help."

Two days later a Dallas nightclub owner named Jack Ruby shot and killed Lee Harvey Oswald at the Dallas police headquarters, and it was the Hunts who needed help. Ruby had been outraged by the tone of LIFE LINE's attacks on Kennedy, and police found two LIFE LINE scripts in his coat pocket when he was arrested, along with Lamar Hunt's telephone number. Across the country, editorial writers, though careful not to criticize Hunt by name, lambasted LIFE LINE for fostering the "climate of hate" in Texas that many

Americans came to believe had at least contributed to the president's death. Hunt began receiving death threats. His phones rang in the middle of the night. At one point, someone even fired shots at Mount Vernon.

By Christmas the Hunts became targets of the federal investigation that would be subsumed by the Warren Commission. An FBI man interviewed Lamar, who denied knowing Jack Ruby; apparently Ruby had intended to call him but never had. Agents questioned Bunker about the "Welcome Mr. Kennedy" ad the following spring. Several people, including a convicted con man, came forward to say they had seen Ruby and H. L. Hunt together on more than one occasion, but Hunt himself was apparently never interrogated. Hunt, however, remained nervous about the Warren Commission probe, and directed Paul Rothermel to canvas his FBI and CIA contacts in an effort to monitor its work. In the ensuing months Rothermel's memos were so thorough that, in many cases, Hunt was briefed on the commission's findings before Earl Warren himself.

When issued in September 1964 the commission's report mentioned H.L., Bunker, and Lamar by name, but effectively cleared them of wrongdoing or any substantive ties to either Oswald or Ruby. During the investigation Hunt ignored all threats to his security, keeping his home number listed in the telephone book and appearing on downtown streets without a bodyguard. But in both his public and private statements he underwent a complete reversal on Kennedy, lavishing praise on the dead president and excoriating his killer. "Thinking people," he wrote in one internal memo, "know patriots had nothing to gain by this vile deed." In a newspaper column he titled "The Assassination Must Not Be Forgotten," Hunt wrote: "The assassination of President Kennedy was the greatest blow ever suffered by the cause of freedom. . . . This is part of the legacy left Americans by his Marxist assassin, and being a Marxist he would have wanted it that way."[1]

While clearing the Hunts, the Warren Commission report did little to dissuade other investigators, private and professional, from reexamining possible ties between the family and either Ruby or Oswald. The most serious of these probes was launched by an ambitious district attorney in New Orleans, Jim Garrison, who in early 1967 suddenly opened his own investigation into the assassination, which Garrison claimed had been planned by

Oswald during his time living in New Orleans. Garrison quickly arrested a local businessman named Clay Shaw, naming him as one of several shadowy conspirators behind Oswald. Those conspirators, Garrison remarked more than once, might have included certain unnamed Texas oilmen. Everyone knew whom he meant.

Once again Hunt sent in Paul Rothermel. Rothermel's work tracking Garrison's probe was even more thorough than before. He advised Hunt to avoid visiting New Orleans for fear of arrest, at which point Hunt canceled a meeting he had planned there with Senator Russell Long. At one point Rothermel procured a hand-drawn diagram in which Garrison's people laid out what appeared to be their theory of the assassination. It consisted of a series of circles and boxes connected with dotted lines and arrows. At the very top was the name "H. L. Hunt"; below was the cryptic notation "screened three times by Paul Rothermel." Below it a series of lines led to boxes containing the names of Ruby, Oswald, the Dallas police, and a host of bit players.[2] Hunt, it appeared, was squarely in Garrison's crosshairs. In fact, he had little to fear. When Garrison finally put Clay Shaw on trial in 1969, Hunt's name was never mentioned. Shaw was acquitted of all charges, ending the probe.

If Hunt thought that would squelch further inquiries, he was sorely mistaken. The late 1960s, in fact, began the heyday of Kennedy-assassination conspiracy literature, and many speculated openly of Hunt's involvement. Starting with a book called *Farewell America* in 1969, authors of every political hue speculated how and why Hunt, among others, might have killed the president; the theories, each wilder than the last, persist to this day and have actually multiplied over the years, leading to an entire subgenre of oilmen-killed-Kennedy books and, in recent years, a profusion of Web sites. Many of the latter-day theories attempt to draw in even more oilmen, notably Clint Murchison, never mind that in 1963 the elderly Murchison could barely answer the telephone without a nurse's help. In 2004's *The Radical Right and the Murder of John F. Kennedy*, author Harrison Livingstone argued that Murchison, Hunt, D. H. Byrd, and the Toddie Lee Wynne family conspired with the CIA and the Mafia to kill Kennedy. Other books, including 1991's *The Texas Connection* by Craig I. Zerbel and *Blood, Money & Power* by Barr McClellan, advanced the case that Texas oilmen killed Kennedy to hasten Lyndon Johnson's entry to the White House.

Much as it had when the Big Four dallied with Joe McCarthy, this kind of talk redounded in popular culture. Within months of the Kennedy assassination a new cultural stereotype began to appear, the Evil Texas Oilman. Powered by the new counterculture's suspicion of all things corporate and powerful, this new image proved far darker than anything seen during the 1950s. The 1973 film *Executive Action*, based on a novel co-authored by the conspiracy theorist Mark Lane, imagined that Kennedy was killed by a cabal of steely Dallas oilmen portrayed by Burt Lancaster, Robert Ryan, and Will Geer. This idea has endured through the years, surfacing again in the granddaddy of all Kennedy-assassination movies, Oliver Stone's 1991 *JFK*. In the movie a group of unnamed Texas oilmen speculate openly of the benefits of removing Kennedy.

But the Kennedy assassination was only the beginning. The notion that there existed a cabal of nefarious Texas oilmen plotting a right-wing takeover of America surfaced in radical literature soon after the assassination and has remained a staple of popular culture ever since. Starting with a series of books and films in the mid-1960s, Texas oilmen began appearing as frightening right-wing nut jobs or James Bond–style supervillains intent on taking over the world. One of the first such portrayals came in Stanley Kubrick's 1964 cold war satire, *Doctor Strangelove*, which featured the actor Slim Pickens, sporting a cowboy hat and a thick Texas drawl, as a B-52 commander so eager to bomb "the Russkies" that, at the movie's end, he rides a nuclear bomb, whooping and waving his hat, as it falls toward Russia.

Perhaps the archetypal right-wing Texas villain arrived in a 1967 British film, *Billion Dollar Brain*, based on a Len Deighton novel of the same name. In both, the villain is a Texas oilman named General Midwinter, "a raving right-wing maniac," in one critic's words, "a cardboard caricature of the Texan as an anti-Communist nut."[3] Midwinter wears a string tie and Stetson, runs an organization called Crusade for Freedom—clearly modeled on Facts Forum and LIFE LINE—and, in a delightfully convoluted plotline, schemes to use his vast computer system to conquer the Soviet Union. He meets his fate when his invasion force falls through the ice of a frozen Latvian lake.

H. L. Hunt, alas, lived on, and watched in amazement as his name was dragged into any number of popular conspiracy theories. The 1968 killings of Robert F. Kennedy and Martin Luther King Jr. led to a new round of death threats against him. At one point, someone actually sent a pair of trained dogs to kill the deer at Mount Vernon. Hunt recoiled from the controversies, all but begging people to calm down. In a January 1968 column titled "Less Hate in '68," he wrote, "The freedom fight is a joyous and constructive crusade. . . . It has no place for destructive hate." At one point Hunt even endorsed the liberal Edward Kennedy for president. For the first time he began sitting for press interviews, including a long interview in *Playboy*, trying to explain how he really wasn't a real-life General Midwinter.

If the national press still viewed Hunt as a noteworthy opinion leader, many Texans, schooled on *Alpaca* and LIFE LINE and years of watching Hunt up close, viewed him with a mix of amusement and disgust. As a *Dallas Times-Herald* editorial writer, A. C. Greene, put it, "If he had more flair and imagination, if he wasn't basically such a damned hick, H. L. Hunt could be the most dangerous man in America."

II.

In retrospect the Kennedy assassination served as a harbinger for the dark days that plagued the Big Rich during the 1960s. For many in the state's wealthiest families, the Swinging Sixties constituted a kind of return to Imperial Rome, a time of baccanalia as their industry withered, of cheering for helmeted gladiators while friends and neighbors went bankrupt, of family quarrels that led to courtroom fights, recriminations, scandals and, eventually, killings.

Texas was booming, with northerners continuing to stream into the suburbs ringing Houston and Dallas, but Texas Oil was slowly dying. By the mid-1960s the two forces that had been squeezing the industry for a decade—rising costs and competition from low-priced Middle Eastern oil—were killing the remaining producers. To find new oil—and most drillers now sought natural gas—meant drilling deeper and deeper, which was expensive. By the 1960s, it cost three times more to drill to twelve thousand feet as it did to five thousand; as a result, between 1959 and 1972 oil field production costs rose by nearly two-thirds. Given the cost, deep wells needed to bring in huge production just to break even.

In 1965, for example, Dallas independent Jake Hamon and a group of partners spent $2.75 million and more than eighteen months to complete a discovery well in the Coyanosa Gas Field in West Texas; the development wells alone cost $1.75 million apiece. Small independents couldn't even think of this kind of outlay. For larger independents, it made every deep well a financial risk—one that more and more began to lose. Of thirty-one independent producers listed on the New York Stock Exchange in 1952, fifteen were bought out by 1962; between 1963 and 1965 more than 150 independents were sold or went out of business. It was the same across the country. As one producer told a West Texas professor named Roger Olien: "You can lose your fucking shirt on gas."

It was only a matter of time before the new realities struck down one of the Big Rich. It was Big John Mecom. By the mid-1960s Mecom and his son John Jr. rivaled the Murchisons as the second most visible clan of Texas oilmen after the Hunts. Like Bunker Hunt, Big John had pushed past the borders of Texas and Louisiana to drill for oil in spots as far afield as Honduras and Jordan. With a net worth estimated at between four and five hundred million dollars in 1964, he had branched into real estate, buying Houston's Warwick Hotel, plus hotels in Peru and San Francisco, fish-meal factories, and an assortment of land and ranches. Like Lamar Hunt, John Jr. was using the family fortune to make his mark in sports. His racing team won the Indianapolis 500 in 1966, the same year he was awarded an NFL's franchise, the New Orleans Saints. Of professional football's twenty-four teams, four were now owned by Texas oilmen.

In December 1965 Big John struck his biggest deal to date, agreeing to buy the *Houston Chronicle* and other holdings, including a 30 percent stake in the giant Texas Commerce Bank, for eighty-five million dollars. In short order Mecom's name appeared on the newspaper's masthead as president and publisher; the famed "Mecom blue" carpet was laid in the lobbies of both the *Chronicle* and the bank. But then, barely six months later, Mecom's name suddenly disappeared from the masthead. Though neither the *Chronicle* nor the rival *Houston Post* explained what had transpired, it was apparent the whole deal had been called off.

What happened, it turned out, was that Mecom couldn't scrape together the purchase price. He had actually been turned down for loans everywhere he looked. In desperation he had tried to cobble together a deal to sell the *Chronicle* to an out-of-state buyer—keeping the bank stock and other assets—but the *Chronicle*'s owners balked. It was the first sign that Big John was no longer so big. His empire, in fact, was drowning in debt. For the rest of the 1960s Mecom sold off various real estate and stock holdings, but nothing he did could right his ship.

In 1970 Big John Mecom filed for bankruptcy. To oilmen across the state, it had never been more clear: No one was safe. Not anymore.

III.

More than one account of John Mecom's fall drew parallels to that of the only other major Texas wildcatter to go under, Glenn McCarthy. Oilmen who read the name shuddered. Because McCarthy's life after the Shamrock was every oilman's nightmare. *There but for the grace of God.* . . .

It was not a pretty story. McCarthy had dragged himself back from Bolivia in 1957, bruised, battered, and, if not exactly penniless, no longer a rich man; unable to build a pipeline to transport the natural gas he had discovered, he sold his Bolivian interests to a group of American companies for $1.5 million, much of which he used to repay debts. On his return to Houston he sat sullenly, wearing his trademark dark glasses, for a network television interview with Mike Wallace, in which he spat the name Edna Ferber and talked about suing her.

He could still look for oil, but that cost money, and he no longer had the will to search overseas. His only asset of value was the Shamrock's aging night club, his beloved Cork Club, which was owned by its members. In a stab at reclaiming his glory days, McCarthy removed the club from the Shamrock and reopened it atop a downtown skyscraper. There he became a nightclub impresario, hosting glittering floor shows packed with dancing girls and Las Vegas–style entertainers such as Soupy Sales and Mel Torme. McCarthy's real passion, though, appeared to be bourbon. Every few months the papers carried a new item about a fistfight or car accident. In 1960 McCarthy engaged in a wild brawl with his son-in-law at a charity benefit, the Bill Williams Capon dinner; police had to break it up. Two years later a cabbie sued him after another fistfight, this one triggered by the cabbie's refusal to move when McCarthy needed to park. McCarthy won. The city, meanwhile, sued him for back taxes, leading to litigation that lasted for years.

By 1964 the Cork Club was in decline. Membership, once near six thousand people, had fallen by two-thirds. It was losing money. McCarthy fired the dancing girls.[4] In the coming years the club's reputation grew increasingly sordid. In 1967 the Houston vice squad launched an investigation after a guest claimed McCarthy had hosted an "orgy" at the club, complete with live sex acts atop the tabletops; nothing came of it. By 1971 it was over. Membership was down to almost nothing. McCarthy announced he was closing the club. Houston shrugged. In 1972 McCarthy sold his beloved mansion, the one he and his wife, Faustine, built in 1937, to a developer, who demolished it and put up apartments. Quietly, so quietly no one noticed, the McCarthys moved into a house in the suburb of La Porte. The story of Glenn McCarthy, once Houston's greatest wildcatter, was all but over.

IV.

The Kennedy assassination was the most traumatic blow to the Texas psyche since the Alamo. No one felt the world's condemnation more keenly than the people of Dallas, which found itself portrayed as a metropolis teeming with violent right-wing extremists—the "city of hate," as one writer put it. The deluge of criticism triggered an unprecedented bout of soul-searching. Many felt the city was being treated unfairly. Others, such as University of Texas professor Robert McGee, writing in The Nation, found in Dallas an explosive combination of Texacentric factors—potent local right-wingers, a near-total absence of a Radical Left, the institutionalization of personal violence, and the widespread ownership of guns—that made the president's killing seem almost preordained. "Barring the probability of Mississippi," McGee concluded, "in a doomed and fated way it had to be Texas, and, in Texas, Dallas."[5]

In the mid-1960s the people of Dallas searched for something—anything—to make them feel good about themselves again. They found it in Clint Murchison Jr.'s Dallas Cowboys. City fathers, looking for anything clean and new to burnish the city's image, began promoting the Cowboys at every turn, and though still a so-so team—in 1964 they won five games—the Cotton Bowl began to fill. In 1965, when the charismatic SMU graduate "Dandy" Don Meredith led the team on a seven-game season-ending streak of victories, pushing the Cowboys into the postseason for the first time, attendance leaped 45 percent. A November game against the Cleveland Browns actually sold out. Suddenly the Cowboys were the hottest tickets in Dallas. Thousands of fans began appearing at Love Field to welcome them back from road games.

A psychologist would have smiled; here was the one thing that could draw all of Dallas's dejected citizens together in a carefree, guilt-free celebration of civic pride. In 1966 the Cowboys won ten games and surged to the NFC championship game, losing by a touchdown to the Green Bay Packers in the final seconds when one of Meredith's desperation heaves was intercepted in the end zone. The outpouring of grief matched anything in Texas sports history. "If ever a team attained tremendous status in defeat," one sportswriter remarked, "it was Dallas." Clint Jr., though heartbroken, tossed off one of the quips Texans were beginning to expect from him. "Oh well," he said with a shrug, "we didn't want to give 'em too much too soon."

Dallas's love affair with the Cowboys, however, didn't erase its ambivalence toward the Big Rich, as Murchison discovered when he attempted to leverage the team's popularity into a new stadium. The Cotton Bowl was a wreck, filthy, and without air-conditioned locker rooms, lined with ancient wooden seats that had housewives tweezing splinters from their rears. Worse, it sat in Fair Park amid the city's worst black slums. Cars in the parking lots were routinely vandalized, hubcaps and aerials yanked off, while over the years scores of fans complained of being mugged after night games. A 1967 contest against the Packers was marred by a postgame rampage in which dozens of patrons were held up and robbed outside the stadium, one man was shot, his wife beaten, and another man stabbed. Murchison pleaded with the City Council to build a new facility, but local politicians, acutely aware of how little the Murchisons and Hunts had done for the city, adamantly refused. "To hell with him," one snapped after a meeting with Clint Jr. "What have the Murchisons ever done for Dallas?"

Matters came to a head in 1966, when Clint Jr. demanded to know whether he was trapped in the Cotton Bowl forever. The city's headstrong mayor, Erik Jonsson, a northern-born founder of the new Texas Instruments electronics conglomerate, didn't care much for oil-family heirs who frolicked in gargantuan mansions and private islands but always seemed too busy to donate to civic causes. He proposed renovating the Cotton Bowl. Clint Jr. said it wasn't good enough. Instead, later that year, he snapped up a ninety-acre parcel at the intersection of three expressways in the western suburb of Irving. In January 1967 he announced he planned to build "the finest football stadium to date in the world" if Irving could just pay for it.

Many thought he was bluffing. But the City Council stirred, placing a referendum to raise twenty-nine million dollars for the Cotton Bowl's renovation on a July ballot. Mayor Jonsson summoned Clint Jr. for what amounted to one last plea for the Cowboys to stay in Dallas.

"If you take the Cowboys out of Fair Park," he asked at one point, "what are you suggesting we put there instead?"

Murchison just smirked: "How about an electronics plant?"

It was no bluff. On Christmas Eve, 1967, the city of Irving announced it could sell enough revenue bonds to build a structure Murchison was to name Texas Stadium. It was to be like no other sports venue in America, and Clint Jr. designed it himself, this time without delay. He wanted Cowboy football played outdoors, as it should be, but Texas's broiling sun was hell on spectators. His answer was a roof with a vast hole over the playing field; the football would be run and passed in the heat, rain, and snow, while fans remained sheltered. Because television was increasingly the key to a team's national popularity, Murchison arrayed seating in the most telegenic manner possible, on a steep incline, which had the added benefit of improving a fan's sight lines, long a complaint of Cotton Bowl visitors.

But the invention that would revolutionize stadium design was Clint Jr.'s introduction of luxury suites, known in Texas Stadium as "circle suites" because they circled the field. The 178 suites, each sixteen feet square, were marketed directly to the Big Rich and their corporate peers. "Your personalized penthouse at Texas Stadium," the sales literature read, "similar to a second residence, like a lake home or a ranch." Each concrete cubicle cost fifty thousand dollars plus twelve season tickets a year, a price tag that typically tripled as well-heeled fans added carpet, furniture, wet bars, stereos, and televisions. In time the suites became a massive source of revenue for the Cowboys, one that over the next three decades led to the demolition and construction of scores of stadiums across America as owners in every American team sport began demanding stadiums with luxury suites of their own.

When finished, Texas Stadium cost more than twenty-five million dollars, a quarter of it from Clint Jr.'s pocket. To recoup costs, ticket prices skyrocketed; counting the stadium bonds every season ticket holder was required to buy, a family seeking four decent tickets would pay forty-two hundred dollars, roughly twelve thousand dollars in today's dollars. Fans were horrified; some took to calling Texas Stadium "Millionaires' Meadows." *Esquire* captured the prevailing reaction: "Wanna buy two tickets to the Dallas Cowboys? Struck oil lately?" Despite pleas from longtime fans and newspaper editorialists, Clint Jr. refused to lower prices, and rightly so.

He knew Dallas. After its opening in October 1971, Texas Stadium was a wall-to-wall sellout every Sunday. The circle suites became must-have trophies for scores of wealthy Texans and their wives, many of whom, as national television cameras captured again and again that first year, came to the games in mink and diamonds. Clint Jr. reserved the choicest suite on the fifty-yard-line for himself, where that first game Lyndon and Lady Bird Johnson sat alongside him nibbling that reknowned Texas delicacy, peanut butter and jelly sandwiches. Murchison surveyed the scene with undistinguished glee. Everything about the stadium was perfect, simply perfect. When someone asked why he left the roof open, he answered, "So God can watch His team play."

Luckily for Murchison, the opening of Texas Stadium coincided with the Cowboys' emergence as a top team in the National Football League. Led by reliable stars such as defensive tackle Bob Lilly and a tobacco-chewing fullback named Walt Garrison, Tom Landry's squad had forged standout seasons in each of its last three years in the Cotton Bowl, but all ended in postseason defeat, the best showing a 16-13 loss to the Baltimore Colts in the team's first Super Bowl appearance, in January 1971. Don Meredith had retired, to be replaced by a veteran named Craig Morton, then, at the outset of that 1971 season, the dashing Heisman Trophy winner Roger Staubach. It was Staubach, backed by an angry young running back named Duane Thomas, who led the Cowboys to nine straight victories that year, followed by a postseason rampage that made Dallas the favorites to win Super Bowl VI against the Miami Dolphins in New Orleans. In the climax of a storybook year, the Cowboys overwhelmed the Dolphins 24–3, a victory that began their decadelong run as the NFL's most popular team.

The Cowboys represented more than just football excellence. With their clean-cut quarterback and their pneumatic cheerleaders and their sparkling new stadium and television ratings no other franchise could touch, they became, by wide acclaim, "America's Team," a red, white, and blue antidote not just for Texas's pain but, in the era of Vietnam and drug-addled hippies and Charles Manson, the nation's. Gone, washed away by a glistening river of victories, was the City of Hate, the Texas of General Midwinter and Pappy O'Daniel and Stetsoned right-wing demagogues. In its place was a new-new Texas, brought to you by the man who now emerged as a beloved Lone Star icon, Clint Murchison Jr. What Glenn McCarthy had sought with the first great symbol of the "new" Texas, the Shamrock Hotel, Clint Jr. accomplished with its spiritual reincarnation, the Dallas Cowboys. He had become not only the folksy new face of the Big Rich, he had become a de facto King of all Texans.

Few had any inkling how unsteadily his crown was worn. For while McCarthy had envisioned using his status to build a corporate empire, Clint Jr. had long grown bored with business. What he wanted most—and this was the tragedy of his life—was to have a good time.

V.

Much as his father had done for the first generation of Big Rich, Clint Murchison Jr. set the standard for a new generation of Texas playboys eager to experience the Swinging Sixties. It was a time of transformation, of "letting your hair down," both for America and an entire class of heirs who grew up in the 1950s beneath their fathers' cossetting wings. The classic case was Fort Worth's Cullen Davis, heir to the four-billion-dollar oil-equipment fortune his father Kenneth "Stinky" Davis built over the years thanks to clients like Roy Cullen, for whom Cullen was named. Cullen Davis had been a little-known John Murchison manqué, a stiff young executive in his father's company—that is, until his divorce in 1968. Practically overnight Cullen took up with a fast-living twenty-seven-year-old young divorcée named Priscilla Wilborn, and in no time the couple emerged as Fort Worth's white-trash Nick and Nora Charles, jetting off to weekends in London or Paris or Rio or Miami, wherever the spirit and the credit card moved them.

Fort Worth, whose buttoned-down upper class was symbolized by the proper Perry Bass family, had never seen anything quite like Cullen and Priscilla. From the Ridglea Country Club, where Priscilla played tennis, to the Petroleum Club, it was all Fort Worth's wealthy could talk about: the black-leather suit Priscilla had bought Cullen, the impromptu appetizer of oral sex she had supposedly given him in the middle of a Dallas restaurant. At one point, Priscilla shaved her pubic hair into the shape of a heart and died it pink. Cullen and Priscilla scandalized Fort Worth off and on until 1975, when they separated and a masked man stormed into the Davis mansion and shot Priscilla's new boyfriend dead. Cullen went on trial for the murder; to this day, in fact, he is the wealthiest American ever tried for a capital offense. After a trial that made national headlines, he was acquitted, and spent the rest of his years as a born-again Christian.

In terms of Texas-style hedonism, though, the Cullen Davises of the world were no match for Clint Jr. All through the '60s he piled on the homes and toys, a three-bedroom hideaway in New Orleans' French Quarter; a five-story Manhattan town house once owned by Aristotle Onassis, where his wife, Jane, stayed when shopping; a four-million-dollar Park Avenue penthouse where Clint and his cronies bedded scores of young women; and a Century City penthouse where they continued their romps while in Los Angeles. He flew to Spanish Cay now in a new Gulfstream jet, its interior done in genuine zebra skin. If friends came along, and they always did, there was a separate jet or two for their luggage. At Spanish Cay he added an eighty-five-foot teakwood yacht, *The Morning After*, which Tom Landry and his coaches used for fishing expeditions when they weren't diagramming plays beneath the island's towering palms. Years later, when he tired of it, Clint sold the yacht to Frank Sinatra.

It was a good life, a Texas life. When Clint Jr.'s Jaguar broke down one day driving to Texas Stadium, he wandered into a roadside warehouse, telephoned his office, ordered up a helicopter, and had the pilot fly him the rest of the way. At his New York penthouse he hired a formal butler, Harry Hughes, who cringed when Murchison and his pals hollered for him by his first name. Hughes was a gourmet cook, yet Clint Jr. disdained his offerings, preferring to grill steaks on one of the balconies. In New York he ate lunch every day at '21', eventually persuading the owner to put chili on the menu. When he tired of waiting for someone to open a decent barbecue joint in Manhattan, Murchison did it himself, opening The Dallas Cowboy on Forty-ninth Street; he reluctantly abbreviated the name to The Cowboy after informed that he had broken NFL rules. The restaurant served his favorites, tamale pie, black bean soup, nachos, and chili, all routinely ruined by the Italian chef Clint inexplicably hired. The Cowboy was blessedly replaced a few years later by an outpost of the barbecued-ribs eatery Tony Roma's, which Clint discovered in Miami, bought, and expanded into a nationwide chain.

As principal owner of America's Team, Murchison was now in a position to befriend everyone who was anyone, from President Nixon to Clint Eastwood. What distinguished him from his father's generation was that he chose not to. Not for him J. Edgar Hoover or John McCloy or Howard Hughes. Clint Jr. preferred an earthier class of crony, the kind of man who could match him drink for drink, who told jokes so he didn't have to, who knew how to find women willing to please. Some called them sycophants; his wife, Jane, who longed for the society friends John and Lupe Murchison favored, dismissed them as trash. Bob Thompson of Tecon was still around, always ready to dance on a tabletop or drive a girl home. In Los Angeles a onetime advertising executive named Bill Dunagan became Clint Jr.'s inseparable friend, squiring him to Hollywood parties and introducing him to Dennis Hopper and other stars. In New York Clint Jr.'s running buddy was Spencer "Spinny" Martin, whose jobs included acting as the beard when Murchison took a girl to '21'.

In the backslapping embrace of these and other cronies Clint Jr.'s personality underwent a transformation. While he remained brusque around strangers, he was a different man with his pals, delighting in endless pranks and silliness. When Bob Thompson left Dallas on a business trip, Clint Jr. brought in a Tecon crane to plop a forty-foot yacht in Thompson's swimming pool. When another pal left his station wagon at Love Field, he returned to find inside a snarling black panther, a surprise from Clint Jr. The Dallas sportswriter Blackie Sherrod, having penned a critical column about the Cowboys, was stunned to return home one evening to find a live goat tethered to his staircase, a note attached from Murchison chiding him for "getting my goat." Occasionally his cronies retaliated. New York's mayor Robert Wagner became a friend, and when Clint Jr. passed out after a boozy lunch at "21," Thompson hired an ambulance, slid Murchison into a coffin, and drove him to Gracie Mansion. When Wagner returned later, he found Clint Jr., a black wreath on his head, asleep in his office.[6]

At other times Murchison's vast appetite for liquor manifested itself in a bizarre penchant for acrobatic stunts. At '21' he would flop to the floor and turn somersaults between the tables, popping to his feet to show the other patrons he hadn't spilled the drink in his hand. Aboard the *Flying Ginny* he drew applause, and the ire of his pilots, by performing a tumbling routine during even the steepest takeoffs. One evening when he and Jane were staying overnight at the White House, Lyndon Johnson was startled to find Clint Jr. standing on his head outside the Lincoln Bedroom. "Clint," the president sputtered, "what the hell you doin'?" Clint Jr. replied that he had wanted to stand on his head in the White House since he was a child.[7] And that was how the King of all Texans often came across, as an overgrown child. He loved costume parties, donning a variety of getups over the years that never failed to have his pals in stitches: a hippie for his forty-eighth birthday party, at other times a cowboy, an aviator, or W. C. Fields.

Some of Murchison's most infamous gags were played on other NFL owners. The rival Redskins took pride in their halftime shows, and for a December 1961 game in which Santa Claus was to appear, Clint Jr. and Bob Thompson had the entire field covered with chicken feed the night before, in anticipation of loosing two hundred chickens into the middle of Santa's show. The scheme was foiled when Redskin executives caught wind of it and, in Clint Jr.'s words, "had the chickens arrested." When the owner, George Marshall, complained to Commissioner Pete Rozelle, he found himself bombarded with strange phone calls. Each time he picked up the receiver, all the angry Marshall heard were clucking noises. Marshall exploded, ordering Rozelle to have the calls stopped. They did, but the next time the Cowboys visited Washington, Clint Jr. hired a man in a giant chicken suit to parade around the playing field.

Many in Dallas found such exploits endearing, especially local sportswriters, who carried on a prolonged love affair with Clint Jr., their ardor fueled in part by languid weekends on Spanish Cay. Clint, in turn, envied their easy way with words. He was always a deeper man than some believed; at Spanish Cay he could stare at the ocean for hours or lose himself in an engineering journal or, to the dismay of his hard-drinking pals, the poetry of Shelley or Byron. He became the kind of man who, too shy to speak clever remarks aloud, loved to put them in writing, firing off thousands of short, pithy notes, many only a single sentence. Most were warm and good-hearted, feelings he found difficult to express in person. Others were playful or teasing.

To a Nevada senator who complained of limestone dust emitted by a Murchison factory: "The biggest cow I ever saw was raised on a diet of limestone dust; of course, her teats broke off the day she calved." To a sportswriter who described him as a 130-pound prep-school halfback: "With reference to your recent column on the Dallas Cowboys, you are full of shit. I weigh 142 pounds." To a television-industry executive who pitched him new business while suing another NFL team: "You must be out of your goddamned mind." Clint Jr. and Senator Edward Kennedy carried on a long correspondence, every word of it in Gaelic.

The repartee, the practical jokes, the love for everything Texan—whatever you thought of him, Murchison projected an image of a man in thrall with life. But nothing, not the pranks, the forty-three-thousand-foot stone mansion, the private island, not even the Cowboys, thrilled Clint Jr. like women. He preferred them young, dumb, and beautiful, and by the 1960s he was pouring an enormous amount of energy into an endless litany of one-night stands. Many of the girls were procured by Bob Thompson in Dallas, Bill Dunagan in Los Angeles, and Spinny Martin in New York, but Clint Jr. turned the pursuit of sex into a daily regimen.

The Cowboys flew to road games aboard Dallas-based Braniff Airways, whose stewardesses became one of his obsessions. In the early '60s Clint actually began attending their graduate ceremonies, sitting in a back row eyeing his would-be conquests. "After the ceremony," one pal told the author Jane Wolfe years later, "he'd point out a few girls and say, 'Go get their phone numbers.'" That he owned the Cowboys was usually enough to get him what he wanted. He spent endless hours deciding which stewardesses would work Cowboy charters. The Cleveland Browns owner, Art Modell, once watched a college football game with Clint Jr. and his pal Bedford Wynne, and was impressed that they spent much of the contest scribbling down notes, only to find the names they were writing were not players but Braniff stewardesses.[8]

Murchison's sexual appetite was insatiable. "There was not a stewardess that Clint didn't want to take to bed," one crony told Jane Wolfe. "He was so in love with these girls he stopped flying on his private planes and flew commercial." And it didn't end there. When Clint attended a Miss America Pageant, he had a pal get phone numbers; if he saw a pretty actress on television, he wanted her number; in the years before they became nationally known, he rutted his way through an entire squad or two of Cowboy cheerleaders. Much as Tex Schramm employed scouts to track and analyze football players, Clint Jr. used the Bob Thompsons, Spinny Martins, and Bob Dunagans of the world to track girls. "Mr. Dunagan," a Murchison secretary remembered, "called Mr. Murchison practically every five minutes in Dallas with news of one girl or another." Almost every night and many afternoons, especially when on the road, there was a new girl in Clint Jr.'s bed, and he didn't like being interrupted. The Murchisons were longtime backers of Richard Nixon's, and when Nixon entered the White House in 1968, he and Clint Jr. spoke once a week or more. One mistress told Wolfe of being in bed with Clint Jr. in his Century City penthouse when Nixon telephoned. "Tell him I'm coming," Clint Jr. hollered, then put the president of the free world on hold while he finished his romp.

His wife, Jane, had known of her husband's infidelities for years. As long as he remained discreet, she tolerated them. His brother John, a stalwart family man, deplored Clint's indiscretions. But the more famous he grew, the better the Cowboys played, the less Clint Jr. cared what anyone else thought. Over the years dozens of the couple's friends saw him with other women; one of Jane's girlfriends spotted him with one in an airport, and in turn was spotted by one of Clint Jr.'s cronies, who told him he had been seen. "Why don't you tell her to mind her own business?" Murchison snapped. For years Jane buried her pain in charity work, working with local arts groups, but every gala she threw triggered an argument over whether Clint Jr. would attend. When he did, he couldn't be bothered to speak more than a sentence or two, and then only if prompted. In time his behavior grew so boorish that Jane began attending functions alone, leading to the inevitable question from her sympathetic friends: "Where's Clint?"

By the late 1960s Jane's anger was manifesting itself in epic shopping sprees. A petite, attractive blonde now in her forties, she collected shoes like her husband collected women. She kept a $100,000 revolving account at Neiman Marcus and was known to top it, forcing Clint Jr. to send an accountant downtown with a check, once for $168,000. Jane amassed more than four hundred pairs of shoes, and thought nothing of walking into New York's Bonwit Teller store and plunking down $50,000 to buy a new designer's entire line. Clint Jr., whose daily wardrobe consisted of little more than dozens of identical powder-blue suits—Cowboy colors—and short-sleeved white shirts, couldn't fathom it. He tried throwing out her charge cards, but nothing worked. "Jane," he quipped, "has a black belt in shopping."

Finally not even shopping helped. In the fall of 1972, barely ten months after the Cowboys' Super Bowl victory, a Chicago acquaintance called Jane to complain that Clint Jr. had stolen his girlfriend. It was the final straw. Jane demanded a divorce. "I can understand a few women here and there, Clint," she told him, "but thousands of women, no." Murchison, while surprised, let his wife go without a fight. In their January 1973 settlement he gave Jane a reported ten million dollars in cash and the town house on Sutton Place in Manhattan. For the newly crowned King of all Texans, it was the beginning of troubled times.

VI.

During the 1960s and early 1970s the Murchison brothers symbolized the twin faces of the Big Rich, the flamboyant and the staid, the high-minded and the low-rent. In reality there were far more Johns than Clint Jrs. The Big Rich were now "modest and calm," the writer David Nevin argued in the umpteenth book analyzing the state, 1968's *The Texans*. "They have made outmoded and unfashionable the gaucheries that once marked Texas; the time is over for the importing of 50,000 camellias to decorate the wedding lawn, of Harold Byrd's brassy parties after the Texas-Oklahoma football game every October, of entire trains taken to Hollywood for capers with film figures."

It wasn't entirely true, as Clint Jr.'s behavior illustrated. Much of the mainstream press, in fact, clung to the fading myth of freewheeling millionaires dancing in oil's black rain, but for every Texas playboy there were dozens of oil families who lived tasteful, quiet lives. Few of the Big Rich were better-behaved than Roy Cullen's offspring. Since their beloved "Gampa's" death his three loyal daughters and their husbands had lived quietly in their River Oaks homes, donated generously, and joined the boards of museums and hospitals. The most visible of the son-in-laws, Corbin Robertson, remained active at the University of Houston, which named its football stadium after him. It was all the notice he wanted. "Just leave us alone," Robertson grumped in a rare interview during the 1960s.[9]

The Cullens, in fact, were probably the last Texas family anyone would suspect of spawning a playboy. But they harbored a secret, and to their dismay, it arrived on their doorsteps in 1964. His name was Baron Enrico di Portanova. The elder son of Roy Cullen's "lost" daughter, Lillie, and the Italian baron she had married in Los Angeles in 1932, "Ricky" di Portanova had grown up in Italy after his parents divorced. Suave, tall, and slender, with luxurious black hair and pencil-thin mustache, di Portanova had worked as a jeweler in Rome but mostly lived *la dolce vita*, chasing actresses, driving Maseratis, and spending every cent of the five thousand dollars a month he received from the Cullen estate. For years Roy Cullen had fretted about his aimless life. "Find some legitimate business in which Enrico can become active," he admonished his father in a 1950s-era letter rejecting an increase in Ricky's allowance. "He cannot have his mind too much on playing society regardless of how many aristocratic friends he may have."

Growing up in Rome, never having met any of his aunts, uncles, and cousins in far-off Texas, Ricky di Portanova had only the dimmest understanding of what the Cullen fortune involved. In 1961, with Italian friends suggesting his "allowance" was a pittance, he made his first trip to Houston to find out. He left with nothing more than a condescending lecture from a Quintana accountant. "They treated me like a foreigner, which really bugged me," di Portanova told a writer years later. "I'm a Cullen, too, not some adopted cousin."[10]

In 1964 the thirty-two-year-old baron returned to Houston, this time to live, and determined to get an accounting from the Cullens that detailed how much money they had, what Quintana was worth, and how he could secure his rightful share. With him came his beautiful, Valentino-draped wife, Ljuba, a onetime star on the Yugoslavian national basketball team who enjoyed a late-1950s moment as a starlet in Italian films. The couple, accustomed to life along the Italian Riviera, was appalled by Houston—muggy, oppressively hot, angry freeways lined with grimy shopping plazas. They moved into an exclusive high-rise in River Oaks, Inwood Manor, but nothing overcame the banality of a city where a night out consisted of barbecue and Lone Star beer. Desperate for distraction, Ricky bought Ljuba a monkey. Neighbors at Inwood Manor heard what sounded like someone dribbling a basketball late into the night, accompanied by simian squeals.

The Cullens did not open their arms for the di Portanovas; in fact, after a strained welcoming dinner party, they had little contact. "Mom had a party for them, I was about fourteen at the time," Wilhelmina Robertson's daughter Beth remembers. "We were nice to them, you know, but there wasn't much we had in common." Convinced he deserved a greater share of the Cullen fortune, Ricky bombarded Quintana with letters and calls; most went unreturned. Quintana executives refused the baron any meaningful information about the company or the estate. In time di Portanova opened a downtown office, a war room for the legal assault he foresaw. Before he could act, however, the Cullens voluntarily mailed him a check for $841,425. If the family thought this would dissuade the baron, they were wrong. If anything, it had the opposite effect. Ricky considered the money a bribe, one he was happy to take, and evidence that there was more to come.

Before he could move against the Cullens, the baron needed to set his own affairs in order. A lawyer suggested the first thing to do was amend his mother's will. As matters now stood, once Lillie Cranz Cullen di Portanova died, all her assets would be returned to the Cullens. Ricky and his brother, Ugo, would be penniless. The baron sighed. This, he knew, would not be an easy thing.

Little is known of Lillie di Portanova's life. After her marriage she disappeared from Texas, and from her family's lives. Some in Houston thought she was dead. Roy Cullen had hired a private detective to keep track of her. In fact, since 1955, Lillie had been living off her trust fund in New York City, in the Times Square Motor Hotel, a tidy hostelry in the heart of the city's seediest area. Most days she could be glimpsed trudging the streets in a black overcoat, black hat, and heavy black boots, shopping bags beneath each arm. She purchased coats at Bergdorf Goodman, snipped off the buttons and for some reason replaced them with safety pins. The hotel staff looked after her with care, in part, one suspects, because of the thousand-dollar tips she was prone to hand out. Most of their conversations took place through her locked door. Lillie's diet appeared to consist almost exclusively of Coca-Cola and sweet cream. In 1965, when her son Ricky reestablished contact with his mother after several years, she weighed more than four hundred pounds and had running sores on her legs.

When di Portanova served notice to change his mother's will, the Cullens hired lawyers who indicated they would try to stop him by challenging Lillie's sanity. Legal wrangling stretched on until August 1966, when the two sides agreed that Lillie would be examined by three psychiatrists in a confer-ence room at a New York bank. The doctors found her of sound mind, and immediately after the hearing Lillie signed documents that created trusts for Ricky, and for his younger brother, Ugo, who still lived in Italy. Four months afterward, Lillie died. She left her sons an estate valued at $5.2 million.

Ricky, however, was only warming up. His next task involved his brother. Three years younger than Ricky, Ugo di Portanova had taken after his mother. Morbidly obese, with long stringy hair and a heavy black beard, he lived with his father in Sorrento, where he spent his days lying in bed between attempts at painting. If Ricky could persuade an Italian court to make him Ugo's ward, he would control twice the money he was able to pry from the Cullens. To do that, however, just as his mother had to be declared sane, Ugo had to be declared insane. An Italian judge, accompanied by a court attorney and a psychiatrist, visited Ugo at his villa that October, two months after Lillie's hearing in New York. He found Roy Cullen's grandson barefoot, dressed in a bathrobe, wandering a room piled high with books, boxes filled with trash, record players, cameras, and arts supplies, includ-ing chisels, pliers, and tongs. Ugo announced he was making "a Christ." He went on to discuss his philosophical interests in Hegel before denouncing the Bible as immoral. The judge ruled him insane.

His preparations complete, Ricky launched his long-planned legal assault in early 1967. That February his attorneys, having persuaded a Houston judge to make Ricky and his father Ugo's guardians, asked the Cullens to pay Ugo $120,000 a year; they got it. One month later, the baron's attorneys demanded that Ugo be paid the same $841,425 the baron had received. A week after that, the di Portanova attorneys served notice on the Cullen-appointed trustees of Ugo's estate, demanding a full and itemized account-ing of the Cullen fortune.

His legal broadside fired, Ricky and Ljuba relaxed and began to celebrate. They had been evicted from Inwood Manor—the monkey had apparently taken its toll—but with new money flowing in, they could now live in style. They bought a home on two wooded acres at 8828 Sandringham in Hous-ton's Memorial section, just blocks from George and Barbara Bush. A full staff was hired, including an Italian groom, Franco Necci, to take care of the horses. None of it, however, made Ljuba any happier in Houston—"this hell-hole," she called it—and as time wore on, Ricky spent more time with a new friend, John Blaffer, the rumpled son of first-generation oil money; Blaffer's father, Robert Lee Blaffer, had been one of Humble Oil's founders.

John Blaffer knew oil, and his insights were invaluable to Ricky's cause; if his mother's estate was being charged for drilling fees, Blaffer pointed out, then the estate—and Ricky—should be receiving royalties as well. But Blaffer's true value to the baron was introducing him to Texas-style hedonism. Blaffer had a wife, but he also owned an apartment complex on South Post Oak Road, and had installed his mistress to run it. More than a dozen oilmen, in fact, had girlfriends living there. Sheppard King—he of the Egyptian belly-dancer marriage—had an ex-wife living there, too, Gloria King, who fell into Ricky's circle.

This was the heart of the Swinging Sixties, and the apartments on South Post Oak swung hard. Parties started every afternoon at five and were often still going strong at dawn. A Texas version of the Riviera, the action drew a hard-living, cosmopolitan crowd, and Baron Enrico di Portanova fit right in. When they needed a break, Blaffer took Ricky on his private plane for overnight hunting trips and tequila-fueled weekends in Acapulco. Ricky enjoyed himself so much that he purchased his own Cessna. If a morning dawned with no imminent agenda, he and Blaffer would fly over to Nuevo Laredo and hit the Cadillac Bar for a long, boozy lunch.

Such escapades did little to strengthen the di Portanovas' marriage. After a brief separation and reconciliation in early 1967, Ricky took Ljuba to Monte Carlo that summer, only to discover her in bed with another man. Ljuba stormed out of their hotel onto the Hollywood producer Sam Spiegel's yacht; Spiegel and a group of friends including the actor Kirk Douglas were heading to Capri. From the yacht, Ljuba telephoned Ricky in tears, then put Douglas on the line to try to persuade the baron to take her back. "Kirk, you stay out of it," the baron said. "I don't think we can be together anymore."

Di Portanova returned to Houston that September to file for divorce and await the Cullens' reply to his legal filings. If all went well—and he had won every skirmish to date—he would soon take delivery of a fortune that might run to the tens of millions of dollars. While he waited, the house on Sandringham was lonely, so he began hosting nightly gatherings of the South Post Oak crowd. One attendee was his new mistress, a pretty River Oaks girl named Sandy Hovas, so bountifully endowed that her classmates at Lamar High School had nicknamed her "Buckets." Before long, Ricky and Sandy were an item.

Then, suddenly, matters took a dark turn. Early on the evening of October 28, 1967, Gloria King and a friend named Norma Clark were visiting Ricky to discuss arrangements for an upcoming dinner party. When Gloria left on an errand, Norma Clark remained behind, sitting with Ricky in the living room. Without warning, a gunshot echoed from the kitchen. The baron's Italian groom, Franco Necci, staggered into the living room, bleeding heavily from a chest wound, and collapsed on the floor. A tall man in a brown jacket strode in behind, waving a .45 automatic. "I'm going to kill all you S.O.B.'s!" he shouted.

Di Portanova had just begun to beg for his life when Gloria King suddenly walked in through the front door. The man with the gun motioned her toward di Portanova and Norma Clark, produced a set of handcuffs, and chained the women together, then ordered the baron to open his three wall safes. There was little inside. Frustrated and cursing, the man with the gun took $350 from the baron's wallet and a 6-carat diamond ring from Norma Clark. Then he ran out.

Police surmised this was something more than a robbery gone bad; while the locations of two of the baron's safes were known to friends, the robber had known of a third that was secret. Clearly, the crime had been planned in advance. A police detective named Paul Nix uncovered a snitch who told them the robber was acting under orders to "kill the Italian." Nix felt certain that Franco Necci, who was Italian, had been murdered by mistake, that the intruder had intended to kill the baron. Few said aloud what some in Houston suspected: there was only one family who would want Baron Enrico di Portanova dead.

The Cullens, meanwhile, remained silent, refusing all interviews as they simmered in their River Oaks homes. No one reached out to di Portanova, much less socialized with him. "You don't tend to have much to do with someone," Beth Robertson observed, "when they're suing you."

Rumors of a murder-for-hire were still swirling in mid-November when Detective Nix arrested an ex-convict and heroin addict named Carl Thomas Preston. The baron, Gloria King, and Norma Clark identified him in a lineup as the man who had robbed them. Preston was indicted for Franco Necci's murder and, while awaiting trial, was convicted of heroin possession, for which he drew a life sentence; then as now, Texas drug laws were nothing to trifle with. The baron offered Preston a hundred thousand dollars if he would identify who had arranged the murder. Preston refused, however, saying his life wouldn't be worth a penny if he squealed. With Preston keeping silent, the theories being floated about who, if any one, was behind the killing remained nothing but unproved suspicions.

There matters lay when the baron's divorce suit suddenly heated up. He was already paying Ljuba five thousand dollars a month in temporary alimony and had made a settlement offer—twenty-five thousand dollars in cash, three thousand dollars a month, and a 1966 Mustang—but Ljuba, keenly aware of the fortune Ricky stood to inherit, held out for more. Rather than face a countersuit, the baron skipped the country in April 1968, taking Sandy Hovas to Rome, where they rented a luxurious apartment with two terraces and a seventy-five-foot-long living room they packed with antiques. He sent condolences to his business manager, Edward Condon, and his secretary, Vivian Flynn, when Condon wrapped the baron's beloved Maserati around a tree.

"Received the tragic news about the Maserati," Ricky wrote Condon in October 1968, ". . . that bunch of stupid American peasants raced my beautiful machine over those goddam flat, ugly, unprepared roads. . . . You allow a Texan to drive a Maserati it is like allowing a baboon to play a Stradivarius. . . . Please have Vivian pack my mink lined trench coat as the weather is getting colder. If you would like to wear it Sweetie go right ahead. All of us girls should be draped in mink after 40."

Money was now pouring into di Portanova's accounts, from his mother's estate and from wise investments Condon made on his behalf. Even as his divorce proceedings dragged on, the baron embarked on a shopping spree of epic proportions. He bought a mansion in Acapulco—it had a name, "Arabesque"—another home in Palm Springs, and a farm outside Rome; a new plane, a King Air Beechcraft, a helicopter, and a speedboat; two Maseratis, a Lamborghini, and a Rolls-Royce; plus five racehorses in England and four more in Rome. There were servants at every house, a secretary in Houston, two full-time pilots, and a captain for the speedboat. "I hope that in the near future," he wrote Condon in March 1969, "I [will] be in a position as my father and think of nothing other than the best things in life—sun, sex and spaghetti."

Di Portanova was having so much fun he didn't bother to return to Houston for the murder trial of Carl Thomas Preston. Despite Gloria King's testimony against him, defense attorneys broadly hinted that someone else had shot Franco Necci. In the end, jurors believed Preston's alibi that he had been in Arizona the night of the murder. He was acquitted; the crime has never been solved, nor has any light ever been shed on the mystery of who—if anyone—hired Preston.

Ricky didn't especially care. Life was good—so good, in fact, that he all but gave up his efforts to pry more money from the Cullens, who had done their best to tie up his pleadings in court. The family's attorneys had succeeded in pushing the matter from a Harris County court into a state court. Rather than wait years to jump-start the proceedings there, and clearly dreading the time in Houston it would require, the baron accepted an out-of-court settlement in 1969. The Cullens allowed di Portanova to become a trustee of Ugo's estate, granting him a measure of control, but refused to produce any kind of accounting. The baron, diverted by his racehorses and Maseratis, by a wonderful life of sun, sex, and spaghetti, ran up the white flag.

For the moment.

FIFTEEN

Watergate, Texas-style

I.

The Hunt family, for all its oddities, harbored no playboys or serious inter-family squabbling during the 1960s—that was yet to come. Hunt him-self, who turned seventy-nine in 1968, was still hailed as the world's richest man, although he almost certainly wasn't. That year *Forbes* estimated him to be only one of the six richest Americans, and for the first time placed Bunker in the top ten, pegging his worth at between three and five hundred million dollars. Outside Dallas H.L. "has become symbolic of the lusty Texas tycoon who flashes $1000 bills, drapes his women in mink, and turns in his Cadillacs when they get dirty," the journalist Jack Anderson wrote in 1969. But there was a wide gap between image and reality.

In fact, Hunt lived a spartan life, still driving himself to work six morn-ings a week, his lunch beside him in a brown paper bag. At Mount Vernon in the evenings he fed carrots to his deer and sat on the veranda holding hands with Ruth, often singing their favorite song, "Just Plain Folks." In 1965 he had moved all the family offices into the new First National Bank skyscraper downtown. Hunt Oil took the twenty-ninth floor. Hunt moved his old furni-ture into a corner office; next door was an unoccupied room whose door still bore the nameplate "Hassie Hunt."

By the 1960s Hunt had given up on oil; even as Placid Oil, which remained controlled by the first family's trusts, discovered the massive Black Lake oil field in southern Louisiana—a strike that probably made the first family's six children billionaires for the first time—Hunt Oil found no significant new reserves in the 1960s. Instead Hunt focused much of his energy on writing, politics, and faddish health products, especially the plant derivative aloe vera, which he believed had all-but-mystical healing qualities. None of this did anything to increase his stature around Dallas. A case in point was the interview Hunt gave to a young *Morning News* reporter named Rena Pederson about his diet. At first everything appeared normal. Hunt, with Ruth at his side, sat in Mount Vernon's dining room, chatting amiably as he gobbled down, and sang the praises of, apricots, dates, and pecans between sips of orange juice and bouillon. "I have lots of money," Hunt mused. "So they call me the Billionaire Health Crank. Heh, heh, heh."

Suddenly, without warning, Hunt dropped to the floor and began furiously crawling around the table. He made one lap around, then two, before the startled Pederson asked what he was doing. Hunt replied that he was indulging in his favorite exercise activity, "creeping." "I'm a crank about creeping!" he blurted, still circling the table.

"Don't go too fast," Ruth pleaded.

"Yes, please slow down," Pederson's photographer said. "I want to get your picture." And that was the photo that appeared in the *Morning News*, the "world's richest man" on his hands and knees, creeping around his dining room. "Yahoo!" Hunt yelled as he finished.

What Hunt-the-philosopher lacked in credibility he made up for in literary output. All through the 1960s he continued to fire off letters to newspapers, almost all on tried-and-true ultraconservative themes; noting the rise of the hippie counterculture, he titled one "We CAN Turn Back the Clock." He began writing a syndicated column, where he engaged in running one-way feuds with everyone from William F. Buckley to Senator William Fulbright, handed out political endorsements no one much wanted, and continued churning out new ideas and terms to describe them; he saw himself as a "freedomist," all who opposed him "anti-freedomists," and dubbed eastern elites "Fabians." At their peak Hunt's columns were carried in thirty-six daily newspapers and twenty-two weeklies, mostly in the South. Between 1964 and 1970 he followed *Alpaca* with ten more self-published books, several of them collections of his columns, but also a family history called *Hunt Heritage* and a second stab at the *Alpaca* story, *Alpaca Revisited*. None exactly roiled literary circles. LIFE LINE, meanwhile, kept chugging along. Despite an ongoing feud with several congressmen over its tax-exempt status, it continued broadcasting its mix of religion and right-wing propaganda all through the years of Woodstock and My Lai; in 1969 it was still carried on five hundred radio stations.

Despite the stream of messages to the outside world, as the years wore on Hunt increasingly withdrew into a world of his own, disdaining the advice of his children in favor of two trusted security men, Paul Rothermel and John Currington, whose offices flanked his own. Rothermel's investigation into the various Kennedy-conspiracy probes made him Hunt's one indispensable aide; over the years he and Currington emerged as Hunt's all-purpose fixers, supervising trust disbursements to all three families, running Hunt's food company, HLH Products, even trying to collect old gambling debts. Whether it was his advancing age, the changing political climate, or the torrents of criticism he endured in the wake of Kennedy's death, by the mid-1960s Hunt began to evidence a mounting paranoia. As the decade wore on, Hunt began seeing vague threats everywhere, from hippies, left-wing politicians, even Hunt Oil employees. At one point, when Libya was threatening to nationalize Bunker's oil fields, Hunt had Rothermel infiltrate a group of local Libyan exchange students, worried they might assassinate him. The real threat, however, came from within.

II.

For much of the 1960s the Hunt family's most irksome headache was Hunt's food business. Between its inception in 1960 and 1968 HLH Products posted losses totalling thirty million dollars, and was increasingly a point of contention between H.L. and the children of his first family, a relationship that had never fully recovered from H.L.'s decision to marry Ruth and adopt her children. Bunker and Herbert tried to persuade their father to close HLH any number of times, but he refused to listen. In time things got so bad that Herbert and his father briefly stopped speaking.

Matters finally came to a head in March 1969, when Hunt Oil convened its annual meeting in a twenty-ninth-floor conference room. It appeared HLH would lose another twelve million dollars that year, and with Hunt Oil finding no new oil, its income from existing reserves had fallen to one million dollars a month, barely enough to cover HLH's losses. If something wasn't done, Hunt Oil's chief financial officer, John Goodson, warned the family that day, HLH threatened to bankrupt Hunt Oil. "It is essential," Goodson said, "that we stop these losses immediately."[1]

While Hunt himself appeared unconcerned, Lamar and Herbert were. As Hunt Oil board members, they pushed to form a "Committee of Four" to review HLH's problems. As the committee's point man they chose cousin Tom Hunt, who had been at Hunt Oil since the 1940s and, after H.L.'s rupture with Herbert, was closer to Hunt than his sons. That trust would be crucial; everyone involved knew HLH was a sensitive topic. All Hunt's political activities, including LIFE LINE, ran through it, and the company was overseen by Hunt's top men, Rothermel and Currington. At first, reviewing HLH's financial documents, Tom couldn't see any obvious problem. The unit operated fifteen factories, and all appeared to be running smoothly. Still, he decided to launch an inspection tour, and what he found startled him. Visiting a plant in Oxnard, California, he found nothing. Not a production line. Not a factory at all. Just an empty warehouse. He found one or two more just like it in coming weeks. Even several plants that appeared capable of production were operating below capacity, the victim of obsolete equipment or, as with a plant in Pennsylvania, lack of a sewer line.

Back on the twenty-ninth floor, Tom reviewed his findings with the board and urged the closings of the inoperable factories. But Hunt, assured by Rothermel and Currington that his nephew was overreacting, refused. By that point Tom sensed something darker afoot. As he delved more deeply into HLH's financials, he noticed a series of strange deals in which Rothermel and Currington had taken sales commissions of some sort. There were dozens of transactions involving companies Tom had never heard of. He didn't understand it all, but the more he studied, the more he felt he was staring at a massive embezzlement. Once Herbert and Lamar were briefed, they began asking questions themselves.

What they learned triggered a rare full-family intervention. Late one afternoon in June 1969, the five active first-family children—Bunker, Herbert, Lamar, Margaret, and Caroline—gathered in Lamar's twenty-ninth-floor office. Waiting until 5:30, when most employees had left for the day, they filed into their father's office. Herbert served as spokesman. HLH's losses, he told his father, were now so bad that the only way Hunt Oil could meet payroll was to take a loan. Herbert had talked to First National about it. To his astonishment, the bank had turned him down—the first time in thirty-five years the Hunts had been rejected for a loan. It was humiliating—enough so that Herbert felt the embarrassment might rouse their father. "Obviously the food company is grossly mismanaged," Herbert went on. "But the problem can't be all mismanagement. There has to be theft involved. We have to look into things and bring the situation under control."

Hunt asked who Herbert suspected. "We don't have any concrete proof," Herbert said. "But we think that Rothermel and Currington, being the two closest to the operation, have to be involved in it."

"Well, you'll have to prove it to me," Hunt said. "I owe my life to Paul Rothermel. There is just no way he could be involved."[2]

In the ensuing weeks Tom Hunt burrowed deeper into HLH's finances, but the complexity of the scheme he suspected was simply beyond his ability to unravel. Worse, Rothermel and Currington complained incessantly to H.L., who eventually told Tom to stay out of HLH operations and limit his investigation to paperwork. Tom appealed to Bunker and Herbert, but both were busy with their own affairs. Finally Bunker suggested they hire the Burns detective agency. "Burns is a national outfit," he told Tom. "They can spot-check the plants in a hurry, and maybe get this thing to a head in a hurry, too."[3]

Once Burns was hired, Tom put HLH aside, joining Bunker and Herbert in preparations for Hunt Oil's bid for acreage on Alaska's northern slope. They were part of a five-company consortium that included Getty Oil and Amerada Hess; all five companies were to contribute $250 million toward the bids, but Hunt Oil was so strapped for cash that Herbert and Bunker were obliged to dip into their trusts to come up with the family's share. Then, on the eve of bidding that autumn, Tom phoned Herbert with a startling discovery. He had found an extra line on his telephone. It appeared to be a "jumper line" that fed into another office. Tom was convinced someone was bugging his phone to stay abreast of the HLH investigation.

Then on Friday, November 14, just as Bunker and Herbert were discussing what to do, Rothermel and Currington suddenly handed in letters of resignation. Later, Hunt attorneys would claim that boxes of HLH papers went missing at around the same time. When Tom found out, he scrambled to the twenty-ninth floor. Other than a single computer printout, almost all of HLH's financial documents were simply gone. The following Wednesday an aide who worked closely with Rothermel and Currington, John Brown, telephoned H.L. and said he, too, would resign "if you people up there, and particularly Tom, don't quit interfering." Tom told Hunt to let Brown resign and he did.

The three resignations, coupled with the vanished documents, meant it would be that much harder to ferret out an embezzlement if there was any there in the first place. The Burns agency had been working the case for several months by then, and had yet to find anything concrete. Bunker realized it was time to get serious. He remembered a talk with his old friend Bud Adams, Lamar's partner in the AFL. Adams had talked of uncovering a kickback scheme in his own organization by hiring a private detective to wiretap the suspect's telephones. A call to Adams produced the detective's name, Clyde Wilson. Wilson headed Houston's top private-detective agency.

A week later the president of Wilson's firm, a onetime East Texas deputy sheriff named W. J. Everett, came to Dallas and listened as Bunker outlined the problem in the living room of his Highland Park home. Bunker said he wanted taps on all telephones being used by Rothermel, Currington, John Brown, and three other employees. Everett, a squat, no-nonsense man who wore his hair in a crew cut, suggested there might be an easier way to solve the case. No, Bunker said, he wanted wiretaps. The only way his father would ever believe these men had betrayed him was if he heard it in their own words. Everett nodded. There was just one problem. Bud Adams had used wiretaps when they were still legal. But the Omnibus Crime Act of 1968, passed into law that January, made wiretapping a federal crime punishable by up to four years in prison. By his own admission, Everett didn't tell

Bunker this. What he later claimed to have said was, "Mr. Hunt, if we do this and get caught, we could all be in trouble, criminally and civilly."

According to Everett, Bunker said the family could handle any legal repercussions. According to Bunker, this entire exchange never took place. Whatever the case, W. J. Everett left Bunker's living room promising to talk it over with Clyde Wilson and get back to him. A few days later Everett returned to Dallas with one of his electronics men, and the two cruised the streets near Rothermel's and Currington's suburban homes to get a sense of what the job involved. It would be risky and technically complex—"a little ticklish," he told Bunker—but they would do it. On Bunker's okay, Everett brought a three-man team to Dallas to begin work in late November 1969. They swung by Herbert's house to pick up a two-thousand-dollar down payment for their services.

While a series of detectives rotated through roles on the Hunt job in the coming weeks, the lead wiretappers remained two of the Wilson agency's best men: Patrick McCann, a twentysomething electronics whiz, and his assistant, Jon Kelly. McCann was the brains of the operation. To do the job, he rigged up tiny transmitters that, when clipped onto a phone line, would broadcast to a nearby receiver. Installation was no problem: the detectives simply climbed to the tops of nearby telephone polls and snapped the bugs into place. The tricky part was the Blaupunkt FM receivers. McCann paired them with tape recorders he packed into briefcases painted in camouflage colors. He and Kelly then painted their faces with shoe polish and, working in the dead of night, dropped the suitcases into bushes near each of the homes. "It was just like a movie," Kelly would recall. "We really got into it."[4]

The equipment required daily maintenance, mostly to replace batteries and tapes, so the detectives remained in the area, checking into nearby hotels and moving frequently to avoid suspicion. Everything went smoothly at first. After a week, McCann began sharing the recordings with Herbert. An initial batch turned up little of interest. But in the second week they began hearing terms that sounded promising, something about a "leg breaker," then mention of a "big daddy." Herbert gave the men more money and told them to keep at it.

By the third week, in mid-December, Bunker became involved, and he showed far more enthusiasm than his brother, his ears perking up when the tapes carried a mention of gold bullion hidden in Mexico. The detectives repeated their warnings that this was all very risky. They had already picked up rumors that Rothermel had asked someone in the Houston police department about the Wilson agency. When Bunker asked Jon Kelly what he would do in the event of discovery, Kelly told him he kept a bucket of acid in his car for destroying evidence. "I was lying to him about the bucket of acid," Kelly admitted later, "but it sounded like a good story, and Bunker really seemed to like it."

They broke for Christmas. Afterward, W. J. Everett told Bunker he was reluctant to continue the operation; there were enough rumors flying around police circles that he felt certain Rothermel was on to them. But Bunker insisted. Everett prevailed upon him to narrow the number of wiretaps from six to three, just Rothermel, Curry, and John Brown. Pat McCann, meanwhile, thought it wise to stop hiding briefcases in bushes. Instead, he recommended parking rental cars near the target homes and stowing the briefcases in the floorboards beneath newspapers. They would rotate the cars regularly in hopes of avoiding suspicion.

Once the new setup was in place, in early January 1970, McCann returned to Houston, leaving twenty-five-year-old Jon Kelly in charge with another detective. Kelly constantly shuttled among the three target homes, sliding into the rental cars to retrieve the tapes. Everything was going smoothly enough, but for some reason Kelly found himself growing increasingly worried. At one point, he noticed a white chalk mark on one of the cars' tires, suggesting someone was keeping tabs on its movements. When he checked in with Herbert—he used the code name "Kirkland" on these calls—he recommended backing off awhile. Herbert ignored him, telling him it was probably just a traffic cop.

Then, late on the afternoon of January 16, 1970, his partner dropped Kelly on a sidestreet near Paul Rothermel's home in suburban Richardson to retrieve a rented Mustang they had left behind. After checking for chalk marks, Kelly slid into the driver's seat, started the engine—and saw the police car. It was parked in a driveway behind him. Inside, a patrolman appeared to be lighting a cigarette.

When Kelly pulled away, careful to drive the twenty-five-mile-an-hour speed limit, the patrolman followed. Just around the corner, there was a stop sign. The officer switched on his flashing lights. The patrolman, George Taylor, got out and approached the car. Kelly stepped out to meet him. As would become clear months later, Officer Taylor's curiosity was no accident. One of Rothermel's neighbors had noticed the Hunt team's car-switching routine and called the Richardson police, who had been through enough domestic disputes to guess this was a private-detective surveillance. "We've been watching this car," Officer Taylor told Kelly, "and we would like to know what's going on."

Rather than lie, Kelly said nothing. Officer Taylor reached inside the Mustang, lifted a pile of newspapers on the backseat, and drew out a briefcase. Inside, he found a speaker and a tape machine.

"Have you been doing some wiretapping?" he asked.

Kelly said nothing.

"Are you a private investigator?"

"Yes."

"Are you out working on a divorce case or something?"

"I think I'd better talk to my attorney."[5]

It was then that Officer Taylor escorted Jon Kelly to the Richardson city jail, where he was booked as a suspicious person. The matter might have ended there, but the Richardson police weren't stupid. They knew wiretapping was now a federal crime, and when they listened to the tapes found in Kelly's car, they heard a mention of H. L. Hunt's name—and called the FBI. When Herbert heard the news, he called Bunker in New Zealand, where he was bidding on some offshore acreage. "Well, that doesn't concern us," Bunker remarked.

"Apparently it's more serious than you think," Herbert said. "There might be a violation on our part."

What the Hunt brothers did next would become the focus of a federal investigation that foreshadowed the Watergate scandal three years later: first a wiretapping, now a cover-up. The U.S. government later alleged that Herbert first destroyed a batch of tapes. Then, after Kelly's release from jail, Herbert and a Hunt Oil attorney sat down with Kelly and his attorney and, as the two men quoted Herbert later, told them they had "nothing to worry about," that things had been "taken care of." The Hunts had a friend at the Justice Department. All Kelly had to do, Herbert said, was hire an attorney named Charles Tessmer. The Hunts would pay for everything. Tessmer and their friend at Justice would handle it from there. Kelly agreed. In March 1970, two months after his arrest, he appeared in a Richardson court, where Tessmer succeeded in having the suspicious-persons charge dismissed. The Hunts later admitted paying Tessmer's bill, but have always denied that it was part of a cover-up.

The FBI began an investigation, but months passed with no word of its progress. None of the detectives cooperated, at least not initially. The Hunts, meanwhile, increased the pressure on the Rothermel group. After more than a year of work, a New York detective agency had finally assembled what it felt was a strong case against them, alleging a massive embezzlement scheme. In November 1970, ten months after Kelly's arrest, the Hunts filed suit against Rothermel, Currington, and John Brown. Rothermel fired right back, directing his wife to file suit against Herbert for invasion of privacy. Then Rothermel played his trump card. He had been at H. L. Hunt's side for almost fifteen years by then. He knew the secrets. And one by one, in interviews with reporters for the *Houston Chronicle* and other Texas newspapers, he began parceling out bits of what he knew:

Hassie's mental problems. Frania Tye. The gambling. Stories of tainted HLH food sold in ghetto stores. Rothermel insisted all his "side deals" had been explicitly approved by H.L. himself. The entire wiretapping operation, Rothermel alleged, had nothing to do with any embezzlement scheme. Rather, he charged, it was about H. L. Hunt's inheritance. Rothermel said he had prevailed upon H.L. to alter his will to give Ruth's branch of the family a greater share. The crusade against him, he said, was Bunker and Herbert's revenge. A blizzard of allegations true, false, and somewhere in between, Rothermel's charges achieved exactly what he intended: they forced the Hunts to the peace table. In a May 1971 settlement, everyone involved agreed to drop their lawsuits. Everyone involved agreed never to discuss it publicly again. All the legal documents were sealed.

It was then, just as the Hunts thought the whole ugly episode might melt away, that the FBI struck. On May 12, 1971, a Dallas grand jury indicted Jon

Kelly and Pat McCann on federal wiretapping charges. Both men remained represented by top-flight Dallas attorneys, according to Kelly, paid by the Hunts themselves. Neither man told the grand jury a thing. When they came up for trial that August, neither man testified. A jury found them guilty, and the following month sentenced both to three years in prison. Through it all, neither Kelly nor McCann said a word.

What bought their silence, Kelly later claimed, was a promise from Hunt attorneys that the two would receive $1,250 for every month they spent in prison—provided they kept quiet. There was just one problem: Kelly decided he didn't want to go to prison. If he could cut a deal with prosecutors, he realized he might yet go free. Kelly, in effect, put himself up for auction. Whoever produced the best offer—the Hunts or the government—could have his full cooperation. The Hunts pursued him like a hound, arranging for one of H.L.'s friends, a Houston businessman named E. J. Hudson, to give Kelly a $700-a-month security-guard job while he remained free pending his appeal.

Federal prosecutors saw what was happening. In December 1971 they granted Kelly immunity from further prosecution and prevailed upon the sentencing judge, Robert M. Hill, to order him to testify. Kelly, secure in his Hunt-arranged job and represented by a Hunt-paid attorney, refused. Judge Hill cited him for contempt of court and strongly urged Kelly to get a new attorney. Kelly got the message, firing Charles Tessmer and replacing him with the canny dean of the Houston bar, Percy Foreman. Foreman, best known for representing Martin Luther King's assassin, James Earl Ray, was equally well known in Texas for a keen eye for value, taking jewelry, artwork, cars, houses, and even speedboats if a client didn't have cash. Jon Kelly had none of this. To his delight, Percy Foreman said he didn't care; Kelly could pay what he could afford. After taking a thousand-dollar down payment Kelly borrowed from friends, Foreman phoned Kelly's employer, E. J. Hudson, and said he might need a little more, say, fifty thousand dollars. The Hunts paid. For the moment, Jon Kelly had no idea what game his attorney was playing.

Prosecutors, meanwhile, kept upping the pressure. On January 14, 1972, W. J. Everett was indicted; five days later, Pat McCann cut an immunity deal and handed over a tape of Bunker discussing the wiretaps. Once again Percy Foreman offered to help. He telephoned E. J. Hudson and promised he could "control" both Kelly and Everett in return for another fifty thousand dollars. Hudson relayed the message to Bunker, who agreed to pay. He scrawled out a hundred-thousand-dollar IOU to Hudson on the back of a piece of stationery; Hudson had a cashier's check cut and got it to Foreman. "Purpose in paying Foreman," Hudson wrote in a diary he kept, "is to avoid indictments of Bunker and Herbert. . . . Kelly safe but McCann out of control." Percy Foreman proved as good as his word. When Kelly asked whether it might be best to cut a deal with the government, he told him the Hunts had the means to hire a Mafia assassin. "The government," Foreman told his gullible client, "can't help if you're dead."

Controlling Jon Kelly, however, wouldn't kill the FBI investigation, Bunker realized. Only the most powerful men in Washington could do that. In 1972 Texas Oil had two good friends in the capital, John Connally, now Richard Nixons's Treasury secretary, and the president himself, who was up for reelection that fall. On a memorable evening that April, Bunker and his father sought help from both.

III.

One after the other, the airplanes appeared as distant glints in the late afternoon gloaming, slowly coming into focus as they descended over the Bermuda fields east of San Antonio, then dropping from the sky to skid into roaring stops on the great ranch's airstrip. The Murchison brothers, Clint and John, arrived in their father's aging *Flying Ginny*, a white, propeller-driven fossil next to the other oilmen's sparkling new private jets. Bunker and H. L. Hunt came in a rented Lear, on loan from a hopeful aviation salesman. Finally, just at sunset, came the telltale *whop-whop-whop* of a helicopter bearing the presidential seal.

This was Picosa, John Connally's sprawling ranch outside the town of Floresville. He had bought it years before with the money Sid Richardson paid him, and over the years had expanded it to three thousand acres, a good chunk of it bought from the Murchisons. That evening strings of light glowed in the live oaks around the big house. Heavy pots sagging with white chrysanthemums hung all around the patio. Connally, wearing a western shirt and bolo tie, met Nixon and his wife, Pat, with a chuckwagon, then rolled them up to the great fieldstone house where his Mexican servants waited with the barbecue, great slabs of tenderloin grilled in a hand-dug pit beside ears of roasted corn. The Nixons moved easily through the crowd, the president lingering over handshakes, slapping an oilman or two on the back. Everyone was there. White-haired George Brown of Brown & Root smiled as he shook the president's hand. Perry Bass was there, down from Fort Worth. Even the Cullen family's Corbin Robertson, freed from the headaches of his Italian cousins, had come from Houston. There were bankers and newspaper publishers and lawyers as well—Glenn McCarthy's old counsel, Leon Jaworski, stood to one side—and all had come together in an effort to keep Nixon in the White House another term.

Though on its face little more than a fund-raiser, that warm April evening was freighted with symbolism. Texas Oil's onetime champion, Lyndon Johnson, was nowhere to be seen. The former president was recuperating from a heart attack but wouldn't have felt welcome anyway; his embrace of mainstream Democratic values, as evidenced in his Great Society welfare programs, had long since alienated his oilmen pals. In his place had risen Connally, the man who first made his name ferrying bags of cash from St. Joe's, who had served as governor and now Treasury secretary, a lawyer Texas Oil hired when he was out of government and counted on when he wasn't. As head of Democrats for Nixon, Connally represented the rear guard of Texas wealth that had fled the Democratic Party in the last twenty years to become conservative Republicans; he would soon join them. This was a kind of coming-out for the silver-haired old pol. There were already whispers he might replace the unpopular Spiro Agnew on the Republican ticket, and more than a few that evening hoped he would run himself in 1976.

That night at Picosa constituted a perfect snapshot of a state in transition, the remnants of Old Texas—Hunt, George Brown, the King Ranch's Bob Kleberg, the onetime governor Allan Shivers in an awful green-checked jacket—mixing freely with the New: H. Ross Perot, Erik Jonssen of Texas Instruments, the Murchison brothers fresh off their Super Bowl victory three months before. With the airplanes lining the ranch's runway, it had all the trappings of Edna Ferber's Texas. *All Texas was flying to Jett Rink's party. All Texas, that is, possessed of more than ten million in cash or cattle or cotton or wheat or oil . . .*

Yet there was no rush for tables, no malfunctioning PA systems, no one punched or kicked or stampeded. Everyone was perfectly behaved. As the barbecue simmered, the oilmen and executives in their tailored Brooks Brothers suits plucked up flutes of Moët & Chandon, took their seats around the lawn, and lobbed soft questions to the president, about Vietnam and price controls. Afterward Connally walked Nixon down to a pasture fence, where ranch hands on palominos gave the president a riding demonstration. Nixon was relaxed and ruminative. "This is a big country," he mused, "and it produces big men." He took a long look over the fields. "Now I feel I've seen what Texas is supposed to be."

That night at Picosa was redolent of so many past gatherings of Texas Oil power. Some mentioned the Shamrock. But in truth it had more in common with an evening almost no one remembered, because almost no one knew of it, the Fort Worth barbecue in 1937 where Sid Richardson and Clint Murchison in all likelihood cut their first deal with Franklin Roosevelt. Richard Nixon, who had known the men of Texas Oil since those long-ago evenings sipping bourbon with Clint and Sid around the Del Charro pool, would raise millions from them in coming months, and was far more open to suggestions than Roosevelt. At some point that evening, in fact, and this would never be confirmed, one or two people would say they spied Bunker Hunt murmuring into Nixon's ear. Maybe it didn't happen. Maybe it happened in Washington back channels.

But however it happened, a meeting was arranged between Bunker and an assistant attorney general, Richard Kleindeinst. It took place at the Mississippi ranch of another Hunt ally, Senator James Eastland.* Bunker wanted to cut a deal. He felt he had information the Justice Department could use. The Israeli prime minister, Golda Meir, was due to visit New York, which

---

*It would later be alleged that Senator Eastland accepted a bribe of either fifty or sixty thousand dollars to intercede on Bunker's behalf; a Dallas grand jury, however, declined to indict him. Bunker strongly denied doing anything improper.

had the FBI anxious. Bunker's success in Libya had produced death threats against him, including some, it was said, from the Arab terrorist group Al Fatah. In the course of his normal business, Bunker had built a large file on Fatah operatives. He informed the Justice Department—allegedly via Kleindeinst—that he could furnish a list of Fatah operatives in the United States if the Nixon administration would only drop the wiretap case. Bunker would later claim a deal had been struck, but that Nixon himself had reneged on it as his own wiretapping troubles—the Watergate scandal—grew.

Whatever happened, the federal case against Bunker and Herbert marched on. Soon after the Picosa barbecue, disaster struck. While working his sweetheart job as a night watchman at E. J. Hudson's offices, Jon Kelly found the businessman's diary. Its pages laid bare how Kelly's own lawyer, Percy Foreman, was conspiring with Bunker to keep Kelly "under control." Outraged, Kelly immediately fired Foreman, went to the government and cut a deal for immunity.

On June 24, 1972, he and Pat McCann began telling a Dallas grand jury everything they knew. It took another nine months for federal prosecutors— no doubt mindful of Bunker's ties to the Nixon White House—to complete their work. Finally, on March 2, 1973, a full three years after Kelly's arrest on a quiet suburban street, the grand jury indicted Bunker and Herbert on federal wiretapping charges. If convicted, H. L. Hunt's sons now faced up to four years in prison.

IV.

By the early 1970s, while intrigues spawned by HLH Products drew him more deeply into escalating family dramas, Bunker Hunt was beginning to cut a profile of his own. The 1961 discovery of the Sarir Field in the Libyan desert had made him one of the world's richest men, but it had taken six long years to see any profit. Bunker had endlessly pestered his partner, British Petroleum, to build a pipeline to the field, but BP executives, concerned that release of Sarir's giant reserves would drive down world oil prices, had dragged their heels, not finishing the 320-mile pipeline until December 1965.

A tax squabble with the Libyan government delayed its opening another year, but finally, in January 1967, oil began to flow from the desert.

Bunker's headaches, however, were only beginning. Under BP's supervision, the pipeline pumped barely 100,000 barrels of oil a day, a rate Bunker felt was scarcely a quarter of what was possible. To his exasperation, BP executives continually downplayed the field's potential, at one point telling a trade publication Sarir "would never do more than 150,000 barrels a day." Determined to pry more oil from the Sahara, Bunker leaked figures to the *Oil & Gas Journal* proving that the field's capacity was more than three times BP's public estimates. When BP, which was half-owned by the British government, found itself the target of scathing criticism in the House of Lords, the ploy worked exactly as Bunker hoped. Within months BP nearly tripled production, to more than 300,000 barrels a day. Bunker pressed for more and got it, eventually pushing production to 470,000 barrels a day, more than twice the output of all other Hunt-owned oil fields in the world. By 1970 Bunker was drawing thirty million dollars in annual profits from the desert, and with the foreign-tax credit, every last cent of it was tax-free.

Even so, he led a leisurely life, rolling out of bed in his Highland Park home most mornings around ten. The house, a subdued French colonial set on Turtle Creek, was decorated with English horse drawings. Around eleven Bunker would head downtown for lunch, usually with Herbert at the Petroleum Club. He remained quietly active in right-wing political circles, donating to George Wallace's 1968 presidential campaign as well as John Birch causes. After Sarir, however, Bunker began living large. He started buying land, first ranches in Oklahoma, Texas, and Montana, then millions of acres of cattle stations in the Australian Outback, eventually more than five million acres in all; he populated the land with twenty thousand head of cattle, making him the world's largest breeder of Charolais cattle. In a bet on Alaska's North Slope, he began buying real estate in downtown Anchorage, becoming the city's largest landlord. These, however, were mere investments. Bunker's real passion was thoroughbred horses.

He had dabbled in horseracing since 1955, but after Sarir went on line he took the plunge, purchasing stables in Kentucky, Ireland, France, and New Zealand, hiring top trainers and bidding on horses at auctions around the world. Racing brought out the very best in Bunker Hunt. He knew blood-lines and track statistics the way a Wall Streeter knew stocks, and while a rank outsider who could be socially awkward around racing's blueblooded owners—his rich Texas accent was a source of some ribbing—Bunker was a natural around the stable, chatting easily with jockeys, trainers, and stable boys. What distinguished him as a horse owner was his decision early on to concentrate on European championships, avoiding the Kentucky Derby and other American events altogether. Europe remained the easiest place from which to supervise his Libyan operations, and as the years went by Bunker spent increasing amounts of time in London, Paris, and Zurich.

His first champion, Gazala II, won the French 1,000 Guineas and French Oaks races just as Sarir came on line in 1967. At the same time, Bunker laid the ground for future success by paying a then-record $342,000 for half-interest in a colt called Vaguely Noble, which went on to win four major European races, including the Arc de Triomphe, and was named 1968 European Horse of the Year. Among the colts Vaguely Noble sired was Bunker's greatest horse, Dahlia, which won the 1973 English and French championships, becoming the first filly to top $1 million in winnings. Bunker's stables would keep producing champi-ons for years, including Empery, winner of the 1976 Epsom Derby; Exceller, a 1978 champion; and the filly Trillion, a 1979 European turf champion.

In 1969, however, when Bunker turned forty-three, he was still little-known to anyone inside or outside Dallas. That May, finally ready to take his place on the international stage, he and his wife, Caroline, threw a lav-ish party for five hundred guests at Claridge's hotel in London. The affair featured three bands, including the Woody Herman orchestra, and was fea-tured prominently in the British gossip pages. As it turned out, Bunker's celebration was premature. Barely three months later, on September 1, 1969, a twenty-seven-year-old army colonel led a coup that overthrew King Idris of Libya. His name was Muammar Gadhafi, and his ascension to power spelled trouble for Western oil companies operating in the country. Gadhafi was a bellicose nationalist, a Socialist, and a tad unstable, and within months he began demanding a greater share of oil profits. Over his every word hung the unspoken threat that was every oilman's worst nightmare: nationalization.

Overnight, Bunker found himself enveloped in a maze of international intrigues. His worldview, shaped by John Birch doctrine, had always tilted toward conspiracies, whether involving Communists, Jews, the Trilateral Commission, the Rothschilds, or, a particular bugaboo he shared with his father, the Rockefellers. But it wasn't all in Bunker's head. To his list of imagined enemies Bunker now added the very real threat posed by Libyan security agents and, thanks to Gadhafi's pronouncements of pan-Arab pride, Middle Eastern terrorists of every stripe. He began receiving death threats from Palestinian groups and he took them seriously, hiring security guards and private-detective agencies to watch his back. When he traveled commercial, he began booking multiple reservations on multiple flights, sometimes using false names. Given the turmoil in Libya, it was only a matter of time before the Central Intelligence Agency came calling, asking to plant a spy in his offices, but Bunker blocked the move, he later said, three different times. It was a rejection he would come to rue.

For all his ranting about the evils of Western oil companies, Gadhafi moved against them in slow motion. Expecting the worst, the oil companies began circling the wagons. In January 1971 Sid Richardson's old banker, John McCloy, now working for the Rockefellers, drafted a defense pact called the Libyan Producers Agreement, known as "the safety net." In it, all oil companies working in Libya, including all seven major oil companies, agreed to negotiate any demands from Gadhafi as a united front. If the worst happened and one of their number was nationalized, the signees agreed to help offset the stricken company's losses by supplying oil from other sources. Bunker, though skeptical of anything involving the Rockefellers and New York lawyers, reluctantly signed on. He thought of it as insurance.

Finally, after two years of posturing and threats, Gadhafi struck, seizing the Libyan operations of Bunker's partner, British Petroleum, in December 1971; BP's sin, Gadhafi announced, was the British government's support for Iran in a border dispute with Gadhafi's ally, Iraq. Gadhafi ordered Bunker to take over BP's operations and market its oil. Bunker, hewing to the safety net, refused. Gadhafi retaliated by expelling Bunker's drilling technicians and installing native Libyans in their place. In November 1972 Gadhafi increased the pressure yet again, demanding that all foreign companies turn over 51 percent of their fields to the state. At the Libyans' mercy, Bunker tried to cut a deal, sending an attorney to Tripoli to propose that Gadhafi buy him out in return for two years of oil production. Gadhafi thundered that he wouldn't pay a cent for his own country's oil; a month later, he shut down Bunker's production.

Bunker might yet have saved himself. But unlike the other oil companies, he had never tried to appease Gadhafi. Where Mobil and Occidental spent portions of their profits building schools and hospitals in Libyan cities, Bunker had resolutely refused, keeping all his profits for himself—a position that earned him no friends in Tripoli. When Gadhafi moved against his next target, nationalizing 51 percent of the Italian national oil company, ENK, he repeated his demand to Bunker: Turn over 51 percent, a telegram from Gadhafi ordered, or be nationalized. Bunker ignored the threat.

Instead he hired John Connally, who by then had resigned his position as Richard Nixon's Treasury secretary, and sent him in to negotiate. Connally's status as a confidante to Presidents Kennedy, Johnson, and Nixon was enough to persuade Gadhafi to free up Bunker's wells. The reprieve, however, was only temporary. On May 17, 1973, two months after Bunker and Herbert were indicted on wiretapping charges in Dallas, Connally resigned as Bunker's attorney to become an adviser to the Nixon White House. A week later Bunker's wells were again shut in. Finally, on June 11, with Anwar Sadat of Egypt and Idi Amin of Uganda at his side, Gadhafi held a press conference to announce he was nationalizing Bunker's production. He characterized the move as "a slap on America's cool, arrogant face." Bunker took the call in his office. When he put down the receiver, all he said was "Fuck."[6]*

Gone, in a matter of hours, was a field throwing off $30 million a year in tax-free cash. Counting future production, Bunker put his losses at $4.2 billion. He had no one to blame but himself; if he had only tried to placate Gadhafi, he might have survived. The other Western oil companies did exactly that. After pledging to stand united against Gadhafi, the others handed over half their holdings and continued operating in Libya for years. Bunker was outraged, at Gadhafi, but also at seemingly everyone else;

---

*A measure of Bunker's anonymity was the *New York Times* headline announcing his nationalization: "Bunker Hill Nationalization Will Cause $4 Billion Loss."

the American secretary of state, Henry Kissinger, for refusing to intervene on his behalf; the other oil companies, who he felt had weakened his bargaining position by cutting side deals; and, bizarrely, the Rockefellers, whose secret hand he suspected was behind it all. Bunker's ejection from Libya led to an orgy of litigation that took years to unravel. The centerpiece was the largest antitrust lawsuit in history to that point. Filed in the spring of 1974, Bunker's suit demanded $13 billion in damages from Texaco, Mobil, Shell, Gulf, and nine other international oil companies. BP, in turn, slapped Bunker with a $76 million lawsuit for payments it was due.*

Bunker had lost one of the world's largest oil fields, a field far greater than the one his father had purchased in East Texas. Even so, he remained one of the world's ten richest men, probably wealthier than his father at that point. With a net worth approaching two billion dollars, his only real decision was what to do with all his cash. It was then that Nelson Bunker Hunt made a fateful decision. He began buying silver.

V.

In February 1974 H. L. Hunt turned eighty-five. His health was failing. Chronic back pain forced him into a wheelchair, and his eyes were so clouded, he consented to be driven to his office by a chauffeur. He began to think about dying, and his sins. He believed that upon his death, God would judge him. At a reception for the singer Pat Boone, who Hunt had long regarded as a paragon of American virtue, the entertainer leaned down to hear something Hunt was trying to whisper.

"Pray for me, Pat," Hunt said.[7]

On September 13, 1974, the old man collapsed at his desk. He was rushed to Baylor Hospital. After a week of tests, the doctors gathered everyone from both Dallas branches of the family into a single room to give them the diagnosis: advanced cancer, of the liver. Hunt never left the hospital. As he

---

*Bunker did manage to recoup twenty million dollars in costs from the Libyan government for the oil field equipment he was forced to abandon.

weakened, many of his children came to his bedside, taking turns holding his hand. He lost the ability to talk, bobbing in and out of consciousness. Finally, on November 29, the day after Thanksgiving, his heart gave out, and he died.

News of Hunt's death prompted an outpouring of something approaching affection. For days lines of cars inched by Mount Vernon. All the Hunts were bombarded with flowers and phone calls. The national obituaries tended to be dismissive; the *New York Times* termed Hunt "a militant anti-Communist . . . and ultraconservative." In his column William F. Buckley lamented "the damage Hunt did to the conservative movement" with his "silly books" and "simplistic literature." Hunt, Buckley wrote, gave "capitalism a bad name, not, goodness knows, by frenzies of extravagance, but by his eccentric understanding of public affairs, his yahoo bigotry and his appallingly bad manners."

After years of gleefully pillorying Hunt, the Texas press turned winsome at his death. "He was many men in one, multitudinous and contradictory," a *Texas Monthly* writer judged. "Good and bad, but on a larger scale, right out of Ayn Rand. In an age of midgets and conformists, he was a rogue who broke rules and cut a large swath and then, at last, lay down with a smile and allowed the ubiquitous and unctuous preachers to make him a monument to nobility." This and similar eulogies were an early sign of a developing Texas nostalgia, a harking to the days when giants walked the oil fields, when men like Hunt and Clint Murchison and Sid Richardson and Roy Cullen helped build something unique in midcentury Texas—an image and culture loud, boisterous, money-hungry and a bit silly to condescending northerners, but proud and independent to many Texans. With H. L. Hunt's passing the last of the greatest Lone Star wildcatters was finally gone, and with him, one sensed, a foundation of Texas culture was eroding.

On December 2, 1974, more than eighteen hundred people filed by Hunt's open casket at First Baptist Church. Each of Hunt's remaining thirteen children were there. Heads turned as fifty-six-year-old Hassie Hunt, the spitting image of his father, paused before the casket and cried. The hymns and eulogies seemed to go on forever, until they slid the coffin into a limousine, then drove it to Hillcrest Memorial Park, where it was lowered into a hole in the dirt beside Hunt's first wife, Lyda. It was then the troubles began.

SIXTEEN

The Last Boom

I.

By 1973 Texas Oil remained mired in its second decade of doldrums, an afterthought in a world of petroleum now ruled by the Ay-rabs. Its unlikely savior, in fact the last man on earth anyone in Dallas or Houston expected to come to the rescue, turned out to be Bunker Hunt's nemesis, the Libyan dictator Muammar Gadhafi. In October 1973, when Syria and Egypt attacked Israel in what came to be known as the Yom Kippur War, the United States and Netherlands supported Israel. The militant Gadhafi implored the organization of oil-producing Arab states, OPEC, to boycott both countries, and it did. Within six weeks, the price of Arab oil rose from seventeen cents a barrel to $5.40.

The Arab oil embargo was a nightmare for ordinary Americans. Prices on every conceivable oil-dependent product, from airplane flights to plastic bags, skyrocketed. In a matter of weeks gasoline, a product people thought as available as oxygen, was declared in short supply; millions of Americans simmered in around-the-block lines of Fords and Chevys and Mustangs waiting to fill their tanks. For Texas, however, the embargo represented a second coming. Suddenly all anyone was talking about was finding more American oil. Just as happened with the shortages after both world wars, the Arab oil embargo triggered a massive drilling boom across the country, and in Texas.

In West Texas wildcat activity leaped 22 percent in 1974 alone. Across the state, well completions rose by a third. All the majors poured money into exploration, spending three times more in 1976 than just two years before.

Competition drove land-leasing prices into orbit; the University of Texas, which owned thousands of acres in West Texas, reported its bids rose 900 percent. Higher prices for oil suddenly made offshore exploration far more economical; all across the shallow waters of the Gulf of Mexico, drilling platforms rose from the waves. All through the mid-seventies, would-be oil finders poured into Texas. Some, like a young Harvard Business School graduate named George W. Bush, were sons of Texas oilmen who never expected to follow in their father's footsteps; Bush formed a wildcatting outfit named Arbusto and began sinking holes around his native Midland.

In Washington, politicians browbeat the executives of Exxon, Mobil, and other major oil companies, demanding to know why their profits were soaring while Americans waited in gas lines; Congressman Henry "Scoop" Jackson coined the phrase of the day—"obscene profits." But for once, no one got angry at Texas Oil. Everyone understood it was the majors who sold gasoline; Texas oilmen had been down so long they were now considered "the little guy." Even when the Murchisons and a handful of other oilmen wandered into the margins of the Watergate scandal—Pennzoil's Hugh Liedtke was caught flying a planeload of cash to Washington for Nixon's reelection—no one blamed Texas Oil.

It was as if the Big Rich of Texas had come full circle, back to the honeymoon days of 1948 to 1953, when they were viewed as harmless, nouveau riche eccentrics. As the 1970s wore on, in fact, Texas oilmen would increasingly be seen not as dangerous but as entertaining, a function not just of the Hunts' various soap operas but of a new television drama loosely based on the family, *Dallas*. The show, billed as the story of "dramatic feuds in the land of the big rich," became the most watched series in America. As far as the outside world was concerned, the new face of Texas Oil was J. R. Ewing.

For the state's actual oilmen, the party that began with the 1973 embargo ran for five solid years. Drilling boomed. Profits mushroomed. Then, in 1978, things got even better. The fall of the shah of Iran led to dramatic disruptions in Arabian oil exports, driving the price of oil into the stratosphere, to an average of $12.64 a barrel in 1979, then $21.59, then $30, then, finally, amazingly, to $34 a barrel in 1980. In just seven short years prices rose 2,000 percent.

The world had never seen anything like this—prices and demand and drilling and profits, all at historic, unpredecented highs. And neither had Texas. Suddenly, everyone wanted into the oil game. Geologists fled the majors to become wildcatters. Doctors and dentists pored money into discovery wells. In Houston, Dallas, and Midland new skyscrapers grew like grass. It was the '50s all over again.

For the first time in twenty years, new millionaire oilmen began popping up across the state. There was Clayton Williams, the Midland oilman who flew Texas A&M flags over his rigs and hosted black-tie cattle auctions. And Sybil Harrington, the Amarillo heir who became the Metropolitan Opera's preeminent backer, bankrolling more than a dozen major performances at Lincoln Center. Everywhere one listened, there were echoes of the golden age. Out in Abilene, an oilman named "Cadillac" Jack Grimm—a moniker bestowed by his friend Bunker Hunt—followed in Tom Slick's footsteps by funding expeditions in search of Bigfoot and the Loch Ness Monster. A garrulous type who placed second in the 1976 World Series of Poker, Grimm made world headlines mounting three expeditions in search of the *Titanic*; his 1981 effort, using deep-sea submersibles, returned with a photo Grimm swore showed the sunken ship's propeller. (It didn't.) Another expedition, this one to find evidence of Noah's Ark on Turkey's Mount Ararat, returned with a sliver of wood Grimm swore was part of the ark. (It wasn't.) Still later he tried, in vain, to find Atlantis.

II.

Texas Oil was back, but the days had long passed when any of the Big Four families depended on oil to put food on the table. By the early 1970s the Hunts and especially the Murchisons were thoroughly diversified. John and Clint had vacated their father's old 1201 Main headquarters in 1965, taking the twenty-third floor of the new fifty-story First National Bank skyscraper, six floors below Hunt Oil. For the first time members of the two families could ride elevators together and exchange daily pleasantries. The Dallas

Petroleum Club, where Herbert Hunt served as president in 1968, took the building's top floor.

Few people in Dallas knew what the Murchisons were up to. In the years following the Alleghany debacle, the Murchison brothers all but disappeared from public view. While Clint's Cowboys were famous around the world, the operator at Murchison Brothers answered calls with a polite, "Seven four one six oh three one." The fact was, other than the Cowboys, there wasn't much to talk about. For all the chatter around Dallas of how brilliant John and Clint were, their assets hadn't changed much over the years: life insurance, construction, and real estate still formed the empire's foundation, as they had for years.

Clint's 1972 divorce signaled an era of unsettling changes for the Murchison family. He had built a shiny new headquarters for the Cowboys but, as with all his investments, he kept his hands out of day-to-day management, allowing Tex Schramm, personnel chief Gil Brandt, and coach Tom Landry to work their magic, and they did; Clint's smiling face, usually displayed alongside a trophy or a sweaty Bob Lilly or Roger Staubach, became a regular feature of Dallas sports pages. What readers didn't know was that as the Cowboys marched to championship after championship, their owner's personal life was spiraling out of control. Clint thought nothing of downing a half dozen vodkas-and-crushed-ice in a sitting, and in the mid-1970s, egged on by the fast crowd he saw in California, he began using recreational drugs. What began with an occasional marijuana joint turned into a keen appetite for cocaine. "It got to the point where he needed drugs in order to perform in bed," one of his mistresses recalled. "Without cocaine, sex was impossible for him."[1]

Clint knew he was in trouble, and told more than one friend he needed to reduce his drinking. He began cutting back on women as well—affairs weren't as exciting without someone to cheat on, he found—focusing his attention on a handful of mistresses; it was during this period that he bought the one in Los Angeles a Jaguar and a home in Beverly Hills. Then, in 1974, came the promise of salvation. Her name was Anne Brandt, and she was the freshly divorced wife of the Cowboys' director of player personnel, Gil Brandt. Smart and focused, a petite brunette who kept her hair short, Anne

had lived a hard life. Raised in Oklahoma, the daughter of a traveling salesman and his alcoholic wife, she ran away at sixteen and by twenty-one had
two children and two ex-husbands. She made a career as a legal secretary,
married and divorced Brandt, and was still only thirty-four when she began
seeing Clint.

Eyebrows had barely had time to raise when the couple was suddenly
married, in a small wedding at the Murchison mansion in June 1975; Anne
had been divorced less than a year. When Dallas society finally digested
what had happened, the new Mrs. Murchison was met with a resounding
thumbs-down. The ladies of Dallas remained loyal to Jane, who had remarried one of her decorators and settled in New York, judging Anne far too
coarse for their tastes. Anne, it became clear, was no Jane. Among other
things, she could be intensely jealous; anything that took Clint from her
side, from his mistresses to the Cowboys, she disdained. Some of his oldest
cronies now found themselves frozen out of Clint's life. Worse, Anne had a
violent streak, and when Clint did manage to sneak away for a weekend, the
consequences could be hair-raising. Anne fought, struck, and even bit Clint
during screaming tirades that sometimes went on for days.

Her behavior, Anne knew, was rooted in a deep-seated insecurity. At
every party or Cowboys function they attended, she spent hours worrying
over what to wear. She sank into depression, and when a girlfriend suggested
she come to a meeting of the Dallas Christian Women's Club, Anne went
along, and that was it. Not a year into her new marriage, she pledged to
become a born-again Christian, plunging into nine months of intensive
therapy from which she emerged determined to devote her life—and her
marriage—to God. Out the mansion door went the bottles of vodka and the
vials of cocaine. Banished, at least from the Forest Lane mansion, were her
husband's oldest friends. While Clint met the onslaught of religious fervor
with sighs and rolled eyes, Anne dived headlong into a new career speaking
to religious groups all across Dallas, where she implored churchgoers to give
themselves over to Christ. "I let Jesus handle even the littlest things," she
told *People* magazine. "Like I pray for parking spaces."

The turmoil in his private life did nothing to concentrate Clint's focus
on business at Murchison Brothers. As the 1970s wore on, he could get

interested in little beyond the most complex—and often riskiest—real estate deals. After a half century of successful family investments, he had no real sense that anything could go seriously wrong. Where his brother John remained the careful, studied keel of the Murchison Brothers ship, Clint sank millions into deals on handshakes, on napkins, at urinals, risking vast amounts on investments he seldom took time to study. Worse, just as he had peopled his private life with hard-living cronies, his business life was increasingly crowded with fast-talking sycophants who promised him the moon, which was all Clint wanted to hear. A solid 8 or 10 percent return bored him. By the mid-1970s, he simply couldn't be bothered with any investment that didn't promise tripling his return or more.

There was the ten million dollars he threw away on an Oklahoma plant that was to convert cattle manure into natural gas. Clint named it the Calorific Reclamation Anaerobic Process, CRAP for short. It never worked. He poured more than fifty million dollars into a computer company called Optimum Systems that had expanded into health care software after beginning its life as a program to rank and analyze college football players for the Cowboys. Murchison thought he could build it into an integrated software-management company to rival Ross Perot's crosstown colossus, EDS. He never would. When Tex Schramm bought a marina in Key West, Clint agreed to invest twenty-one million dollars to turn it into a five-star resort, six hundred condominiums around a lush golf course. After Clint, who like his father was chronically short on cash, missed an interest payment, construction stalled, and lenders eventually foreclosed.

What real estate developments did get finished were marred by shoddy workmanship, the kind of detail Clint had long since fobbed off on subordinates. A case in point was the Hillandale project in the Georgetown section of Washington, D.C., the brainchild of a new pal named Lou Farris Jr. Barely thirty of the planned two hundred homes were built; the first residents warned everyone within earshot that the roofs leaked and the doors jammed. Still, Farris and other cronies kept plying Murchison Brothers with vouchers for more and ever larger expenditures; when accountants questioned what they were for, Clint just scrawled out the word "Pay." "Clint Junior," the family's longtime attorney, Henry Gilchrist, once observed, "just threw money at people he perceived to be his friends. He was doing this out of sight of the family and family advisers and there was no way of stopping him."

John Murchison was not amused. John had become everything his brother was not, mature, worldly, personable. He and Lupe were now respected members of Dallas society, hosts of the city's most elegant parties; John's thoughtful insights, so unlike Clint's, were sought by anyone weighing major new projects, especially those in the world of arts, museums, and civic boards. Uppercrust Dallas had been frankly dismayed in the early 1960s when John began uncrating the Miros and Rauchenbergs he and Lupe bought in New York and hung throughout the Big House. But as Dallas matured, John's taste in art came to be seen not as bizarre but cutting edge, and he emerged as a driving force behind the Dallas Museum of Art, serving as chairman and then president from 1972 to 1978. "Everyone deferred to him because he had superb judgment," one Dallas civic leader recalled. "He would say something at a park board meeting and suddenly everyone would say, 'Yes, that's the answer. Why didn't I think of that? It was so simple.'"[2]

By the early 1970s John's judgment was telling him to steer clear of his brother's investments. For the first time John began venturing out on his own. North Dallas and its surrounding suburbs were booming, and in 1972 he and another developer set aside two hundred acres of pasturelands Big Clint had used for grazing cattle and began work on a $250 million residential, office, and retail project they called Bent Tree; it was the first major project John had attempted without Clint, and it was a resounding success. Yet in his own way John was no more enamored of day-to-day work at Murchison Brothers than his brother. Their companies ran themselves, and as the years wore on he and Lupe began traveling more and more. Summers they spent in La Jolla, autumns at a new plantation they bought in Georgia, while winter always found them in Vail. In between they spent long periods at a sparkling new home they had built on the harbor in Sydney, Australia, jetting off to Paris, fishing in Iceland, and photographing elephants in Kenya. Their four children were usually left behind, in the care of governesses.

Whether in Texas, Vail, or Sydney, they entertained lavishly, parties that drew the crème of not just Texas but the nation. President Ford skied with

John and Lupe in Vail, while Texas senator Lloyd Bentsen and other politicians were regulars at Big House affairs. Their gatherings, like a Gladoaks pigeon hunt they hosted in the mid-1970s, fused East Coast sophistication with Texas country life. Private jets disgorged hundreds of guests for the Gladoaks shoot, men with antique English shotguns, wives in rustic fashionables, all assembling beneath a dozen giant tents for a catered luncheon of fried bass, string beans, corn bread, and peach cobbler. Afterward the men struck off into the brush to blast pigeons released for their aim, returning to the tents at dusk for an outdoor banquet that lasted to the wee hours.

John might have overlooked Clint's string of bad investments were it not for the debt Clint was piling onto the Murchison Brothers balance sheet. His brother, raised on their father's ability to get rich using other people's money, never met a loan he wouldn't take. In fact, Clint chased new loans with much the same fervor he chased women, always believing he could engineer a return higher than the interest rate. "Borrowing money was a game and a challenge to Clint," one of his attorneys recalled. "He was always trying to see how much he could borrow. Most businessmen would get an idea for a deal and then go out and get the financing for it. Clint did the opposite. He'd pledge a Murchison Brothers asset, get a twenty-million-dollar loan for it, and then he'd look for a place to put the twenty million. Most people thought it was ludicrous, but it worked for him for years."[3]

When the lending game changed during the 1970s, Clint didn't. Where Big Clint had built his fortune on interest rates of 2 or 3 percent, soaring oil prices in the '70s brought soaring inflation, and with it soaring interest rates—by 1977 as high as 18 or sometimes 20 percent a year. Higher rates threatened Murchison Brothers from every angle. Not only did loans cost more to pay off, higher interest rates meant higher mortgage rates, so high many Americans couldn't afford a new home. The resulting slowdown in the housing market struck Murchison Brothers a hard blow; several of Clint's biggest projects, including the ones in Washington and Key West, stalled. His cash flow slowed. Yet the banks still had to be paid. Neither John nor Clint, schooled on Big Clint's lectures to spread money like manure, kept much cash on hand. By 1977 their aides were scrambling almost every month to scare up enough to pay the banks.

John's lawyers repeatedly warned that in a nightmare scenario, Clint could drag them both under; all Clint's deals involved both brothers' money. John was away from the office so much that his brother had long ago given up seeking his approval for new investments, often the deals that were now facing trouble. John warned Clint to stop it. Clint promised he would, then didn't, then like a child caught in a lie would promise that his next deal would make them whole. As a Murchison Brothers executive recalled, "Clint would say, 'John, just give me a little more time. We're going to have a Hail Mary pass. If this next deal works, it will clean up all the bad deals of the whole year.'"⁴

By 1977 Clint's assurances were ringing increasingly hollow. Too many Murchison Brothers investments were going south. Wall Street was stuck in the fourth year of a dreadful bear market, and the handful of public companies John and Clint controlled had been eviscerated. They had taken one of Tecon's subsidiaries, Centex Construction, public in 1969, and the stock had soared. Amid the wreckage on Wall Street its value had plummeted from $146 million to $22 million. The stress, as it had during Alleghany, wore on John; he took up smoking again. At night in the Big House, Lupe watched him stare out the window, lost in thought. Nothing he could say or do would change his brother, John knew that. It was too late anyway. The damage had been done. His chief counsel begged John to give him a single reason he wouldn't dissolve the partnership with Clint. "Well," John said, "he *is* my brother."

More than once John warned Clint he would break up Murchison Brothers. Clint begged him not to. But their investment styles had diverged so thoroughly, they were no longer able to agree on much of anything. The final straw appears to have been John's refusal to plow another cent into Clint's computer company, Optimum Systems. Not long after Clint refused to abandon it, John demanded out. The lawyers began drafting what they called a dissolution agreement. Completed in mid-1978, it called for Clint to take sole control of their riskiest investments, removing their debt from John's balance sheet. It began the extraordinarily complex unraveling of Murchison Brothers, hundreds of investments that each had to go to either John or Clint. The lawyers guessed the process might take three years. In the final agreement both brothers promised to have the partnership dissolved by October 1981.

Afterward they shook hands and promised nothing would change between them. Yet John was wracked with guilt. Big Clint had formed Murchison Brothers for his sons all the way back in 1942, and John and Clint Jr. had operated it as equal partners their entire business lives. John felt he had destroyed his father's greatest creation, his legacy, and he worried that Clint, left to his own devices, would implode. Clint looked forward to the freedom of making his own decisions, but one look at his financial statements told him they wouldn't be much fun. His capital pool had been halved. Of the assets he would take—Optimum Systems, Tony Roma, Tecon—much of it was spiderwebbed with entangling bank liens. He could still wheel and deal, but the game was no longer about fun. It was about survival.

The breakup of Murchison Brothers should have eased the family crisis. Instead it provoked another. It arose from, of all places, the third generation, John's and Clint's children, who had grown tired of waiting for their inheritances. Their money was held in trusts. Beginning in 1949, Big Clint had created three funds for seven of his eight grandchildren. John and Clint, as the executors, had stuffed the trusts with stock, 80 percent of it in their Centex construction company, worth $150 million before the stock's collapse and still valued in the tens of millions. The trusts were to begin throwing off cash to each of the grandchildren as they turned twenty-five. But Clint's oldest son, Clint III, hadn't gotten anything when he turned twenty-five in 1971, and neither had John's oldest, John Dabney Jr. Six of the seven grandchildren were now twenty-five or older, and not one had seen a cent from their trusts.

The problem, it turned out, was that Murchison Brothers had been using the trusts as a piggy bank. Any time John or Clint needed a loan, he dipped into the trusts for Centex stock to use as collateral—an ethically questionable practice, and perhaps illegal. When the firm's attorneys had realized what was happening—a review was performed when Clint III reached twenty-five—they sought to cover the Murchison brothers by drafting a legal paper all the grandchildren were obliged to sign. Broadly speaking, the document said what John and Clint had done was okay with them. With the matter cleared up, at least in their own minds, John and Clint kept using stock in the trusts as collateral for loans all through the 1970s.

If John and Lupe had been more attentive parents, the question of the trusts might have worked itself out. But they weren't. For twenty years they had spent far more time traveling the world than raising their four children. The children resented it—deeply. "John had a very Victorian view of his children," a friend recalled. "He thought of them as the absolute and complete responsibility of governesses and he did not want his life disturbed in any way by them."[5] No one resented John more than his eldest son, John Jr., a tightly strung young man. Like his siblings, John wasn't even sure his father loved him; he never said so. "When my parents were home, which was seldom, they never wanted to be alone with us," he recalled. "They always had to have a lot of friends and famous people around, people who would entertain them constantly."

John Jr.'s simmering anger grew when, after graduating from the University of Colorado, he returned to work at Murchison Brothers in 1975. If he thought this would draw him closer to his father, he was mistaken. "Basically John put his son in a room and said, 'Here, look over these files and see what businesses we're involved in,'" recalled a Murchison Brothers executive. "That was about all the business instruction John Junior got." While his father lived at the Big House and toured the world, John Jr. was forced to make do on a salary of only eighteen thousand dollars. He talked it over with his cousin, Clint III, and in 1977 the two simultaneously sat their fathers down, politely pointed out that they were misusing the trusts, and asked for their inheritances. When John Jr. finished his presentation, John simply rose from his chair and walked out of the room. He abhorred family conflict, and couldn't understand why his children couldn't do as he and Clint had done, and follow their parents' wishes. Clint had a more visceral reaction; he thought the boys were ingrates and said so, loudly. For the moment, John Jr. and Clint III were powerless to change things.

All this—the faltering businesses, the mushrooming debt, the breakup of Murchison Brothers, the challenge from the children—was still unresolved on the evening of June 14, 1979, when John, who had returned to Dallas just that day from a week with Lupe at the Paris Air Show, suffered an asthma attack while delivering a speech during a Boy Scouts fund-raiser at the home of Texas governor William Clements. Coughing and choking, he stepped away from the podium. As he began gasping for air, the governor guided him into a state patrolman's car and asked the officer to take him to the emergency room. When the attack worsened en route, John climbed into the backseat to lie down—just as the officer ran a red light and was struck broadside by an onrushing car. John suffered a heart attack and lost consciousness. An ambulance rushed him to a nearby hospital. An hour later he was dead.

He was fifty-seven. In his last months John had been studying to convert to Catholicism, and after being baptized on his deathbed, he was buried in a Catholic ceremony beside his father in Athens. Every single obituary referred to him as the brother of the man who owned the Dallas Cowboys; not one mentioned what few in Dallas knew, that John owned exactly as much of the team as Clint. No one wrote, because no one really understood, that John had been the cement holding together the family empire, and that without him the Murchisons were doomed.

III.

If the Murchisons were Exhibit A in any discussion of how *not* to diversify, a Texas businessman didn't have to look far to find a shining example of diversification done correctly. It loomed just to the west, in Fort Worth, where Sid Richardson's nephew Perry Bass and his sons had begun transforming their smallish fortune into something special. The 1960s had been grim for the Basses. Perry, who lived quietly with his wife, Nancy Lee, in the suburb of Westover Hills, had left the reading of his uncle's will with a 25 percent share of all the oil fields they had found since 1939, including the massive Cox Bay and Pointe a la Hache fields south of New Orleans. An internal financial statement Perry generated in 1968 put his net worth at twenty-five million dollars, less than 5 percent that of the Murchison brothers. Perry's four sons had shares of the fields as well, valued in total around twenty-five million dollars, making the Basses worth maybe fifty million dollars—barely half "a unit," as wealthy Texans termed a hundred million dollars. They were certainly wealthy, but by Texas standards they were no longer Big Rich.

By the 1960s, while Perry Bass had lost interest in oil, he had found something of himself. Coming of age in Sid Richardson's shadow, he had grown up a quiet, serious, inarticulate man, but after Richardson's death he began to relax and open up. His sons, returning from Andover and Yale, found him a new man, one who was now pleasant, comfortable with small talk, and had even begun giving speeches to local charity groups. "I had no idea how important it was to be glib," Perry remarked over dinner one night on St. Joe's, which had become the Bass family's private retreat. Perry and Nancy Lee enjoyed riding out every major storm in the seaside home Perry had built so many years before, including Hurricane Carla in 1961; the only serious damage Uncle Sid's house ever sustained, he was proud to say, was the time high winds blew a seagull through one of the windows.

Perry discovered no significant oil reserves during the 1960s, but according to family associates, he laid the groundwork for what came later with a single deal. All of Sid Richardson's oil reserves—the Keystone Field, sundry West Texas wells, plus the big Louisiana fields—now rested in the control of the Sid Richardson Foundation, which doled out an occasional scholarship but little else. Perry, who had never been pleased with his inheritance, badly wanted his Uncle Sid's oil fields back. In the mid-1960s the family's bank, Chase Manhattan, proposed loaning him the money he needed to buy them, in the range of fifty million dollars; the reserves themselves would be the collateral. Perry sat on the foundation's board—in fact, he was its dominant figure—and any transaction between the family and the foundation, especially one so large, risked being viewed as blatant self-dealing. Perry put the proposal to his fellow board members, found them amenable, then sought a fairness opinion from the Texas attorney general. Once the attorney general approved the deal, Perry, backed by Chase's cash, purchased all of Sid Richardson's oil fields. And not a minute too soon. Several years later, in 1970, Congress passed a law outlawing just this kind of insider dealing.

Paying off the Chase debt, however, took every last cent of the Bass family's cash flow during the late 1960s. By April 1969, when Perry's eldest son, twenty-seven-year-old Sid, arrived back in Fort Worth after graduating from Stanford Business School, the debt had finally been paid off. It was then that Perry called Sid into his corner office on the twelfth floor of the new Fort Worth National Bank Building. Everything they had, Perry told his son, was now his to manage. The oil company, renamed Bass Brothers Enterprises, was worth $120 million. With the Chase debt repaid, it was throwing off $1 million in pretax cash flow. "Do what you want," Perry told Sid. "I'm going sailing." And he did exactly that, serving as navigator the next several years on his friend Ted Turner's attempts to win the America's Cup. Never again would Perry Bass have a significant say in the family's finances.

On paper, Sid Richardson Bass was hardly the rough-and-tumble oilman his great-uncle Sid Richardson might have selected to guide the family fortune. Slender, blond, and intellectual, what Sid Bass liked to do with oil was paint; a number of his canvasses hung beside the Monets in his parents' home. ("That one?" Perry liked to josh. "Oh, that's a Sid Bass.") But Sid's second passion was the stock market; he had begun trading stocks as a boy, and he was certain that the kind of modern securities investing he studied at Stanford held the key to the family's future. His conviction deepened when he finished his first task, a top-to-bottom review of the Bass oil business. It was a godforsaken time to be a Texas oilman. Low-priced Middle Eastern crude was killing them. Bass Brothers, which had five hundred employees in West Texas and Louisiana, was barely breaking even.

If he was going to invest, Sid knew he needed talented people around him; he envisioned a group of five advisers swapping ideas around a conference table. One of his first hires was an energetic Stanford classmate, a twenty-five-year-old Fort Worth kid named Richard Rainwater. Rainwater had grown up middle-class in the city's Lebanese community; his father had Indian blood, hence the name. Rainwater was hawking stocks for Goldman Sachs in Dallas when Bass hired him. It was the single smartest decision Sid Bass ever made. Rainwater, he soon realized, was the ultimate stock-market nerd, an investing dynamo brimming with ideas. "I wanted to hire five guys,"

Sid joked to friends. "But after bringing in Richard, it was pretty clear I wouldn't need the other four."

They became a mismatched team, working in adjoining offices and sharing a secretary. Sid, trim and erudite, wore Saville Row suits; Rainwater, tall, swarthy, and painfully blunt, favored weatherbeaten work shirts and collected baseball caps with funny sayings. Sid was uncomfortably aware of both their responsibility and their inexperience. They didn't know how to do a business deal. They didn't know how to negotiate. They chose another Stanford classmate, John Scully, to handle their stock trading, shoveling him roughly twenty million dollars in those first few years; their largest investment, several million dollars, was in the Church's Fried Chicken chain.

"I want to invest in solid companies, with solid management, and stay for the long haul," Bass told Rainwater.

"You sound like Warren Buffett," Rainwater remarked.

"Who's Warren Buffett?" Bass replied. Rainwater had been collecting the little-known Omaha investor's writings for years, and shared them with Bass, who all but memorized them. Soon the two began reading the works of Benjamin Graham and other Wall Street greats. In terms of an investment philosophy, Sid disagreed with Wall Street wisdom in only one important way. Conventional wisdom held that a modern portfolio should be thoroughly diversified, with a mix of stocks, bonds, and perhaps real estate, the better to weather downturns in any one area. Sid, however, wanted to make fewer investments, but make them large. "You only have so many good ideas in life," he mused to Rainwater. "Why not put your money in what you really believe?"

Sid's second idea was venture capital. The biggest returns, he reasoned, would come from putting the family money into tiny start-ups. In their first year together, Bass and Rainwater searched for promising local companies in which to invest. They selected three, including an Italian restaurant with a frozen food business and a Dallas outfit that made telephone equipment. Within months, however, all three companies were faltering. Rainwater, meanwhile, served as Bass's link to Wall Street; he regularly attended the "road shows" at Goldman Sachs staged by companies touring the country trying to raise money. In 1970 he took in a presentation by a venture capitalist named David Dunn, who was attempting to raise money for a new computer company. Rainwater was so impressed, he invited Dunn to Fort Worth for lunch.

Bass listened, fascinated, as Dunn laid out the way he identified start-up managements in which to invest. Find someone who has been successful before. Don't ask them to do anything they haven't done before. Avoid new products; stick with the tried and true. Three simple principles, but they became Bass's mantra. Afterward, Dunn told Bass he hoped he would invest in his new computer venture. "We're not gonna do that," Sid said, smiling. "Because you've never done that before."

Several months later, Dunn's new venture fell apart when his major investor backed out. Bass hired him on the spot to run the family's new venture capital arm. Dunn's first task was to assess the three companies Bass and Rainwater had already invested in. He came back to the Bass offices shaking his head. When Sid asked whether their investments made sense, Dunn cracked, "I've never seen so much money in the hands of so little talent in all my life." Impressed, Bass and Rainwater gave Dunn eight million dollars to start a venture capital fund they called Idanta Partners, headquartered in San Diego. It was a brilliant move. Over the next decade Dunn would invest Bass money in scores of computer start-ups, including Storage Technology and Prime Computer. By 1980 he would turn that initial eight-million-dollar stake into two hundred million dollars.

For the next four years Bass and Rainwater quietly played the stock market. Their returns were decent but not spectacular. They worked alone, just the two of them; any staff they hired, Bass told Rainwater, would just generate ideas they would have to go over again themselves. Every evening they would convene in one of their offices to see what they had learned. Bit by bit, they were building confidence in themselves. Then, in October 1973, came word of the Middle East oil embargo. On November 2, Bass returned from a trip to Wall Street deeply shaken, convinced that oil shortages would soon send the markets into free fall. He told Rainwater he wanted to sell every stock they owned within five days. Rainwater nearly choked. But they did it, unraveling every position they had save Church's, which had gotten too big.

The day of Bass's decision, the Dow Jones Industrial Index stood at 921.

Within a year it hit 574. He and Rainwater felt like geniuses; their confidence soared. Oil prices, meanwhile, began to rise, quickly quintupling the flow of cash from the Bass-owned oil fields; in the mid-1970s, Bass Brothers threw off between ten and fifteen million dollars in new cash every month for Bass and Rainwater to play with. It was a perfect recipe for investing success: low stock prices, tons of excess cash, and a confidence bordering on the cocky. All through the mid-1970s Bass and Rainwater picked up stocks at bargain prices, soon amassing enough cash to buy entire companies, National Alfalfa in 1976, then the small Pick chain of twenty-two hotels. By 1978 their largest single investment was in Sperry Hutchison, the Wisconsin-based maker of Green Stamps.

Outside Fort Worth, the two young Texans remained unknowns. Slowly, though, Sid was increasing his profile in the city. He and his wife, Anne, a willowy Indianapolis girl he began dating at Yale, joined the Fort Worth Art Museum's board; Anne proved a charity whirlwind, devoting her time to the Junior League, the Van Cliburn Foundation, the Fort Worth Symphony, and the Fort Worth Country Day School, which Perry Bass had helped found so that his grandchildren wouldn't have to go to eastern boarding schools, as his children had. In 1972 Sid and Anne completed their dream home on eight wooded acres in the new Westover neighborhood. Designed by the noted New York architect Paul Rudolph, it was, one writer put it, "one of the most beautiful houses in Texas; serene and white, [rising] out of the treetops in a series of rectangular boxes, one stacked on top of the other, ends jutting out dramatically like cantilevered bridges." Anne spent several more million dollars on the landscaping and formal gardens. Somehow she still found time to give birth to baby girls.

In 1973 Sid's mind began drifting toward philanthropy. The Basses gave millions each year to Fort Worth's three major museums and other causes, but as Sid told his father, what the city really needed was something to bring back its old cow-town pride. The downtown, he saw, was beyond sad, block after block of boarded-up storefronts pockmarked by the odd porn theater. Sid bought up two blocks and was beginning a modest restoration project when Charles Tandy, CEO of Fort Worth's largest non-oil company, Radio Shack, announced plans for a far more ambitious revitalization. The two men soon joined forces, and by the time they had completed work in 1980, Sid had taken over and modernized a wide swath of downtown, complete with a new Worthington Hotel and two immense glass skyscrapers designed by Paul Rudolph, one thirty-eighty stories, the other thirty-four. Downtown Fort Worth's most noticeable landmarks, the glow on their glass skins can be seen for miles. Bass Brothers made the new buildings its new world headquarters, but few others followed suit. For years the buildings stood half empty.

By the late 1970s Sid and Anne had emerged as Fort Worth's first couple. Anne drove a Rolls-Royce Silver Cloud. Sid bought a private jet. Anne was beginning to cut a profile in New York, popping up in fashion magazines and joining the board of the American support group for the Paris Opera. When a New York ballet group performed in Fort Worth, Anne and Sid hosted a glamorous after-party unlike anything the city had ever seen. The world's greatest ballet dancers, Mikhail Baryshnikov, Peter Martins, and Heather Watts, couldn't believe the splendor of the Bass home or its art. "It was incredible," Martins remembered years later. "We knew they were young and rich, but we couldn't believe when we got to their house. When we came in, there was an orchestra playing. It was. . . ." He was at a loss for words.

The Basses' social emergence, however, came with risks neither Sid nor Anne had foreseen. Late one evening in 1973, they were pulling into their garage when Sid spied two black-clad men poised on the roof. As they stepped from the car, the two men appeared behind them, flashing guns and smashing the car's rear windshield. In the kitchen, the couple was ordered to lay on the floor. Sid was frightened, all the more so because his two infant daughters were upstairs with a babysitter—who promptly saved them all by activating the alarm system. The men ran and were never caught. Afterward Sid had a security gate built and manned around the clock. In later years, when reporters wanted to know why he so studiously avoided the public spotlight, friends sometimes pointed to the events of that night to explain.

All through the late 1970s Bass and Richard Rainwater continued piling up their investment profits. In 1978 they broadened their business by hiring a Wall Street whiz kid named Tommy Taylor, who opened a Bass trading floor to perform complex arbitrage trades; in his first full year of operation, Taylor alone brought in eight million dollars in profit. Yet Bass Brothers wasn't without its challenges. The challenges, in fact, were the brothers. The second Bass brother, Ed, had graduated from Yale in 1968 and after a stint in the Coast Guard had moved to New Mexico, where he designed and built adobe houses, made pottery, and became involved with a counterculture commune called the Synergia Ranch; some called it a cult. Of his siblings, Sid was closest to his youngest brother, trim, deferential Lee, born in 1956, who would join Bass Brothers in the early 1980s. The two were fifteen years apart but close in temperament. Sid treated Lee almost like a son.

For Sid, the headache was always the third brother, Robert, known as Bob. Six years his junior and boyishly handsome, with prematurely graying hair and rimless eyeglasses, Bob Bass had married the only girl he ever dated, a Fort Worth accountant's daughter named Anne; around town, everyone called her Little Anne, to differentiate her from Sid's wife. Bob and Little Anne met when his parents threw him a party for his thirteenth birthday and had never parted; so tight were the two that Bob had gotten permission from his parents to marry while still at Yale. Sid always suspected that Bob's problem was a keen sense of entitlement; while the rest of the Basses acted vaguely embarrassed by their fortune, Bob seemed to genuinely love being rich. When he first arrived at Bass Brothers in 1974, Sid tried to give him something to do. As far as he could tell, however, Bob simply sat in his office, noodling around with the occasional real estate deal.

Bob, his brothers felt, deeply resented Sid's leadership position. He seldom if ever invited Sid to his home. Bob was no closer to his parents; when his mother would bring over Christmas presents, she was seldom invited inside. In those first, cautious years together at Bass Brothers in the 1970s, Bob and Sid managed a strained if peaceful coexistence. Their relationship was a rare blemish in what, for Sid at least, was a glorious life. Still, by 1979, everything was beginning to change. The family's long years of anonymity were poised to end. A new decade was dawning, the 1980s, and for Sid Bass, who turned thirty-eight that year, nothing—his marriage, his money, his partners, his fame, his family—would ever be the same again.

IV.

The passing of H. L. Hunt, like that of the Soviet Union or Tito's Yugoslavia, unleashed years of pent-up frustrations among the thirteen children of his three families. It didn't happen overnight, in large part because Hunt's will, filed in a Dallas probate court the day after his funeral, stated that any family member who challenged its tenets would receive nothing. To his first family's dismay, Hunt left the bulk of his assets, including Mount Vernon and his 80 percent stake in Hunt Oil—its stock valued at roughly $125 million—to his wife, Ruth. But that wasn't the shocker. What stunned everyone was that Hunt had named as his sole executor not Bunker or Herbert or Lamar but a new actor in the Hunt drama, Ruth's thirty-one-year-old son, Ray Hunt.

Ray was the baby whose 1943 birth had inaugurated Hunt's secret life with Ruth. Born Ray Wright, he had become Ray Hunt at fourteen when the second family moved into Mount Vernon following Lyda Hunt's death. Blond, handsome, and popular, Ray had been senior class president at Dallas's St. Mark's prep school before heading to SMU, where upon graduation he married and joined Hunt Oil as a vice president in the mid-1960s. He had a tough time of it; while he enjoyed a congenial if distant relationship with his father, Bunker and Herbert shunned him, refusing to invite him into their deals or their homes.

Blocked from anything but a salaried position at Hunt Oil, Ray struck out on his own in 1969, buying a dude ranch in Lewisville, north of Dallas. Like all the Hunts, he thought big. He branched out into commercial real estate, and in 1974 stunned Dallas by proposing the largest development in downtown history, a multiblock collection of office towers, theaters, and a new arena called the Reunion Center, all of it to be topped by a massive geodesic ball. At the time of his father's death, Ray was the acknowledged head of the "third" Hunt family. His older sister, June, had begun what would become a long career as a Christian singer and inspirational speaker. The middle daughter, Helen, was a housewife. The youngest, Swanee, had married a minister and settled in Denver, where she and her husband ran a home for the emotionally disturbed. None had much to do with the first family.

But now, as executor of H. L. Hunt's estate, Ray found himself thrust into the first family's daily lives. Hunt Oil, which he now controlled on his mother's behalf, served as the umbrella organization for everyone's finances, administering the flow of money and paperwork for all the children's trusts and investments; it was the family office. Ray moved into his father's corner suite and began trying to make sense of it all. In those first hectic weeks after the funeral, there were dozens of issues he needed to hash over with Bunker and Herbert, but to his frustration, though his half brothers had offices on the twenty-ninth floor as well, neither would return his phone calls. It was sheer petulance. The first family had ignored Ray for years; they weren't about to acknowledge him as the Hunt family's de facto head.

Bunker and Herbert, meanwhile, wasted little time distancing their interests from Ray's. While Bunker and his siblings still owned 18 percent of Hunt Oil, they controlled the far larger Placid Oil outright, and their trusts controlled Penrod, the drilling subsidiary. That fall, while their father lay in the hospital, Bunker and Herbert had initiated a rare hostile tender offer for the nation's largest sugar refiner, Great Western United, and after a brief legal tussle, managed to win control that January; it was the first time any of the Hunts had taken over a publicly held company, and many on Wall Street wondered why, given the regulatory and media interest it would bring. Then, in January 1975, six weeks after their father's death, Bunker and Herbert took their first step toward severing their businesses from Hunt Oil's, forming a new company, Hunt Energy Corporation, with an aggressive million-dollar exploration budget. They didn't bother to tell Ray, who found out anyway and redoubled his efforts to talk with them.

When one of his messages crossed Bunker's desk in early February, Bunker read it. "Ray wants to see us?" he asked Herbert in mock surprise. "Well, I guess we better go see him then."[6]

There, at a meeting in Ray's corner office, Bunker and Herbert informed their half brother that they were switching all the first family's administrative affairs from Hunt Oil to the new Hunt Energy, which would take over

Placid's old space on the twenty-fifth floor; Bunker and Herbert were moving their offices downstairs as well. It was by all accounts a civil meeting. Bunker and Herbert had no problem with Ray assuming their father's position as president of Hunt Oil; they agreed to maintain seats on its board. Otherwise, everything would go on as it had before. They even promised to return each other's phone calls.

But what truly drew the two families together—at least for display in public—were preparations for Bunker and Herbert's wiretapping trial. It was set for September 1975 in the windblown West Texas city of Lubbock, home of Texas Tech University. The good news was that two Hunt aides, including John Currington, had been convicted of mail fraud—the embezzlement scheme—that March, which would tend to make the wiretaps appear, if not legal, reasonable. The bad news came in July, when Bunker and Herbert, along with four of their attorneys, were indicted on additional charges of obstruction of justice in connection with their alleged attempts to silence the wiretappers.

The real problem, though, was that the Hunts had no serious defense. They would argue that they hadn't known wiretapping was illegal, which doesn't matter in criminal trials. Their sole remaining defense, and this was the line their attorney, Phil Hirschkop, reluctantly began feeding skeptical reporters, was that Bunker and Herbert had been set up by the CIA. It was all revenge, Hirschkop argued, for Bunker's refusal to help the agency in Libya. From the beginning, though, Hirschkop sensed the trial would be more about form than substance. H.L.'s security man, Paul Rothermel, had fed reporters a theory that the first and second families were at war, that the wiretaps were part of a scheme to cheat Ray's family out of its inheritance.

Hirschkop needed the jury to believe the wiretaps were about ferreting out an embezzlement, and he made sure Ray, his mother, Ruth, and his three sisters would appear at the trial to present a united front. At the same time, Hirschkop badly needed Bunker and Herbert to emerge from their long years in the shadows and present themselves as people a West Texas jury could embrace, good ol' boys in cowboy boots who drove battered Chevys and rooted for the Cowboys on Sundays. The brothers did their part, easily charming a series of reporters who produced exactly the kind of aw-shucks puff pieces the defense needed. Herbert smiled and talked about raising chickens as a boy. Bunker joked about his weight. The *Dallas Morning News* reporter, whose feature appeared on the eve of the trial, was impressed by their "country charm and easy humor. . . . It would be hard to guess from their casual demeanor that these are two of the wealthiest men in the world."[7]

When jury selection began on September 17, 1975, there were almost as many Hunts in the courtroom as prospective jurors. Brothers, half brothers, cousins, wives, children—everyone who could smile for a camera came to Lubbock. H.L.'s widow, Ruth, described by one reporter as a "kindly grandmother," told newsmen she was certain her stepsons were innocent; Hirschkop asked and received permission for her to sit in the courtroom. At breaks Ray smiled and chatted with Bunker, Herbert, and Lamar, who made sure to smile and chat with courtroom personnel and always wore their cowboy boots; the *New York Times* man, presumably not a Texan, called them motorcycle boots. Meanwhile, a line of Hunt wives beamed as flashbulbs popped; the picture appeared in the *Morning News* beneath the headline "Hunt Wiretap Trial Becomes a Family Affair." "They sure don't act like rich people," one observer murmured.

On the witness stand Bunker and Herbert made no effort to deny the wiretapping. One of the private eyes, W. J. Everett, insisted he had told the brothers it was illegal, but Bunker and Herbert denied it. Herbert was the star, actually tearing up as he explained he had only done it to protect his beloved father. The character witnesses, especially two Dallas ministers and E. J. Holub, a mammoth Texas Tech alumnus who played for Lamar's Kansas City Chiefs, charmed everyone. After a week of testimony the jury took just three hours to reach its verdict: not guilty on all counts. Bunker emerged from the courtroom arm in arm with Herbert, telling reporters how sorry he felt for ordinary Texans who couldn't afford the fine lawyers the Hunts could.*

It was a shining moment for the Hunts, their most significant legal victory since besting Dad Joiner forty years before, and proof that the two Dallas

---

*The Hunts' victory gave them the upper hand in settlement talks that ensued over the federal obstruction of justice charges. Eight months later, prosecutors agreed to an exceedingly generous resolution. Charges against Herbert were dropped. Bunker pled no contest and paid a thousand-dollar fine.

branches of the family could live together in peace. But the warm glow was not to last. H. L. Hunt had once had another family, born to a woman most of the Hunt children only dimly remembered, and six weeks later, in the fall of 1975, she emerged from the mists and sued.

V.

It had been a long time coming. H. L. Hunt's all-but-forgotten second "wife," Frania Tye, now Frania Tye Lee, had withdrawn to Atlanta after World War II upon marrying one of Hunt's men, John W. Lee. The couple had used the money from Frania's 1942 separation agreement to buy an estate in suburban Chamliss called Flowerland, where Frania, a pretty blonde then in her forties, threw elegant parties in an effort to penetrate Atlanta society; they later added a 747-acre plantation. Frania's social ambitions were stifled, however, by nagging rumors, apparently started by a woman who knew her in Long Island before the war, that she had been Hunt's mistress. She and Lee had divorced in 1957, and Frania moved into a Tudor home in Atlanta's Ansley Park neighborhood. She never remarried. Friends believed she never got over the idea that she would one day "remarry" Hunt.

Frania had raised four of Hunt's children to adulthood. In 1975, when they suddenly sprang from the wings of the Hunt family stage, none were exactly thriving. The eldest son, Howard, who turned forty-eight that year, was a self-employed mechanical engineer. His sister Haroldina, forty-six, had married a doctor and given birth to seven children. However, like her half brother Hassie Hunt, she had been overtaken by mental illness, believed to be a form of schizophrenia, and had undergone electroshock treatments. The second daughter, Helen, one of H.L.'s favorites, had moved to New York and attempted to become an actress. Disappointed by her prospects, she had returned to Atlanta, married a real estate developer, bore him a son, and became active in local theater circles, founding the Theatre Atlanta Women's Guild. In 1962, while on a monthlong tour of Europe, she and her husband were killed in an airplane crash along with 126 others outside Paris.

Given their tangled parentage, all of Frania's children suffered to some extent from identity issues, especially the youngest, Hugh. Born Hugh Richard Hunt, he had changed his name over the years to first Hue R. Lee, then Hue Richard Lee, then finally to Hugh Lee Hunt. A broad-shouldered, intelligent, if insecure young man, Hugh made it into Harvard Business School only to drop out and return to Atlanta, where he blew much of what little money he had attempting to become a real estate developer. All four children had Hunt-money trusts, but they were a pittance, dribbling out barely six thousand dollars a year. Howard and Hugh had had sporadic contact with H.L. over the years, usually to borrow money. When Helen graduated high school, H.L. sent her a mink coat. But that was the extent of Hunt's support. When the first family mentioned Frania's brood at all, they called them "the Lee people."

By all accounts, it was Hugh Hunt who persuaded his mother to explore the idea of some kind of lawsuit. The family had discussed it over the years. Frania's enthusiasm waxed and waned. She was seventy now. Years before, Hunt had told her all records of their marriage had been destroyed. She and John Lee had actually gone to Tampa once to check, and found it was true. Still, Hugh pressed. Maybe they looked in the wrong place. Lyda's sons were billionaires now. Why couldn't Frania's children have money, too? Hugh took his mother once more to the courthouses in Tampa, but they were unable to find any record of the marriage. It didn't matter, Hugh said. They had lived as man and wife.

Frania hired an attorney. On November 11, 1975, fifty years to the day since she had exchanged vows with "Major Franklyn Hunt" at the little bungalow in Tampa's Latin quarter, her family filed two suits against the Hunt estate in Louisiana. In the first, filed in Shreveport, Frania asked to be declared Hunt's "putative" wife. She asked for half of everything Hunt had earned during their nine years together—the period where he took control of the East Texas field—and all the income that flowed from it, an amount that could run into the billions. In a separate suit filed in Baton Rouge, Hugh asked a court to declare him and his siblings H. L. Hunt's legitimate heirs.

For the Hunts, the only good news was that no one noticed. Other than a wire-service story in the *Morning News*, the press ignored it.

VI.

By 1976, with the sole exception of Bunker's reversal in Libya, the children of H. L. Hunt's two Dallas families, and the corporations they controlled, were thriving as never before. Thanks in large part to rising oil prices, the Hunts were probably the world's wealthiest family. They invested together and separately, at their whims, in nearly two hundred different partnerships and companies, everything from coal and sugar to the Shakey's Pizza chain; their real estate holdings alone were valued at $1 billion. Placid Oil, run by professional executives under the guidance of Bunker, Herbert, and Placid's unofficial chairwoman, Margaret Hunt Hill, was the largest single Hunt entity, raking in more than $350 million in annual revenues atop reserves valued at $4 billion or more. The family-owned drilling contractor, Penrod, run by Herbert, operated offshore and onshore rigs for customers all over the world. Its assets topped $500 million. All told, the Hunts employed more than eight thousand people. Their net worth hovered in the $6 billion to $8 billion range, easily topping the next-wealthiest American families, the Mellons and Rockefellers.

The only Hunt who concentrated his moneymaking efforts outside oil was Lamar, who, after cofounding the American Football League, had become one of America's most successful sports entrepreneurs. Lamar still ran the Kansas City Chiefs—they won the Super Bowl for him in 1970—but in 1967 he began to branch into new sports, cofounding the first serious professional tennis league, World Championship Tennis. The WCT revolutionized the sport, then run by amateur associations. Under Lamar's supervision it staged the first big-money tournaments, instituted innovations such as the seven-point tiebreaker, and in short order transformed tennis into the mass-spectator sport it is today. Not all Lamar's ventures worked so well. He started a professional soccer franchise in Dallas, the Tornado, but the league never caught on. He tried to buy the island of Alcatraz from the city of San Francisco, thinking he might turn it into a shopping and tourist attraction, but the project died in the face of the usual Bay Area protests.

Though the first family was loathe to admit it, the most successful Hunt during the 1970s was Ray, who began making over his father's old Hunt Oil Company much as his Reunion project was changing the face of downtown Dallas. Like Sid Bass, Ray hired a crew of outsiders to help modernize and streamline Hunt Oil's corporate structure, consolidating his father's old real estate and farming businesses, then dismantling the scandal-plagued HLH Products. Once his house was in order, Ray concentrated on finding new reserves. In 1976 he snapped up his first acreage in the North Sea, a 15 percent stake in a consortium that began drilling the following year. The discovery well came in strong, revealing a new field whose reserves were estimated as high as five hundred million barrels. Hunt Oil's share was initially valued at two hundred million dollars, then as high as a billion dollars as oil prices rose. As more than one Dallas oilman remarked, Ray had inherited his father's luck.

Buoyed by his success, Ray hired dozens of new land men and geologists and began buying acreage all over the world, concentrating in the North Sea and Gulf of Mexico. In three short years he quadrupled the size of Hunt Oil's staff, increased its offshore leases from a hundred thousand to a million acres, and boosted the value of its reserves from one hundred million dollars to a billion dollars or more. For the first time in years, Hunt Oil was once again a legitimate player in international oil circles.

But Ray was much more than an oilman. He remained active as a real estate developer, accumulating hundreds of acres he intended to build out in the Fort Worth suburbs. Unlike Bunker and Herbert and the rest of the family, the Reunion project, fast nearing completion, had made Ray a public figure in Dallas, a genuine civic leader who accepted invitations to join the boards of the Chamber of Commerce, the Museum of Fine Arts, and the SMU Alumni Association. To the delight of many in Dallas, Ray emerged as everything the rest of the Hunts had never seemed, friendly, approachable, civic-minded, *modern*. Where his father's idea of a media outlet was LIFE LINE, Ray started a new magazine just for Dallas, *D*, a glossy offering immediately embraced by wealthy readers. Haute Dallas simply fell in love with him. Ray was invited to join the best country clubs as well as the exclusive Idlewild, Terpsichorean, and Dervish clubs, all of whom had shunned his half brothers. A 1977 profile in *Time* anointed Ray "the Nice Hunt," and soon *BusinessWeek* and other publications rushed to add compliments of their own. "He's the last out of the elevator [at night]," one peer told *Time*, "and the last one walking down the hall."

If Ray was welcomed into the bosom of greater Dallas, the first family's respect proved far harder to gain. Ray's mother, Ruth, tried her best to unite the two families, throwing debut parties for Herbert's daughters at Mount Vernon and inviting the first family to camp and hunt on H.L.'s Wyoming ranchlands. Her overtures, however, were seldom returned. As one of Ray's sisters described relations between the two branches of the family: "We only see each other at weddings and trials." Ray's ascendancy only made things worse. Bunker clearly resented his popularity. The first family still owned 18 percent of Hunt Oil, and despite its success, they badmouthed Ray's management at every turn. He was too young, they said, too inexperienced. He knew nothing about the oil business. The North Sea field? Beginner's luck.

In early 1977 these festering resentments finally burst into the open, when Bunker and the first family demanded that Ray buy out their stock in Hunt Oil. The catalyst was Reunion. The first family accused Ray of taking a sweetheart loan from Hunt Oil to satisfy a brief cash crunch during construction. Ray denied this, saying the "loan" was a housekeeping matter involving the merger of one of his real estate partnerships into the Hunt Oil portfolio. Whatever the details, Bunker and Ray's attorneys proved unable to come up with a mutually agreeable buyout price. On paper the 18 percent stake could be valued at $100 million or more, money Ray said he simply didn't have at the moment.

The real problem, though, was the IRS. Ever since his father's death Ray's tax attorneys had been negotiating what promised to be a gigantic bill for taxes on the Hunt estate. Just how large the bill would be depended largely on a valuation of H. L. Hunt's oil reserves at the time of his death. Ray's attorneys were pushing for the lowest possible valuation. Selling stock in Hunt Oil, however, would value those reserves in harsh black and white. By pushing for the highest purchase price, the first family could, indirectly, pull the rug from beneath Ray's negotiations with the IRS. Bunker and Herbert didn't especially care; it was Ruth and the estate who had to pay the taxes, not them. To drive home the point, Bunker once again brought in John Connally, who told Ray's men he was prepared to play "hard ball" and go straight to the IRS with higher valuations of Hunt Oil's reserves.

Negotiations between the two sides of the family dragged on all through 1977, pushing relations toward a new low. Finally, in December, a settlement was reached. Ray agreed to transfer a group of assets, including a North Dakota pipeline and some timberlands, into a new subsidiary, which the first family would then buy with its stock. The settlement heralded a final split between the two families' finances. As far as Ray was concerned, Bunker and Herbert and Lamar were now out of his hair once and for all. As if in celebration, that same month brought the symbolic birth of the Reunion project. Just before Christmas, Ray flipped a switch and for the first time lights blazed all around the giant geodesic ball atop the fifty-five-story Reunion Tower.

It was a defining moment for Ray Hunt. After years of work, Dallas had a new skyline, and a new Hunt family leader, one who actually seemed to care about the city and its people. Ray's work, however, was far from over. It was time to deal with Frania Tye.

VII.

As executor of his father's estate, it had fallen to Ray to deal with Frania Tye Lee and her lawyers. No one thought their two lawsuits would ever reach a court trial; the whole story was simply too embarrassing to the Hunt legacy. The question became how much "the Lee people" would accept to go away. Frania's attorneys were angling for a number around a hundred million dollars. Ray and his lawyers just shook their heads.

The trial was set for federal court in Shreveport on January 9, 1978. Ray was expecting the worst, a media circus, teary testimony from the aging Frania, tales of how she had been wined, dined, hidden away, and eventually dumped by the heartless H. L. Hunt, whose Dallas children were billionaires while the Lee children still paid rent. To counter the sympathy she would no doubt attract, Ray badly needed the first family to come to Shreveport and present a united front, just as they had for Bunker and Herbert's wiretapping trial three years earlier. Bunker told him to kiss off. Eventually, though, it dawned on his siblings how much everyone had to lose—not just their father's legacy but a fair chunk of his millions—and they agreed to appear.

On the witness stand Frania was everything the Hunts had feared, a gray-haired, trembling grandmother, her voice quavering as she walked the six jurors through her whirlwind courtship with Major Franklin Hunt in 1925, their "marriage" in Tampa, the years together in Shreveport, her discovery of Hunt's other wife in 1934, then the soap operas that led her to accept Hunt's settlement and flee Texas in 1942. Frania agreed to part with her "husband," she insisted, only after Hunt promised to take care of their children and remember them in his will.

"He promised me that in his will he would name me as wife, that he married me, that they're his children, and that he would leave the same amount of funds that the first family had," Frania said.

She stuck to her story throughout a cross-examination by Hunt attorneys, led by the man who had defended Bunker and Herbert in Lubbock, Phil Hirschkop. Asked why she had signed a settlement agreement that pointedly omitted any mention of their marriage, Frania said, "I signed that statement to protect Mr. Hunt from a bigamy charge. That was my contribution to his life."

"You say you swore to facts that were not true?"

"Yes."

"Doesn't that bother you to do that?"

"At that time I did that for the man I loved. And women in love are not philosophers, nor do they know the law that well."

"They know the truth, don't they?"

"Yes," Frania testified, "but they make a lot of sacrifices. I was thinking not only of him, but of all the children."

It was an immensely sympathetic performance, and Frania's attorneys followed it up with several sharp blows to the Hunt defense. They called in Hunt's onetime aide, John Currington, who testified that H.L. had privately admitted marrying Frania. Then they revealed their secret weapon, a voided entry Frania and her son Hugh had discovered in a Tampa courthouse that clearly listed the marriage in 1925. The faded writing was the source of considerable intrigue. After its discovery—and certification by both sets of lawyers—someone had sliced the original entry from a courthouse ledger. Hunt attorneys brought forth a handwriting expert who argued that the entry was in a different hand from adjacent entries, suggesting it might be a forgery. The Hunts tried to date the ink but couldn't. The damage had been done.

After two days of testimony, Bunker was ready to settle. He and his brothers and sisters lingered outside the courtroom during recesses, smiling and mingling with Ray's side of the family just as they had in Lubbock. But no amount of public backslapping and cowboy boots could offset the impression Frania had made on the jurors. Phil Hirschkop took one of her lawyers aside and suggested the Hunts might consider a settlement. The lawyer, Roger Fritchie, said Frania might be willing to reduce her demands from one hundred million dollars to twenty-five million.

When Frania's team concluded their courtroom presentations at noon on Friday, January 13, Hirschkop made the offer. The Hunts, he said, were willing to pay Frania $3.5 million but would not recognize her as Hunt's wife. Hirschkop gave Frania's attorneys until 5:30 that day to respond, at which point he would withdraw the offer. The appointed hour came and went. Frania's lawyers never responded.

When the court recessed for the weekend, the first family scattered; Lamar took his wife, Norma, to Miami to watch Clint Murchison's Cowboys beat the Denver Broncos in Super Bowl XI. Only Ray remained in Shreveport, studying the week's testimony. The more he read, the worse he felt. Frania's case appeared unbeatable. By Sunday night, when everyone reconvened, Ray was willing to increase the settlement offer to ten million dollars and recognize Frania as his father's wife. The first family, however, would have none of it. They authorized Hirschkop to offer more money, but refused to recognize Frania.

On Monday both sides and their attorneys gathered in Judge Tom Stagg's chambers to negotiate a deal. Frania was willing to settle on favorable terms, in large part because Judge Stagg had issued several rulings in the Hunts' favor. Among other things, the judge ruled that the jury couldn't consider

Hunt's supposed promise to include Frania and her children in his will because she could furnish no independent witness; her financial claims, as a result, would be limited to only the period when they lived as a man and wife, from 1925 to 1934.

It was over within hours. Frania agreed to accept $7.5 million, representing half of H. L. Hunt's community property in 1942; the money would come from trusts left both Dallas branches of the family. In return Frania pledged to end both lawsuits. She would not be recognized as H. L. Hunt's wife, even though it was now clear to everyone she had been. Afterward Frania approached Bunker in a hallway. "I just wanted you to know that I don't have any hard feelings about this," she said. "I've always liked you kids."[8] Bunker managed a weak smile.

A month later, Ray returned to Shreveport to deliver the $7.5 million check. When he reached Judge Stagg's chambers, however, he received a rude surprise: Frania didn't want it. Actually, it appeared her son Hugh was the problem. Frania and all the other Hunt children had signed the settlement, but for reasons no one seemed to understand, Hugh refused. It took another ten long months before he reappeared in Shreveport to explain himself. Everyone gathered in the judge's chambers, Ray with a group of lawyers from both sides. Hugh seemed unable to explain his opposition, but in questioning in front of Judge Stagg, the Hunt attorneys insinuated that he wanted to reserve his right to challenge any of his half siblings' wills. The Hunts argued that the judge should simply force Hugh to sign.

And there things lay for months, an inconclusive final act to a sad family drama, when suddenly, with practically no warning, the Hunts found themselves caught up in the financial equivalent of a thermonuclear explosion. In the panic it injected into world markets, in the unprecedented audacity of its scope and aims, in its sheer Texas weirdness, it was unlike anything the world had ever before seen.

It was Bunker and his silver.

SEVENTEEN

The Great Silver Caper

I.

By all rights, Bunker Hunt should have emerged from the 1970s the world's richest man, a position he may have briefly held for a period in the late 1960s. Oil prices had risen to all-time highs. Inflation pushed interest rates to records. All across Texas, new millionaires were once again hatching like mayflies. In Fort Worth, Sid Bass and Richard Rainwater, products of Yale and Stanford, were demonstrating what a modern, diversified investment strategy could do for even the smallest oil fortune. But in the 1970s Bunker and his brother Herbert, who pooled their investment capital, charted a far different course.

Over the years they had slowly diversified, going into sugar and real estate. Bunker was the big-picture man, the one who devised their ideas and strategies; Herbert was the detail man, the one who put his brother's ideas into action. Bunker's ideas, however, were at best unconventional. At worst, they were stupid. By and large, Bunker ignored the advice of Wall Street's best and brightest. The product of a single semester at the University of Texas, a man who fervently believed in Jewish, Communist, and Rockefeller plots to subvert the world, he devised his own unique investment strategy: silver.

Afterward, everyone involved would offer a different story of how it all began. A New York commodities broker named Alvin Brodsky claimed to have been the first to interest Bunker in silver, in 1970, as did one of Bunker's prep-school friends and a pair of Dallas silver brokers. Probably all four made the recommendation; silver was a popular investment at the time and was trading at historic lows. To understand why Bunker listened, it helps to understand his worldview. Bunker, like his father, believed the world was slowly falling apart. The Communists, the Jews, the Rockefellers, the Russians, the Chinese, the hippies—everyone was out to destroy the world and Bunker's position in it. "Bunker," noted a Los Angeles rare-coin dealer named Bruce McNall, who befriended him during the 1970s, "was just obsessed with the idea that the Russians were coming over the Rockies."[1]

It's impossible to know whether Bunker's philosophy was a product of extreme nouveau riche insecurity, his father's harem-scarem politics, or something else altogether. Whatever its cause, it channeled his mind to a place familiar to all the world's worried wealthy, from Middle Eastern Jews to South American dictators, who for centuries have invested their money in tangible, inflation-resistant items: diamonds, silver, and especially gold, anything that could retain its value if a family, a country, or a world economy suddenly collapsed. Bunker was perhaps the ultimate case of the worried wealthy. As the tumult of the 1960s built to a crescendo, he had begun funneling much of his fortune, including his Libyan profits, into tangible investments.

At first, silver was just one of many. Bunker started small in 1970, when the price still hung near all-time lows, $1.50 per ounce. Working with Herbert, the two slowly increased their purchases over the next three years, buying five-thousand-ounce "penny packets" through a Wall Street brokerage, the Bache Group. Silver's price rose steadily as they did, eventually doubling, to $3 per ounce, in 1973. It was then that Herbert read a book, *Silver Profits in the Seventies,* by Jerome A. Smith, an author of several financial newsletters who specialized in gloom-and-doom economics. Smith argued that the world faced an imminent economic and political collapse, and that the only safe hedge was gold and silver. He urged wealthy investors to buy both and store them in Switzerland—just in case Russia invaded the United States.

Bunker didn't actually read the book—he was never much of a reader—but from what Herbert said, it sounded brilliant. In hindsight, in fact, the brothers appeared to follow Smith's strategies to the letter. In mid-1973, working through a variety of brokers to mask their identities, they began buying silver contracts on the New York market. A silver contract entitles its owner to buy the metal at a specific price. No one actually buys the silver itself; a contract is a bet on the movement of prices. It's simple: If you buy a December contract at ten dollars and the price of silver moves to twelve, you make a two-dollar profit. If it falls to eight, you post a two-dollar loss. What was unusual about the Hunts' evolving strategy was the amazing number of contracts they bought—an initial twenty-million-ounce December 1973 contract, followed by several more large orders, until by early 1974 Bunker and Herbert sat atop contracts entitling them to buy fifty-five million ounces of silver—roughly 9 percent of *all the silver on Earth*. No one, not even governments, controlled more.

But what was truly jaw-dropping about the Hunts' purchases—what stultified investigators when the truth eventually came out—was that Bunker and Herbert, unlike almost every other investor on Earth, *actually took the silver*. It came in ingots, and it cost around $175 million, in cash. Years later, one of Bunker's top men told a congressional committee how the operation worked, a story both Bunker and Herbert denied. According to this story, the Hunts selected a dozen cowboys from Bunker's Circle T ranch west of Dallas, handed them rifles, and divided the men among three chartered 707s. The three planes, flying at night, landed in Chicago and New York, where the cowboys stood, rifles at the ready, as lines of armored cars drove up with forty million ounces of silver ingots from the New York and Chicago exchanges. Once loaded, the planes took off, crossed the Atlantic, and landed the next morning in Zurich, where the silver was loaded into still more armored cars. With the Circle T cowboys riding shotgun, the cars ferried the silver to six separate Swiss banks.*

There was another tale about these flights, one so improbable that most investigators later dismissed it altogether. According to this story, Lamar Hunt went along on one of the planes. The silver ingots were distributed evenly around the hold, leaving a large open space in the middle, which the plane's owner decided to fill with, of all things, a cage containing a circus elephant. Over the mid-Atlantic, so the story goes, the plane began to lurch

---

*The remaining fifteen million ounces of silver were said to be stored in warehouses in Chicago and New Jersey.

out of control, at which point it was discovered the elephant had looped its trunk around the wires that controlled the aircraft's flaps. Lamar and an aide, it was said, saved the day by opening the elephant's cage and tossing in a rubber tire, diverting the elephant to a new toy and saving everyone's lives.

As was his habit, Bunker moved into the silver market with stealth and secrecy, but warehousing 9 percent of the world's silver supply couldn't stay secret for long. In the spring of 1974 the rumors finally began to fly. One question spun through the world's silver exchanges: Who the hell was Nelson Bunker Hunt? And what was he doing with all that silver? No one knew, and everyone needed to. If that much silver were sold, prices could crash. For the moment, traders believed the Hunts would buy more. Silver's price rose to six dollars per ounce.

It was then, in April 1974, that Bunker made his dramatic debut on the floor of the New York commodities exchange, known as the Comex. All around the vast room, busy traders stopped and stared—as a paunchy man in a blue suit and thick glasses waddled through the aisles, craning his head, studying this strange new world of commodities. Before Bunker left, a financial reporter from *Barron's* cornered him and asked his intentions. "Just about anything you buy, rather than paper, is better," Bunker said. "You're bound to come out ahead in the long pull. If you don't like gold, use silver. Or diamonds. Or copper. But something. Any damn fool can run a printing press." Few *Barron's* readers had the first clue what Bunker was talking about, and he didn't linger to explain. In fact, it was the last anyone in the silver market would see Bunker for a long time.

In the months to come, the hubbub over the Hunt silver purchases fell away. As it did, silver's price sagged, falling a dollar, then two, before finally settling below four dollars per ounce. In Dallas, Bunker was perplexed. How could they get the price to go back up? He began making quiet trips to Europe and the Middle East, taking leading silver dealers out to lunch and mumbling questions. One of the market's most important suppliers was a sly Abu Dhabi bullion dealer named Haji Ashraf. Driving up silver's price was simple, Ashraf told Bunker over lunch in London. All you needed was for wealthy Arabs to begin buying.

Bunker thought it an idea worth pursuing. Which was how he found himself in March 1975 aboard a commercial airliner descending across the mountains of Iran toward the capital city of Tehran. He had come to see the shah of Iran with a simple proposal: the shah should begin buying silver. If he and Bunker bought in tandem, they could drive silver's price through the roof. The meeting had been arranged by a prince Bunker had met during his prep-school days. When the prince met Bunker at the airport, however, he had bad news: the shah was away. Bunker ended up meeting with the finance minister, a man named Hushang Ansari. Ansari hadn't the faintest idea who Bunker Hunt was. He smiled during Bunker's presentation, shook his hand, waved good-bye, and forgot all about it.

Irked, Bunker left Iran for Paris, where he spent several days inspecting his horses. At a stable, he spied an Arabian sheikh. It gave him an idea: Why not call the Saudis? He phoned a family friend with connections in the kingdom, who promised to arrange a meeting with King Faisal himself. He advised waiting two or three weeks, however, since word of Bunker's Iranian visit was making the rounds. They wouldn't want the Saudi king to feel he was Bunker's second choice. The meeting, however, never came off. Faisal was assassinated several days later.

Batting 0-for-2 with Middle Eastern royalty, Bunker returned to the drawing board. He needed more people buying silver; that much was clear. He couldn't simply do it himself. It was expensive, and that very year commodities markets, after years of unregulated growth, had finally gotten a federal regulator, the Commodity Futures Trading Commission, the CFTC. There were now limits on how much a single entity could buy and sell. Searching for a pliable partner, Bunker and Herbert turned to their newest company, Great Western Sugar, acquired in late 1974. At their direction, one of Great Western's subsidiaries began buying commodities and metals across the board, sugar, copper, gold, and especially silver—lots and lots of silver. In June 1976 alone, Great Western purchased twenty-one million silver contracts. But to Bunker's dismay, silver's price simply wouldn't stay up. It rose to $5.20 that August, but by the end of 1976 it had fallen back to barely $4.

It was then the fledgling CFTC stirred, lobbing a series of questions Bunker's way about how much silver he actually controlled. Bunker explained, lamely, that it wasn't an investment. He was only buying silver to use in a complex barter arrangement with the government of the Philippines; in essence, he claimed they had tried to trade silver for sugar, but the deal never came off. In the process of answering these inquiries, Bunker's attorneys raised a more worrisome concern. Great Western was still a public company. Its shareholders, or even the Securities and Exchange Commission, might not look too kindly at one of its subsidiaries trafficking in massive amounts of silver. Bunker and Herbert simply announced an offer to buy back Great Western's publicly traded shares. Now they could do as they pleased.

Their next move, brought off in the spring of 1977, was to have Great Western launch a rare hostile tender offer for a company named Sunshine Mining, which controlled the nation's largest silver mine, outside Kellogg, Idaho. The Hunts quickly bought up 28 percent of Sunshine's stock for twenty million dollars, at which point its management surrendered, handing over permission for the brothers to acquire the rest of the company's stock. The Sunshine deal not only gave the Hunts control over another thirty million ounces of silver, they now qualified as commercial users of silver, exempting them from most trading limits.

At this point, the Hunts' commodities strategy took a detour. While the Hunts remained fixated on silver, they had continued dabbling in other tangible investments, and in January 1977 Bunker decided to get serious about one of them, soybeans. His angle was the weather; Bunker's decision was heavily influenced by a New Mexico climatologist named Iben Browning, who predicted that climate changes in South America would soon lead to a worldwide soybean shortage. Using this and other predictors, the Hunt brothers began buying unprecedented amounts of soybean futures contracts. Federal regulations set the legal limit on one individual's soybean-contract holdings at three million bushels. By that spring, Bunker and Herbert controlled six million bushels between them, plus an astonishing eighteen million bushels in the names of their children, giving them control over one-third of the total U.S. soybean supply. Their buying drove up prices, to ten dollars a bushel from six dollars, giving them a paper profit of nearly one hundred million dollars.

It was a flagrant flouting of the law, and in April 1977 the CFTC told the Hunts to reduce their holdings to three million bushels—or else. The

Hunts sold off a mere two million, leaving them with twenty-two million. At that point the CFTC went public, suing the Hunts and showing the press what they had done. Bunker was outraged. This was government harassment, he told reporters, pure and simple. Plenty of commodities buyers worked together to get around legal limits, he said, yet the government only pursued the Hunts. "I think the reason, frankly, they jumped on us is that we're sort of a favorite whipping boy," Bunker said. "We're conservatives, and the world is largely socialist and liberal. As long as they want to jump on somebody, they want a name and they want somebody that's on the other side."

The CFTC case against the Hunts went to court in Chicago in September 1977. The Hunts came away with a Pyrrhic victory. While the judge ruled they had clearly violated federal regulations, he refused to order any damages, telling the CFTC it had brought the wrong kind of action: rather than try to prevent the Hunts from buying more soybeans, it should have sued them for attempting to corner the market, a clear violation of the Sherman Antitrust Act. It was a prescient observation to make, because as later investigations would make clear, that was just what the Hunts were about to try to do.

II.

"Cornering" a market is defined as an illegal attempt, typically by a group of investors working together in secret, to buy enough of a particular commodity to be able to dictate its price. It is the rarest of economic crimes, and the chanciest, a scheme that not only requires massive amounts of capital but puts it all at risk. The last known attempt to corner the silver market was masterminded by a Bombay financier named Chunilal Saraya, who between 1907 and 1912 led a group of maharajas that bought enough silver to "squeeze" other buyers; that is, the Saraya group hoarded so much silver that, when others needed to buy it to fulfill margin-loan requirements, it was able to demand and receive sharply inflated prices from the desperate buyers. The British government ultimately intervened to end the scheme.

Modern antitrust laws make cornering a market all but impossible; the mere act of two independent entities secretly buying in tandem is against the law. (Hunt attorneys, in fact, advised Bunker that Great Western's commodities purchases could easily be judged in violation, one reason Great Western faded from the Hunts' plans.) But as Bunker's conduct during the soybean case showed, he didn't think much of federal antitrust laws, federal regulators, or federal trading limits; he considered all of it "red tape" a smart buyer simply worked around.

By 1978 the Hunts had been sitting atop their mountain of silver for five long years. Prices hovered around six dollars per ounce, the same level they had been when Bunker first toured the Comex in 1974. No matter what he tried, no matter what he bought, he couldn't get the price higher. The fact was, he didn't need to. The Hunts could have sold out then and realized a few hundred million dollars in profit. But like Clint Murchison Jr., doubling or even tripling his money bored Bunker. He wanted a true home run, the more intricately mastered the better.

His opportunity came on October 1, 1978, when, at a horse auction outside Paris, Bunker was introduced to a dark, slender thirty-five-year-old named Naji Robert Nahas, a Lebanese-born, Egyptian-raised, Brazil-based businessman who acted as a middleman for several Saudi Arabian sheikhs attempting to invest in South America. One of his clients was a Saudi named Mahmoud Fustok, who sometimes invested alongside Saudi Prince Abdullah Ibn Abdul-Aziz, the stuttering thirty-six-year-old who headed his country's internal security forces; Prince Abdullah also happened to be second in line to the Saudi throne. That day outside Paris, Bunker, Nahas, and Fustok started out talking about horses. They ended up talking about silver. Yes, the two Middle Easterners agreed, it sounded exactly like something that might interest Prince Abdullah.

Bunker began to get excited. He ordered fifty copies of a bullish silver analysis in *Myers' Finance and Energy Report* translated into Arabic and mailed to leading Saudi investors. At roughly the same time, the ubiquitous John Connally managed to introduce Bunker to a second group of well-connected Saudis, who he was helping with their first American investments. One was Khalid bin Mahfouz, a young man in his thirties whose family ran the National Commercial Bank in Jeddah, the largest bank in Saudi Arabia. The second was a seasoned Saudi middleman named Gaith Pharaon. It didn't say so on either man's business card, but both were widely regarded as front men for the Saudi Crown Prince himself, Prince Fahd, a plumpish, freewheeling casino gambler whose many roles included running the Saudi central bank.

What happened next has never been fully explained, given that all the Saudis involved would remain beyond the reach of government subpoenas. But it was the perfect moment for Bunker to gather a silver syndicate. During the winter of 1978–79 silver's price, pushed by inflation and escalating interest rates, finally began to rise, to $6.50 an ounce, then $7, then $8. Was it then that Bunker and the Saudis hatched a conspiracy to corner the market? Everyone involved would always deny it, but the facts were there in black and white: on July 1, 1979, Bunker and Herbert formed a legal partnership on the island of Bermuda with front men representing Khalid Mahfouz and Gaith Pharaon, themselves almost certainly front men for Prince Fahd. Immediately this partnership, called the International Metals Investment Company—IMIC for short—began buying silver futures. In short order IMIC accumulated contracts representing forty-three million ounces of silver. Combined with the fifty-five million ounces the Hunts already owned, these purchases gave Bunker and Herbert indirect control of somewhere between 12 and 15 percent of the world's silver supply.

Then, just as the Hunt-Saudi partnership began buying, Prince Abdullah's front man, Naji Nahas, jumped into the market. Working through several accounts managed by a veteran broker at Conti Commodity Services named Norton Waltuch, Nahas managed to amass contracts representing forty-two million ounces of silver. Nahas's purchases, in turn, were mimicked by a half dozen mysterious Arab trading companies. Word of the developing scheme had spread, it appeared; any number of wealthy Arabs were buying in Bunker's wake in hopes of mirroring his profits.

But if half the Persian Gulf now seemed to know what was in the works, CFTC commissioners in far-off Washington hadn't a clue. In fact, as reflected in minutes of the commission's July 27, 1979, meeting, the agency's top men had only the dimmest clues to the Hunts' intentions.

"Can I ask one general question? The Hunts," a commissioner named David Gartner asked at one point. He went on:

> Every week we see them in something, silver, soybean oil, livestock, whatever. Do you think there's any possibility these guys are just having fun, just horsing around? Like playing Monopoly like you and I might do, or nickel and dime poker. Is this a little game they're going through?
> JOHN MIELKE, CFTC DIRECTOR OF MARKETS: Well, they're playing with some awful big bucks. I was looking at their silver position and on Chicago and New York combined—and I'm talking basically about the two brothers, Bunker and Herbert—their position . . . is worth 475 million dollars.
> GARTNER: That's a lot of money.
> MIELKE: That's a lot of money. . . .
> GARTNER: It just seems to me there are people with a hell of a lot of money and not a lot to do with their time, fiddling around like you and I might play a game of checkers.[2]

One week later, the first reports of the IMIC and Nahas-Waltuch silver purchases began to inch across Wall Street ticker tapes. At first they attracted little notice. But by the third week of August, IMIC's purchases were gaining steam. The price of silver responded, leaping to over $10 per ounce, to $10.61 on August 31. By the time CFTC commissioners returned from the Labor Day holiday, rumors of who was buying all this silver were flying fast and furious. Most traders assumed it was the Hunts. But even the Hunts couldn't buy this much. Someone had to be helping them. The best guess was the Saudis. CFTC attorneys began trying to identify IMIC's owners. Even if they could identify them, however, it was far from clear what the agency should do. At the CFTC's September 7 meeting, commissioners argued whether it was worth doing anything at all. "We're either going to get a nonanswer when we ask or the right answer, but even if we get the right answer—that it's the Hunts or the government of Saudi Arabia—what do we do?" one commissioner asked. To which another responded: "This is economic warfare. Is Saudi Arabia trying to corner the market? They should get out of the market if they don't give us the information."

It took two more weeks for CFTC attorneys to identify IMIC's owners, at which point Herbert Hunt confirmed the family's involvement. Bunker, meanwhile, met twice in those early weeks with Norton Waltuch and his client Naji Nahas in Paris, presumably to align the two groups' purchase strategies. They continued to move in lockstep. All through September, the Hunt and Saudi buying plowed onward. Silver's price rose as they did, hitting $17.88 on October 1. In time the mere sight of Norton Waltuch walking onto the Comex trading floor was enough to drive the price of silver up another fifty cents. By mid-October CFTC officials were beginning to catch on. They identified the two buying consortiums as the "Hunt group" and Norton Waltuch's "Conti group." It was then that officials of the two main silver exchanges, in Chicago and New York, realized they were facing their nightmare scenario:

They were running out of silver.

The Hunts and their allies now controlled 62 percent of all the silver in the Comex warehouse, plus 26 percent of all the silver held by the Chicago Board of Trade, the CBOT—in addition to all the silver they had stored in Zurich. It was the classic recipe for a market "squeeze." If the Hunts and Saudis kept buying, they would soon control enough silver to dictate the world price. In Chicago, Board of Trade officials beseeched Herbert to stop buying. Herbert promised they would. Within days, however, CBOT officials noticed another eruption in purchases. A few calls traced it—not to the Hunts themselves, but to accounts controlled by their children. Bunker, meanwhile, accepted the CFTC's offer to sit down with its staff in Washington to explain what he was doing. He proved the soul of reason. No, no, Bunker said, the Hunts weren't trying to corner the market. Yes, he knew a Saudi or two. But a conspiracy? Oh, no, no, no, no. The last thing he wanted was to cause problems for anyone in the market.

The CFTC was satisfied—for the moment. Not so the Chicago Board of Trade. The CBOT announced it would begin limiting silver holdings to three million ounces; anyone who held more had to reduce their positions by April 1980. Bunker was apoplectic. "You can't do it!" he told a CBOT official. "You wouldn't dare!" Pledging to fight the new limits in court, he told anyone who would listen that it was all one more Eastern Establishment conspiracy. Through it all, Bunker kept buying—and so did the Saudis. Silver hit $20 an ounce, then $25, then $30, before, on the last day of 1979, hitting an incredible $34.45 per ounce. In just four short months the price had more than quintupled.

At that point the CBOT, the Comex, and the CFTC realized something drastic had to be done. The Comex, the world's largest commodities exchange, took the lead, announcing on January 7 that it would soon limit individual trading positions to no more than ten million ounces of silver. In Dallas, Bunker howled. "I am not a speculator, I am not a market squeezer," he told a reporter. "I am just an investor and holder in silver."

Bunker reacted to the Comex news by increasing his silver purchases yet again, ordering a stunning thirty-two million more ounces at a cost of more than $500 million. Prices leaped in response, crossing $40 per ounce, then, on January 17, striking a heretofore unthinkable high: $50. At this level the Hunts' holdings were valued at $4.5 billion. Nearly $3.5 billion of that was profit, a sum larger than all the profits all the original Big Four Texas oilmen had made in their entire lives, H. L. Hunt included. One CFTC commissioner privately estimated that the Hunts and their Saudi allies now controlled 77 percent of all silver in private hands. Soon, another breathed, "they'll have all the silver in the world."

By January 21 the Comex board was in a panic. Dozens of new buyers were flocking into the market; many, it turned out, were Arab friends of Naji Nahas. Facing what had once been unimaginable—handing over control of an entire world commodities market to an unstoppable cartel—the Comex announced unprecedented new restrictions on silver traders. No more silver purchases would be allowed—none—only sales, and sales only to groups the exchange specifically approved. The next day, in Chicago, the CBOT followed suit. In Dallas, Bunker pounded his desk. This was blatantly unfair, he fumed. The powers that be were changing the rules just because he was winning their game.

He was right, but there wasn't a thing he could do about it. Once the Comex's draconian new limits were announced, Bunker and Herbert were all but trapped. They couldn't buy more silver. And if they sold, they risked triggering a panic. The very next day, January 22, silver plummeted to thirty-four dollars per ounce. The price stabilized in the upper thirties for several weeks before sliding again, sagging as low as thirty dollars per ounce in late February. By now the Hunts' silver play was front-page news in the world's business press. Bunker put on a brave face, giving a series of interviews in which he urged caution and predicted silver's imminent rebound. His appearances climaxed with an unlikely interview he gave Barbara Walters. How much silver, Walters asked, did he really own?

"I don't really keep track," Bunker said. "I don't count my money. It's bad luck, and bad taste, I guess."*

Off-camera, Bunker wasn't nearly so sanguine. In fact, he was desperately trying to prop up silver's price. His best hope was a deal he struck with a group of Kuwaiti and Bahraini sheikhs who had bought silver in his wake, cashed out at fifty dollars, and were ready to return to the market, presumably buying in overseas markets. The idea fell apart, though, as silver's slide began to hasten. In early March it sagged below thirty dollars, then tumbled all the way to twenty-one dollars by March 14. In fifty-five short days, Bunker and Herbert had sustained a paper loss of more than two billion dollars. It was at that point that something amazing occurred to the Hunts. They were running out of money.

III.

Carrying costs on all that silver would have already bankrupted lesser men. It cost three million dollars a year just to store it all. But the real cost was the cost of money. No one played the commodities market with cash. Most purchases, and almost all the Hunts', were made "on margin," that is, with money borrowed from the banks and brokerage houses that handled the trades. Bunker and Herbert had used their earliest silver purchases to collateralize loans to pay for their buying campaign with the Saudis. As the price

---

*The Hunts always denied trying to corner the silver or any other markets. They characterized their silver efforts as simply a solid bet on a solid commodity.

of silver rose through the fall, so did the value of their collateral, allowing the Hunts to take more and more loans to buy more and more silver contracts. As long as the price rose, their mountain of debt remained an afterthought. But once the price began to fall, so did the value of that collateral. The further silver fell, the more Bunker and Herbert had to fish into their own pockets for cash to pay the carrying costs on all those loans, a process known as margin calls. By mid-March, margin calls at just one of their brokers, Bache, reached almost ten million dollars *per day*.

But it wasn't just the silver. With the prices of both silver and oil skyrocketing, Bunker and Herbert had both gone on shopping expeditions of epic proportions. They had bought millions of shares of stock in companies such as Global Marine, Penn Central, and the First National Bank of Chicago; after Bunker's chat with Barbara Walters, rumors swept the stock market that the brothers were poised to buy Texaco. Bunker bought racehorses like other men buy cigarettes; by 1979 he owned more than seven hundred thoroughbreds, the largest stable in the world. Bunker alone spent more than fifty million dollars collecting ancient Greek and Roman coins; Herbert bought almost as much, concentrating on Byzantine issues. They bought artwork and more ranches and more real estate.

But then, all through January and February into mid-March 1980, hundreds of millions of dollars slid through Bunker's fingers like so many grains of sand. Every week the banks and brokers demanded more collateral; Bunker produced it by arranging still more loans, including two hundred million dollars from Swiss Bank Corp., in early March. Yet even as silver's price dropped, he needed more and more cash to sustain silver contracts he had arranged months earlier. Bunker might have pulled it all off—Merrill Lynch and other financial institutions were still offering credit—if only he had the time. But on March 14, 1980, the Hunts' time ran out. Once again, the rules were changing. That day the chairman of the Federal Reserve Board, Paul Volcker, in a move to stem an inflation rate that was rising to historic heights, ordered U.S. banks to begin curtailing their loans by instituting special "credit-restraint programs," noting with emphasis: "Special restraint should be applied to financing of speculative holdings of commodities or precious metals." Whether Volcker's demand was directed at the Hunts or not did not matter. The effect was the same. American banks and brokerage houses wouldn't lend the Hunts another cent.

This was the worst possible news: without new money to lubricate its pistons, the entire Hunt silver machine would begin melting away. Bunker didn't panic—at least not at first. Within hours he was on a plane to Europe; he had always been able to secure loans on the Continent, especially in Switzerland. But everywhere he went that week—to Paris, Zurich, Frankfurt, Bern, and Geneva—he heard the same dismaying message. Volcker had spoken; even though they didn't have to, European banks would follow his advice. Their vaults were closed. Even Swiss Bank Corp. turned Bunker down. And that was all it took.

On Monday, March 17, three days after Volcker's pronouncement, the Hunts' cash began to run out. For the first time, they were unable to meet Bache's margin call. Bache's chairman, Harry Jacobs, held off demanding the money another two days, but that Wednesday, March 19, Bunker began giving Bache bullion to satisfy the debt. By that Friday Bunker's situation was growing desperate. If word of their plight got out, silver would crash. He needed a miracle, or a few hundred million dollars, if he was to satisfy a new round of margin calls the following Monday, including a massive payment due the big silver producer, Englehard Industries. Which is how Bunker and Naji Nahas found themselves on a private jet descending across the Red Sea toward the Saudi city of Jeddah that Saturday. Prince Abdullah was his last hope.

That day Bunker met with one of his sons, Prince Faisal, and Saudi bankers. Bunker argued that silver's drop was temporary, that he could weather his cash crunch with a loan of two hundred million dollars or so. But the prince shook his head. His silver positions were now in the red; all the Saudis faced margin calls of their own. They could see Volcker's policy meant Bunker was finished. They would take their medicine, but if the Hunts went under, they would not go with them.

Bunker returned to Paris in defeat. They were unable to pay the margin call on Monday. Bunker spent that day and the next grasping at his last straw, a scheme devised by the Wall Street investment bank Drexel Burnham

Lambert to raise money selling bonds backed by his silver. Bunker was deep into these preparations on Tuesday, March 25, when he got the call from Herbert in Dallas. Another margin call, this one for $135 million, had come in. Bache would pay it, Herbert said, but swore it would pay no more. Bunker thought it over, then sent word back to Herbert via Lamar. The three-word message: "Shut it down."

It was over. The next day Herbert informed Bache the Hunts could no longer satisfy any further margin calls. To raise the needed cash, Bache informed the Hunts it would be necessary to begin selling their bullion. The next day, Wednesday, March 26, it began doing just that, unloading silver valued at one hundred million dollars, at a loss. At the same time, Bache notified the CFTC that the Hunts would likely lose eighty-six million dollars in the next day's trading—money the Hunts didn't have. Neither did Bache.

Suddenly there arose the specter that a major Wall Street investment bank might founder. This was the kind of news that could easily lead to a wider meltdown of world financial markets. In Washington, Paul Volcker and officials from the SEC and Treasury Department rushed into an emergency meeting. As they did, Bunker boarded a flight home from Paris. Before leaving, he had a letter dropped off at the Associated Press office. It contained a two-paragraph press release stating that the Hunts and their partners held more than two hundred million ounces of silver and that a "substantial portion" of it was to be sold in the form of silver-backed bonds. When the news crossed the wires, silver traders immediately viewed it for what it was, a pipe dream born of Bunker's desperation. Silver had closed at $20.20 that day. Wednesday saw a massive sell-off. At the close of business the price of silver had fallen to $15.80.

Bache's chairman, Harry Jacobs, was staring at disaster; every dollar he had lent the Hunts he had borrowed from someone else, and Bache, too, was running out of cash. He pleaded with Comex officials to shut the silver market. They refused. He telephoned Volcker and the CFTC. If silver fell to $9.85, Jacobs warned, Bache's credit agreements allowed its banks to call its loans. Bache would go under. Wall Street would panic. The entire financial system was in jeopardy. Volcker chaired emergency meetings all that day and into the night, but at two A.M. he finally threw up his hands. The markets would open on time. They had no choice. The government's regulatory structure simply wasn't designed to handle a financial crisis on this scale. There were too many chefs, and no recipe for a rescue. Worse, they didn't have enough information about what Bunker had been up to. All they had was hope.

The next day, Thursday, March 27, 1980, would go down in Wall Street lore as "Silver Thursday." At eight A.M., ninety minutes before the market opened, Herbert stepped to a pay phone in New York and called a staff gathering at the CFTC. He said market conditions had grown so bad that the Hunts would no longer pay their debts by selling silver; he urged the CFTC to close the market and take the unprecedented step of forcing holders of all silver contracts to settle at Wednesday's closing price, $15.80. Herbert's idea would allow the Hunts to sell their silver for $2.5 billion; after paying their debts, they might yet see a profit of $1 billion. If the market opened on schedule, Herbert warned, the consequences would be dire. "All the Hunt family will be washed out," he said. "We will go broke."

CFTC commissioners convened an emergency meeting at 9:30, nine people on nine phone lines with nine sets of opinions. It was impossible to tell what was actually happening, whether the Hunts actually faced a financial implosion or not. Maybe this was some sort of convoluted Hunt scheme to corner the market. No one knew. No one knew what would happen if the Hunts really were out of money.

"What about oil wells as collateral?" one commissioner asked.

"The Hunts say they have collaterized everything," a staffer answered.

"We just don't know," said another commissioner.

When trading commenced, every kind of rumor swept Wall Street trading floors. The Hunts had a one-billion-dollar margin call. Bache had a one-billion-dollar margin call. Merrill Lynch and the Hunts' other lenders, including First National of Chicago, might go under. Traders could agree on only one thing: sell. By noon an orgy of panic selling drove the price of silver down to $10.80. The stock market, staring at a worldwide meltdown, dropped like a stone, hitting its lowest level in five years, then, on the strength of rumors that the Hunts would somehow survive, rallied, ending the day almost even. And to everyone's surprise, the price of silver held through the day, closing at $10.80.

It was, all the papers agreed the next day, the wildest single day of trading since the 1929 stock market crash. Wall Street had survived. Bache survived. Somehow, to the CFTC's amazement, the Hunts survived. All around the world, in Washington, in Paris, in the Persian Gulf, traders held their breaths: Was it over? No one knew. The wary silence extended into the Friday trading day. To the relief of everyone, the markets calmed. Silver closed at $12. At the Federal Reserve, Volcker and his staff were just beginning to think the worst was over when, out of the blue, came ominous news. The Hunts, it turned out, owed Englehard $665 million for nineteen million ounces of silver, all due that Monday. If Englehard sued, the Hunts would have no option but bankruptcy. Worse, the Hunts faced another $900 million of silver-related debts in the coming weeks. If the Hunts defaulted, silver would collapse. The panic of Silver Thursday would resume—and spread.

For the moment, everything depended on Englehard and the Hunts. That Saturday Englehard executives flew to Dallas, where they met with Herbert and Lamar, who had quietly purchased a good deal of silver himself; Bunker was off traveling again. The Englehard men, led by CEO Milton Rosenthal, were polite but straightforward: pay on Monday, or be forced into bankruptcy. Herbert walked Rosenthal through a list of every asset the Hunts owned, from oil fields and ranches to all of Bunker's racehorses. All of it, Herbert emphasized, was already collateralized to someone else. There was no free cash. None. The Hunts, he said, were "under water." Bunker appeared the next day, hopelessly jet-lagged. At one point, he heaved a sigh and uttered probably the most famous words to ever escape a Texas oilman's mouth: "A billion dollars," he said, "isn't what it used to be."

The only possible solution, everyone could see, was a loan, a big one, maybe a billion dollars or more. In one small way, they were in luck. That same weekend every major banker in the United States, including all the Hunts' principal lenders, were gathered at the Federal Reserve City Bankers Association annual conference in Boca Raton, Florida. Paul Volcker himself was there, as a scheduled speaker, and he gave his tentative approval for some kind of bailout package. On Sunday both the Englehard and Hunt teams boarded private jets and arrived in Florida in time for meetings to begin that night at 10:00. Before heading to bed, Volcker told the bankers to keep him informed. As talks stretched past midnight, Volcker, clad in a wrinkled pajama top over his suit trousers, stuck his head into the talks a time or two.

By sunrise the outlines of a deal had begun to emerge. The Hunts agreed to give Englehard 8.5 million ounces of silver, plus a 20 percent interest in oil properties they owned in the Beaufort Sea off the northwest coast of Canada. Once Englehard was made whole, the question became what to do about the Hunts' remaining debts. No one involved was eager to make new loans, but without them, Bache and other major Hunt lenders were at risk of collapse. At Volcker's gentle urgings, a bailout package slowly took shape. For the Hunts, it came at a steep price. In return for $1.1 billion to repay their silver debts, the Hunts agreed to mortgage almost all of Placid Oil's assets, including 114 producing properties in North America, its best fields in Louisiana, and all its best tracts in the North Sea. At the time, it was the largest bailout of its kind in U.S. history.

In return, Bunker's older sister, Margaret Hunt Hill, demanded he and Herbert mortgage almost all *their* personal assets to Placid—all the race-horses, all the ranches, all the ancient coins, plus four million acres of oil and gas leases in Bunker's name, seventy thousand head of cattle, just about every last thing he and Herbert owned, down to CB radios, lawn mowers, and a water cooler. It was humiliating. But Bunker had no choice. The press coverage, predictably, was withering. Cartoonists depicted the porcine Bunker as a pig wallowing in silver. *Newsweek* put him on its cover. It and every other magazine leaped to compare him to J. R. Ewing, a comparison Bunker loathed.

Somehow, the Hunts had survived. Thank goodness, everyone agreed, the price of oil had kept rising, keeping the family's far-flung energy businesses healthy despite all the new debt piled on their balance sheets. As long as oil prices kept rising, Bunker and Herbert saw, everything would be okay. As long as oil prices kept rising.

EIGHTEEN

The Bust

"How do you get a Texas oilman out of a tree?"
"Cut the rope."
—Joke making the rounds in mid-1980s Dallas

I.

By 1982 Texas Oil had luxuriated in a solid ten years of record-high oil prices. Life was good—very, very good. The state's economic boom, however, couldn't last. The problems, as usual, began far outside Texas. A decade of skyrocketing oil prices had forced the world's utilities to scramble for cheaper sources of energy, leading to a wholesale recovery of the coal industry and especially the nuclear power industry, both of which by 1980 emerged as prime competitors to oil; oil's share of worldwide energy use began to fall, from 53 percent in 1978 to 43 percent in 1985. Oil demand shrank even further as Western countries begged their consumers to conserve energy—turning off unused lights and the like—while 1975 legislation mandating a doubling of automobile fuel efficiency cut American oil consumption by two million barrels of oil every day. The killer, though, was the 1979–82 recession. Out-of-control inflation forced the Federal Reserve to implement unprecedented restrictions in monetary policy, which drove interest rates to new highs, which in turn choked off government and corporate spending, which plunged the United States and much of Europe into the worst recession in fifty years.

All these factors—competition from other fuels, conservation, a deteriorating world economy—cut deeply into oil demand, which by early

1983 was forcing the world's oil-producing nations to ponder what had once been unthinkable: cutting prices. That February Britain reduced the price of North Sea oil three dollars, to thirty dollars a barrel. The government of Nigeria threatened to reply in kind. A month later OPEC oil ministers, meeting in London, announced the first price cut in their history, cutting prices by 15 percent, to twenty-nine dollars a barrel.

In Texas, drill bits began whirring to a halt. Across the state, wildcatters who had borrowed heavily to drill expensive deep wells saw the value of their collateral fall. Banks began calling in loans; many of the new wildcatters born during the 1970s began to go bankrupt or, like George W. Bush's renamed Bush Exploration Company, limped into shotgun mergers. As oilmen failed, Texas banks weakened. In October 1983, just eight months after the OPEC price reduction, the state's largest independent bank, First National of Midland, collapsed. In the next decade nine of the ten largest Texas banks would follow suit.

Texas Oil, the engine of the state's economy, coughed, then sputtered, then began, with a series of violent backfires, to die.

II.

Two weeks after John Murchison's death in June 1979, his thirty-one-year-old son, John Dabney Jr., moved into his corner office at Murchison Brothers. To those on the twenty-third floor, it was a show of stunning audacity, but John Jr. was just getting started. One of his first acts was sending a memo to his uncle Clint. "I want Murchison Brothers dissolved quickly," it read, "and I want the trusts delivered."

Clint just smirked, then crumpled the memo and lobbed it into a wastebasket. "We'll just call that the Dabney file," he mused.[1]

If John's death was a shock to the Murchison family system, its aftermath was worse. Within days of his funeral, John Jr., as coexecutor of the estate with his mother, Lupe, had taken control of his father's affairs. In the process he ripped away the veil of civility that for years had lain across the family's festering disagreements. As a businessman John Jr. was a rank novice, but he knew enough to realize that, left unchecked, Clint's investments could bankrupt them all. He wanted his family's fortune free from his uncle's speculations, but most of all he wanted his inheritance, and he wanted it yesterday.

Unable or unwilling to confront Clint directly, John Jr. racked his brain for a way to gain leverage. He found it in the steady stream of paperwork that now crossed his desk. Every investment, every loan—just about every significant piece of paper that went out under the Murchison Brothers letterhead—required two signatures, and in the summer of 1979 John Jr. simply stopped signing. In a matter of weeks the surging engine that was Murchison Brothers began to sputter. What John Jr. signed were memos, memo after memo after memo, all marked to his uncle's attention and all demanding that the trusts be released. Clint, it appears, threw every last one into the Dabney file.

To himself, John Jr. suspected Clint would never relinquish the trusts without a fight; too much of their stock was already pledged for Murchison Brothers collateral, and Clint needed every available asset to surmount his continuing cash-flow problems. In John Jr.'s mind, there was only one thing to do. As summer turned to fall, he began canvassing the best attorneys in Dallas, looking for a lawyer brave enough to sue Clint Murchison. To his dismay, he couldn't find a single one. It wasn't just that half the Dallas bar had done work for Murchison Brothers. For anyone planning a career in Dallas, suing the owner of the Cowboys was suicide. After months of frustrated interviewing, John Jr. ended up hiring an obscure country lawyer based in Georgetown, north of Austin, to begin planning his legal assault.

By autumn the atmosphere on the twenty-third floor had turned glacial. Clint and John Jr. had stopped speaking. Just about everyone at Murchison Brothers, from the secretaries to the garrulous Lou Farris, who had emerged as Clint's right-hand man, sided with Clint. To be seen speaking with John Jr., much less entering his office, was viewed as a ticket to the unemployment line. As his nephew's obstructionist campaign continued, Clint, growing angrier by the day, turned to the one person who might be able to control him: his mother, Lupe.

Her son, Clint told Lupe, was strangling Murchison Brothers. Everything, the cash-flow problems, the debt, the question of the trusts—*everything*—could be resolved in time, he promised, if only he had the freedom to maneuver. Clint pleaded with Lupe to make John Jr. see reason. She tried. He wouldn't. For months Clint kept at his sister-in-law, nagging, wheedling, until taking his argument to its logical conclusion. If John Jr. wouldn't see reason, Clint said, he had to be removed as coexecutor of his father's estate. Unless his nephew was evicted from Murchison Brothers, they could lose everything. Gradually Lupe weakened. She knew nothing about the family's finances. She could barely balance a checkbook. But at Clint's insistence, she began talking to lawyers, and to her surprise they agreed with Clint.

The cold war at Murchison Brothers was entering its second year when, in the fall of 1980, Lupe finally acceded to Clint's wishes and, along with her three grown daughters, sued to remove her son as coexecutor of his father's estate. John Jr. was stunned. When the news hit the papers, so was Dallas; it was the first anyone outside the family knew of the turmoil within. Overnight John Jr. found himself shunned not only by his mother, his sisters, his uncle, and his cousins, but by much of haute Dallas. The whispers raced from the family through their friends to anyone who would listen: John Jr. had lost his mind, he was unstable, he was blowing money on cars and strippers and who knew what else. None of it was true, but it didn't matter. When John Jr. tried to have dinner at Brook Hollow, other members actually walked away. In a bit of perverse payback whose humor was lost on the club's members, he returned one evening with an unusual date, a stripper who catwalked through the stately Brook Hollow dining room wearing nothing but a black-leather string bikini. And tassles.

John Jr. was absolutely alone, but he was determined to get justice. He hired the noted Houston criminal attorney, Richard "Racehorse" Haynes, and in February 1981 returned fire, suing Clint as well as his four children, demanding his share of the trusts plus thirty million dollars in damages. It had become, as more than one wag noted, a real-life *Dallas*. Then, much as happened at the height of the Cullens–Di Portanova struggles, things got strange. Lupe's suit against John Jr. was nearing its April trial date when, returning to the Big House one evening, she and a friend were confronted by two masked men holding pistols. The friend was bound and gagged, then thrown into one of the family Mercedes; Lupe was shoved in the trunk.

When her daughter Barbara arrived a bit later, she and two of her friends were bound, gagged, and left in the living room. The men prowled the house for a full four hours before leaving. Afterward Lupe discovered almost nothing had been stolen, just a few pieces of jewelry.

It was an odd home invasion. The two intruders had used a two-way radio to communicate with someone outside the house, as if they were receiving instructions. Though there was no evidence to back their musings, John Jr.'s attorneys speculated privately that Clint had been behind the robbery. Lupe had been having second thoughts about the upcoming trial, and John Jr. came to believe his uncle had staged the whole thing in hopes that Lupe would blame her own "irrational" son, thus renewing her anger. No arrests were ever made.

The robbery left Lupe traumatized, but the trial went forward, opening in a Dallas courtroom in April. John appeared in court, seemingly ready for battle, but was shocked when, on the first day of jury selection, one of Lupe's attorneys launched into an impassioned denunciation of him as an ungrateful son who was jeopardizing the family fortune to get his hands on money he hadn't earned. John later said he expected to be portrayed as naive, even dim-witted, but the vigor of the attack left him shaken. Lupe's attorneys suspected his discomfort had more to do with his dwindling savings; with little more than an eighteen-thousand-dollar salary behind him, it was far from clear he had enough money to pay his lawyers. Whatever the case, John Jr. initiated settlement talks that very night, and by dawn the deal was done. On the surface it amounted to John Jr.'s total capitulation. He resigned as coexecutor of his father's estate. In return Lupe agreed to lend him three million dollars. John Jr. rationalized it as the only way he could raise the money to pursue the fight that really mattered, the lawsuit against Clint to free his trust. He hired new attorneys and got to work.

Clint, while relieved to be rid of his nephew, had little reason to celebrate. In the two years since his brother's death, his financials had continued to deteriorate. Interest rates remained high, housing starts low, and every month was a fight to feed the banks. It didn't help matters that not even Clint could keep straight the hundreds of loans and investments in the Murchison Brothers portfolio; organization had never been his forte. "The only way we knew a loan had come due was a banker would call," remembers one aide. "Then all hell would break loose, trying to find a way to pay it." Clint had borrowed seven million dollars in 1979 to start a new luxury development, the Summit, in Beverly Hills, but two years later what few homes had been finished remained unsold. He was forced to borrow another three million dollars just to pay interest on the first loan, then added a second mortgage.

The crippling blow, though, was the fall in oil prices, which threatened to sever the last leg of Clint's tottering financial table. The value of all his remaining oil properties fell sharply, as did real estate, and lenders began demanding more collateral to offset the drops in value; almost all of Clint's collateral, however, was already pledged to others. For a time his balance sheet was buoyed by a rise in the stock of Kirby Exploration of Houston, a publicly held wildcatter descended from oil companies assembled decades earlier by John Henry Kirby; Big Clint had purchased a controlling stake in the company from Kirby's estate in 1956. On the promise of a new natural gas field, Kirby's shares leaped as high as forty-five dollars in early 1981; Clint wasted no time leveraging the increase into another big loan. But when the field failed to pan out, the stock cratered, within months falling to thirteen dollars. Clint was left staring at a hundred-million-dollar loss, along with another lender angrily demanding more collateral for his loan.

Through it all Clint remained a picture of optimism, assuring his men everything would be fine once interest rates fell. In 1982 he decided to go on the offensive, snaring tens of millions of dollars in new loans to start or finish developments in Palm Springs, New Orleans, and other Sun Belt cities. Almost all this money came from out-of-state lenders who, unlike the Dallas banks, had little sense how deep Clint's troubles actually ran. He began paying quiet visits to lenders in New York and St. Louis and Memphis. His calling card was always the Cowboys. Clint would talk football with goggle-eyed Bucks and Bubbas at some middle-market bank in Atlanta and walk away with another ten or twenty million that by month's end would be used to pay another bank. In time, though, word spread, and Clint began trolling for money in murkier waters. Among his lenders of last resort, his lawyers would later disclose, was a Louisiana banker named Herman K. Beebe, whose known associates included the New Orleans mob boss Carlos Marcello. In 1985 Beebe would be sentenced to a year in prison for bank fraud.

Perhaps unsurprisingly, it was as his financial condition reached a crisis point in early 1981 that Clint found God. Over time, Anne had simply worn him down. He had been attending church with her for several years, and at her insistence had even sworn off alcohol; the days of cocaine and multiple mistresses were now just a fading memory. When Anne began hosting Bible classes at the mansion in 1978, Clint sat on the edges of the group, half listening. In time he moved his chair in a bit and began asking questions. Then, in 1981, Clint went to see Anne's pastor, Olen Griffing. Griffing's church, the Shady Grove Church in Irving, was a thirty-minute drive and a world apart from everything Clint had ever known. Its blue-collar membership was made up of hard-core fundamentalists who shouted and sang and spoke in tongues.

Griffing had seen this coming. As his finances tightened, Clint had occasionally asked him about his beliefs, including a long discussion on the beach at Spanish Cay; he showed a special interest in how a man could become "saved." Now Clint said he was prepared to give his life to Christ. Griffing suspected Anne was behind the approach, but Clint denied it. One day in the fall of 1981 Griffing drove to the mansion. Clint opened the door in his swim trunks. Together the two men waded into one of the swimming pools. When Clint was ready, Griffing dunked him beneath the water and baptized him. The ceremony was followed by a formal baptism, with Anne standing alongside, at the Shady Grove Church one Sunday that December. Follow the word of God, Griffing intoned, and all your sins will be washed away. The pastor had no idea, of course, how many sins there had been.

Clint knew. As he studied the Bible, they weighed heavily on his mind. He became convinced he would need to pay for his sins. And at some point, apparently in mid-1981, the Lord delivered the check. In truth Clint had known something was wrong for at least two years. His walk had grown unsteady. At times he had trouble maintaining his balance. In early 1982 he fell on a Cowboys flight and broke two ribs. When he began to "wobble," as he put it, friends assumed he had fallen off the wagon. Finally Clint had tests run. They revealed he was suffering from a rare degenerative nerve disease called olivopontocerebellar atrophy, or OPCA. Related to Parkinson's, OPCA is one of several diseases known as "Parkinson's Plus" because they attack all across the central nervous system. For now Clint was just "wobbly." In the future, doctors told him, maybe a year, maybe five, he would suffer tremors, mild at first, then worse. Eventually he would lose the ability to speak. And he would die. No one could say how long he had.

Clint, now fifty-eight, took the news with grace. But he knew in his heart why he was sick. His son Robert later recalled a Bible-study session at the mansion not long after Clint's diagnosis. Another attendee suddenly asked Clint if he understood what had caused his illness. "I think," Clint replied, "that it's because of my past sins."[2]

III.

John Jr. would not forgive, and he would not forget. In October 1981, at roughly the same time Clint underwent baptism in his swimming pool, his nephew's attorneys scored their first victory, persuading a Dallas court to make Clint release six million dollars in Optimum Systems stock held in John's trust. That left twenty-four million dollars John still wanted, and in December he attempted to capitalize on his success with a settlement offer. In broad strokes, he offered to drop his claim for thirty million dollars in damages if Clint would just release the last twenty-four million dollars in his trust. John Jr. then revealed his trump card. If Clint didn't settle, his attorneys made clear, his nephew was poised to go public with details of Clint's financial condition. If he did, Clint's attorneys advised, it could initiate a feeding frenzy. Each of Clint's banks knew but one side of the elephant; if they realized how sick the entire animal was, it could release a flood tide of lawsuits as each attempted to lay legal claims before the others. There was no way Murchison Brothers could survive such a run. It would mean bankruptcy.

Clint, distracted by his illness and the demands of feeding the banks, dragged his heels about making a decision for more than a year. In April 1983 he finally gave in, handing over John Jr.'s twenty-four million dollars and shifting the debt encumbering his late brother's assets to his own. He knew it had to be done, but his greatest fear appeared to be the specter of his faltering finances splashed across the newspapers. He could just see it: the vast Murchison empire, Big Clint's legacy, the second-greatest family fortune in all Texas, ruined by his own stupidity. Surrendering to his nephew, however, did nothing to calm family tensions. Once John Jr. got his trust money, his sisters wanted theirs. Lupe, too, began making noises about finally removing her family's assets from the Murchison Brothers partnership. A series of meetings ensued between Lupe's family and Clint's, dozens of attorneys and accountants crammed into conference rooms on the twenty-third floor.

For Clint, it was unbearably sad. By and large the meetings remained civil, but everyone involved understood that if some kind of agreement couldn't be worked out, they were headed for years in court and years of humiliating headlines. "Everyone was threatening to sue everyone else," one person in the room recalled. "It didn't matter that these people were family." By late 1983 the outlines of an omnibus settlement had emerged. All the children would get their trusts. Lupe would gain control of most of John's remaining investments. Clint left the negotiating table with a mound of debt, a collection of companies leveraged to the hilt, a stack of message slips from angry bankers—and the Cowboys. To those closest to him, Clint appeared to take solace that at least he still owned America's Team. As long as he still had the Cowboys, he was still Clint Murchison.

Numbers are cruel things, though, and Clint knew what they said. At first he just mused about what the team was actually worth. In early 1983, just as he was settling with John Jr., he had a friend send out quiet feelers; in no time, he received an offer for forty million dollars. Clint brushed it aside. "Thanks," he said, "but you know, selling the Cowboys would be like selling one of my kids."[3] When rumors of a sale hit the papers, Clint's men reluctantly confirmed them, lamely fibbing that it might be needed to settle John's estate.

By now Clint had grown visibly ill. The tremors had begun. His hands shook. He began using a cane. His businesses were even sicker. Finally, in the spring of 1983, Clint called Tex Schramm to the mansion. The two men had been together twenty-four years; Schramm was the only general manager the Cowboys had ever known, and he was immensely fond of Clint who, unlike so many sports-team owners, had never interfered with his work. The Cowboys, if no longer the goliaths they had been a decade before, were still thriving. They had made it to three more Super Bowls in the 1970s, winning in 1978, but after Roger Staubach's retirement following the 1979 season, it became clear America's Team had peaked. Their decline, in fact, exactly paralleled Clint's. Between 1980 and 1982 Dallas lost three straight NFC championship games. Yet the Cowboys' image as America's foremost sports franchise—their only rivals were the New York Yankees—was as strong as ever. Clint's players still popped up in movies and television shows and loads of commercials, Tom Landry shilling for American Express, Charlie Waters and D. D. Lewis for Lite Beer, Randy White for Dannon Yogurt.

That day at the mansion, Clint broke the news to Schramm: It was time to sell. He blamed his health. Schramm handled the ensuing auction. Offers poured in from across the country, but Clint wanted the Cowboys to remain in Texas hands. In late 1983 Schramm reached a tentative deal with a group of eleven Dallas investors led by the oilman H. R. "Bum" Bright. The offer was sixty-five million dollars for the team, plus twenty-five million dollars for the lease on Texas Stadium; it would be the highest price ever paid for an American sports team. John Jr., whose side of the family still owned half the team, warned Lupe that Clint might try to reduce the team's price tag in favor of more money for the stadium lease, which Clint alone controlled. His concerns slowed the closing for months, but on March 19, 1984, the deal finally went through: the Cowboys, the team that had been Dallas's salvation during the dark days of the 1960s, that had symbolized a new Texas in the 1970s, and that had briefly made Clint Murchison King of all Texans, were no longer his. The Dallas newspapers devoted hundreds of column inches praising Clint's legacy, practically elevating him to Texas sainthood. Not a word, however, was written of how quickly Clint's hungry banks snapped up the sale's proceeds to satisfy his debts.

In those last ugly years, it seemed that every time Clint struck a deal to rescue himself, it spawned more trouble. The dissolution agreement in 1978 begat John Jr.'s attacks; the settlement with John Jr. begat squabbling with Lupe and her daughters. So it was with the sale of the Cowboys. Clint had begun missing debt payments at least a year earlier. The first lawsuits had come in November 1983; a Cleveland bank charged that Clint owed two million dollars on a meager four-million-dollar loan. A week later a bank in Paris sued, seeking four million dollars. In March 1984, just as the Cowboys were sold, an Arkansas savings and loan hit him with a suit demanding twenty million dollars. In short order another sought twenty-five million dollars.

After the Cowboys sale, the legal dam burst. With Clint's last healthy asset gone, dozens of lenders, realizing the Murchison cupboard was all but bare, raced to lay claim to what remained. Citicorp sued. Merrill Lynch sued. Even one of the banks Clint owned, Nevada National, sued. By Labor Day 1984 it seemed there was a new suit every day. Clint was no longer in any shape to fight back. He had taken to a wheelchair. He stopped coming into the office. When the lawyers needed him, they went to the mansion. By Christmas even that was under threat. To their horror, Clint's attorneys discovered that in his scramble for cash Clint had been *personally* guaranteeing loans, meaning that lenders could now foreclose not just on Murchison Brothers assets, but on Clint's home, his cars, his farms—everything. In early December a Fort Worth bank, desperate to retrieve $9.7 million, filed notice it planned to auction off the twenty-four acres of land surrounding the mansion. During a court hearing it came out that it was only one of *sixteen* banks with liens on the land. Only a restraining order arranged by Clint's lawyers allowed him to remain in his home for Christmas.

At the mansion, Clint sat in his wheelchair, staring, waiting. Just as the doctors had predicted, he had been losing his ability to speak, and as a series of winter storms coated Dallas in a silvery coating of ice, his voice disappeared forever. His creditors, their claims now approaching $175 million, began telephoning his doctors at Sloan-Kettering, demanding to know how long he had. But if Clint's body was failing, his mind remained sharp. In January he hired a new lawyer, Philip I. Palmer, and with Palmer's help he devised an audacious bailout plan. It depended on warm memories—those of his father's oldest friend. Palmer placed the call, to a glass skyscraper thirty miles to the west. And then, hoping a deal could be struck, they invited attorneys representing more than thirty of Clint's creditors to a meeting at the mansion on Friday, February 1.

That morning the sun rose on a scene from a chicken-fried *Citizen Kane*. The great man, ailing but still alert, all but alone in his vast mansion; his hundreds of handpicked live oaks and azalea bushes encased in armor of uncaring ice; the dark-suited attorneys in their BMWs and Mercedes creeping up the slickened driveway from Forest Lane, jaws dropping as they caught their first glimpse of Dallas's Shangri-la. All had read of the Big Rich. Few had seen their lives up close. More than one simply shook his head. How had it come to this?

Inside, everyone gathered in the living room. They watched as a servant pushed Clint into the room in his wheelchair. He didn't speak; he couldn't. Philip Palmer did the talking. As the lawyers leaned forward in their chairs, Palmer announced that Clint had tentatively agreed to a bailout package, a massive cash infusion coupled with plans to develop or sell eight of Murchison Brothers' largest real estate developments around the country. His rescuer, Palmer revealed, was none other than Sid Richardson's great-nephew, Bob Bass. It was a moment of surpassing poignancy. Forty-four years earlier it had been Clint Murchison Sr. who gave Sid Richardson the loans he needed to survive the bleakest years of the Depression. Now, Big Clint's son slouched in his wheelchair fully aware that Sid Bass and his brothers had since achieved everything he hadn't, that while the Basses were investing in Wall Street stocks and high-tech start-ups, he had been snorting cocaine.

Now, it appeared, the Basses would return Big Clint's long-ago favors. One of Bob Bass's men was there, and as the lawyers broke into groups, they pressed him how real this rescue package was. From his answers they gathered it wasn't; everything appeared to be in the discussion stages. One of the creditor attorneys argued that Clint should be forced into involuntary bankruptcy. Others objected, fearing the consequences if he somehow managed to recover. Afterward everyone tried to shake Clint's hand, wished him the best, and drove out beneath the ice-covered trees with nothing resolved. Within days, to no one's surprise, the promise of a Bass-family rescue dissolved like a West Texas mirage. A week later a trio of creditors, led by

Citibank, decided enough was enough. They asked a Dallas federal court to place Murchison Brothers in Chapter 7 bankruptcy. After that Clint had no choice. On February 22, 1985, his attorneys asked the court to convert the petition into a voluntary Chapter 11 bankruptcy. It was over.

In the ensuing months the clerk's office at the Dallas courthouse slowly filled with filings that revealed, in harsh black and white, how the once-proud Murchison empire had been overwhelmed by a wave of debt. To many, it was almost incomprehensible how Clint had staved off disaster so long. Total creditor claims, both against Murchison Brothers and Clint personally, eventually rose to more than $1.15 billion. Clint's assets, including Spanish Cay and the mansion, were valued at barely $71 million. His cash on hand: $4,876.66.

For the next year Clint remained in his wheelchair, mute, while the attorneys liquidated his estate. Sales of real estate brought in $300 million or so, the rest maybe $50 million, leaving creditors with barely twenty cents of every dollar the Murchisons had taken. Clint's suite at Texas Stadium alone brought $920,000. The mansion sold for $14 million to a real estate developer, who planned to subdivide the surrounding land into home lots. For Clint, the final humiliation was the October 1986 garage sale he and Anne were forced to endure before moving out. They wheeled him into the auction, where he sat, glassy-eyed, as bidders came up to pat his hand. Several had tears in their eyes. The sentiment disappeared, however, once the bidding began.

Those handling the sale had high hopes for Clint's twelve-foot mahogany dining table with its ivory inlays. "The table," an auctioneer named Perry Burns announced to the crowd, "was built just for the Murchisons and is valued at twenty-five thousand dollars."

The first bid came in at one hundred dollars.

"I can't believe this," Burns said. "Every important person who visited the city of Dallas in the past twenty years has eaten at this table. Let's get some serious bidding at six thousand dollars."

The bidding inched up, to five hundred dollars.

"This is history," Burns pleaded. "We are talking about history." The table sold for twenty-nine hundred dollars.

Clint's art did no better. A collection of jungle-themed paintings and tapestries, valued between four and twelve thousand dollars, sold for an average of five hundred dollars each. Not even his collection of Cowboys memorabilia stirred the crowd. At one point, Burns held up a scrapbook containing Clint's ticket to the 1971 Super Bowl; it had been autographed by many of the players. "Look," Burns said, "there's notes from Roger Staubach, Bob Lilly, Calvin Hill, Tony Fritsch—everybody. You can't pass this up."

Someone bid twenty-five dollars.

"You've got to be kidding," Burns said.

"Fifty dollars," another bid called out.

Burns made a face. "Oh, how I wish the oil was still flowing," he said.[4]

Dallas, a city that had always run on boosterism and hero worship, had little interest in a failure. It was a cold good-bye, much the same as Texas was handing scores of men forced into bankruptcy that year, including John Connally, who had gone into real estate, the Houston developer Harold Farb, and the renowned heart surgeon Denton Cooley. As the Murchison auctions continued that November—they stretched on through the winter, including several days where the public was allowed to roam through the mansion—Clint was finally wheeled out of his home. He and Anne moved across the street into a small tract house not much bigger than the mansion's living room. Clint spent his days there in a tiny bedroom, looked after by a nurse, obliged to ring a little bell when he needed something. He faded quickly. In March 1987 he contracted pneumonia and was admitted to a hospital. By then he was a vegetable. Anne and his children visited regularly, but there was little to do but wait. On Sunday, March 29, John Jr. appeared, holding his uncle's hand and forgiving him. Clint's eyes were open, but he couldn't speak. He died the next evening.

The following day the Morning News ran an editorial thanking Clint for giving Dallas the Cowboys and produced two articles on his life. And that was it. The era of men like Clint Murchison Jr. had passed. Texas was moving on. Three days later there was a strange coda when many of Clint's old pals left his memorial service scratching their heads. They held it at the Sacred Heart Church. Anne had been in charge. Tom Landry and others gave the eulogies, but the stage was dominated by Pastor Olen Griffing, who stood in a white suit before a Plexiglas podium. He kept referring to "Brother Clint." Someone played a guitar solo. As Clint's pals traded glances, a troupe of girls dressed in togas came out and did some kind of interpretive dance.

This was a side of Clint's last years that few had glimpsed; of those in the crowd, few had ever heard him speak of God. Obviously Pastor Griffing had. "I wouldn't be surprised," Griffing said at one point, lifting his eyes toward heaven, "if Brother Clint's up there right now, dancing before the throne and shouting, 'Great is the Lord!'" For those who had watched the onetime King of all Texans turn somersaults between the tables at '21', who had seen him cadge a million-dollar loan at a public urinal, who had known him to bed seven different women in as many days, who had flown to his private island in a zebra-lined jet, it would not be the way they remembered him at all.

IV.

While oil prices soared into the early 1980s, the stock market didn't. As a result, almost every big oil company was sitting atop oil and gas reserves that, at thirty dollars a barrel, were worth far, far more than the total value of their stock. Analysts called this the "value gap." It didn't take long before the savviest American investors realized that the cheapest place to obtain oil reserves was no longer the floor of the North Sea. It was the floor of the New York Stock Exchange.

Among the first to recognize this were Texas oilmen, one in particular. His name was T. Boone Pickens. Born in 1928, making him just two years younger than Bunker Hunt, Pickens was a restless, headstrong geologist who had left hidebound Phillips Petroleum to build an independent oil company in his native Amarillo. He called it Mesa Petroleum. Pickens made his first fortune in Kansas natural gas, but his true genius lay in his grasp of the capital markets. In 1969, long before hostile takeovers were de rigueur, he had completed the takeover of Hugoton, a gas producer far larger than Mesa.

Pickens spent the 1970s drilling for oil and gas around the world, but by 1982 realized, with a start, how thoroughly undervalued American oil companies had become. Pickens launched his first attack on Cities Service, the nation's eighteenth-largest oil company, which happened to be three times Mesa's size. Cities replied with a tender offer of its own—for Mesa. At that point Gulf Oil swooped in with an offer for Cities of its own, forcing Cities executives to sell out to Occidental Petroleum. Pickens walked away with a thirty-million-dollar profit. In the following years Pickens made offers for Phillips, Gulf, and Unocal, never actually buying his quarry, but always making money on his stock. He cloaked his attacks in the populist verbiage of "shareholder rights," and to some he became a kind of folk hero. It wasn't long before he became the latest Texas oilman to adorn the cover of *Time*.

At their glistening new Fort Worth headquarters, its executive offices adorned with the works of Jasper Johns, Sid Bass and Richard Rainwater took notice. They had stayed busy the last few years, buying a set of resort hotels from American Airlines, then accumulating stakes in several public companies that they happily sold back to management for higher and higher profits; they made sixty million dollars alone on a little-noticed deal with Blue Bell. Then, in 1981, Rainwater was leaving a Wall Street banker's office when, just as the elevator doors closed, the banker piped up, "You should look at Marathon!"

Marathon was the country's seventeenth-largest oil company, with annual revenues of $8 billion. Sid knew it well; every Texas oilman did. It was Marathon, then named Mid-Kansas Oil, that codiscovered the massive Yates Field south of Midland in 1926, the strike that opened West Texas to leasing; Marathon had been pumping oil from the field ever since, more than 800 million barrels, with some 1.2 billion barrels still in the ground. "It's just this massive cavern of oil," Bass told Rainwater. "I mean, you just turn on the spigot and the oil comes gushing out." In a good year the Yates produced 25 million barrels plus natural gas; with oil prices at $30 a barrel, that meant the field was generating $750 million in revenues a year, and Marathon had fields all over the world. Yet when Sid did the math, the value of its stock was barely $3.8 billion. "That's absurd," Sid breathed.

They began buying Marathon stock in early 1981. By the time executives at giant Mobil Co., seeing the same disparity in values, began making noises about taking over Marathon that autumn, Bass and Rainwater had accumulated just over 5 percent of Marathon's shares, at a cost of $148 million. At that point federal regulations required Bass to formally notify the SEC of their holdings. Marathon executives, fearing a hostile takeover, panicked. A month later they sold out to U.S. Steel. Once the merger was finalized, Bass and Rainwater walked away with $160 million in profit, more than doubling their investment. It was by far the best deal they had ever attempted. "Wow," Bass said as the enormity of their gain sunk in. "Let's do that again!"

They began looking at every oil company of any size, even as they continued buying and selling shares of dozens of non-oil companies. Takeover fever was pushing the stock market higher and higher, and even their smaller investments were paying handsome dividends; the Basses took home $50 million selling their stock in one company, Amfac, back to its executives, a deal so small the press barely noticed. As the number of takeovers began to mount, Bass and Rainwater found themselves for the first time in the "deal flow," that is, a regular stop for just about any Wall Street deal maker with a hot idea or an open hand.

Their next major investment was Texaco, which found itself the target of rampant takeover speculation in 1984. Bass and Rainwater bought millions of shares, eventually amassing almost 10 percent of Texaco's stock. They ended up selling it back to the company at a profit of four hundred million dollars. By the time they sold, Rainwater had already selected their next play: the Walt Disney Company. Disney was being hounded by the New York investor Saul Steinberg, who had the backing of one of Rainwater's Wall Street pals, the junk-bond guru Michael Milken. Rainwater had picked up a Florida real estate outfit the year before, and when Disney inquired about buying it, the talks led to a deal in which the Basses agreed to swap the company, Arvida, for two hundred million dollars in Disney stock, about 7 percent of its shares. It was a way for Disney executives to get a chunk of their stock in friendly hands.

As Sid studied Disney's financials, he was underwhelmed. It lost money year after year. Its movie studio hadn't had a hit in decades. The theme parks were aging and worn. Still, Rainwater had done his homework. Disney, he swore, was a neglected gem. The stock was stuck in the fifty-five-dollar range. Simply doubling the thirteen-dollar admission price to the theme parks, money that would sink straight to the bottom line, could make the company worth a hundred dollars a share. With proper marketing and advertising, he told Sid, Disney might be worth as much as two hundred dollars a share.

In the summer of 1984, six months after the Basses acquired their first shares, Disney's management bought out Saul Steinberg. But no sooner had Steinberg withdrawn than more wolves attacked. The Wall Street arbitrageur Ivan Boesky—later to be imprisoned in a notorious insider-trading scandal—accumulated a massive stake in Disney stock, as did a predatory investor named Irwin Jacobs. Disney executives, who met and liked Sid, turned to the Basses for help. In a series of lightning maneuvers, Bass and Rainwater used their $400 million profit from Texaco to buy out Boesky and Jacobs, giving them a commanding 18 percent share of Disney shares. They bought still more, raising their stake to nearly 25 percent, at a total cost of $500 million. In barely six months, they had taken effective control of the company. In short order Bass pushed to hire a new chief executive, Michael Eisner, who in the years to come would transform Disney into one of the most powerful entertainment companies on Earth. By the early 1990s the Basses' $500 million stake in Disney would be worth a staggering $2.8 billion.

It was Sid's crowning achievement. In a span of just sixteen years, in one of the greatest investment performances of the twentieth century, he and Rainwater had increased the Bass family's fortune from $50 million to an estimated $5 billion or more. With Disney, Sid had put all his eggs in one sparkling basket; he would remain its largest shareholder for years. Though left unspoken, Disney also marked Sid's retirement from major investing; almost all his capital, in fact, was now wrapped up in Disney stock. Rainwater resigned not long after, moving into offices below Sid's to make deals on his own; he would go on to be one of the most successful investors in Wall Street history.

Sid's "retirement," however, was bittersweet, for even as he took effective control of Disney he was obliged to break up his father's legacy, Bass Brothers Enterprises. Each of the four Bass brothers would go his own way. The breakup was almost all his brother Bob's doing. Bob and Little Anne seemed to go out of their way to separate themselves from the rest of the family. Raised Methodist, Bob became an elder at a Fort Worth Presbyterian church. He quit using the family's security people, preferring his own. Bob and Anne became ardent historical preservationists—Bob later became chairman of the National Trust for Historic Preservation—and led a local fight against a highway extension that the rest of the Basses favored.

For years Bob had been pressing for more responsibilities at Bass Brothers; Sid and Rainwater, however, worked so seamlessly, there was little left for Bob to do. After agitating to manage his own money for several years, Sid had finally agreed to split up Bass Brothers in 1983, at which point Bob established his own company and began making his first major investments. It took the better part of three years, from 1983 to 1985, to finalize the breakup. The problem was taxes. As the laws stood in 1983, the breakup of the far-flung Bass investments would subject each of the brothers to hundreds of millions of dollars in taxes. Sid prevailed upon Texas senator Lloyd Bentsen to draft a bill to change the tax law, but it went nowhere. In the end, however, a favorable IRS ruling achieved the same result. By the time the breakup went through, Sid and Bob were no longer speaking. Each of the four brothers received a quarter of everything, all told, more than one billion dollars apiece.

The youngest, Lee, stayed in Fort Worth to run the oil company. The second brother, Ed, who had returned from New Mexico in 1979 to develop a downtown area of shops and office space he named Sundance Square, stepped up his commitment to environmental causes. By the time of the breakup Ed was already, as the *New York Times* dubbed him, the largest private sponsor of environmental research in the United States. Among his many and varied projects were a 1,042-acre rain forest preserve in Puerto Rico, a backpackers' hotel in Katmandu, an eighty-two-foot ocean-research ship, a twenty-acre research farm in Southern France, nine hundred square miles of ranchlands in the Australian outback, and the twenty-million-dollar Institute of Biospheric Studies at his alma mater, Yale. He sat on the boards of the World Wildlife Fund, the New York Botanical Garden, the African Wildlife Foundation, and the Jane Goodall Institute for Wildlife Research, Education and Conservation.

Over the years Ed was dogged by accusations that he remained under the influence of John Allen, the futurist he had met at New Mexico's Synergia Ranch. Ed always brushed off the accusations, but much of his work furthered Allen's stranger ideas, including a plan to construct self-contained biospheres where humans could restart civilization after a nuclear holocaust. Ed's most visible project was just such an enclosure, called Biosphere II, a sprawling set of glass buildings in the Arizona desert where, in 1989, a set of "bionauts" would live for two years without any contact from the outside world.

Bob Bass, meanwhile, took his share of the Bass fortune and launched a brief career as an aggressive Wall Street investor, launching a series of hostile takeover attempts and leveraged buyouts during the late 1980s. He and Little Anne remained active historical preservationists. They purchased Ulysses S. Grant's onetime Georgetown mansion and made a second home in Washington, D. C. Their relations with the rest of the family, especially with Sid, never recovered.

Sid, meanwhile, began spending even more time in New York, where in June 1983 he had laid out a then-record $5.25 million for a sprawling Fifth Avenue apartment. It was as grand a space as Manhattan would know, decorated with fine French antiques, a row of Monets in the dining room, two giant Rothkos and a series of Matisses in the living room. Sid was named to the Metropolitan Museum of Art's board the following year. Anne became a leading benefactor of the New York City Ballet, until a nasty spat with its creative director. They might have gone on like that for years, traveling the world, collecting the finest art, but in 1986 Sid fell in love with another woman—a married woman, named Mercedes Kellogg. Both left their spouses in a romantic scandal that kept the gossip columnists busy for months.

After the divorce—Sid's settlement with Anne was said to be $450 million—Sid married Mercedes and retreated from the headlines into a cocoon of gilded luxury. White-coated butlers hovered at every home. A Falcon jet always stood ready for weekends in Paris, Rome, Rio, wherever the mood struck. Sid Bass had ascended to a plateau Sid Richardson could not have imagined, as far removed from the scrublands of distant Winkler County as a Texan could get. To many, in fact, his life was a kind of modern-day fantasy, something Walt Disney might have dreamed up, an endless tableau of private jets and presidential dinners and intellectual evenings with New York writers, artists, and politicians. It wasn't perfect, of course—Sid would tell you that—but as endings went, it beat the hell out of Clint Murchison Jr.'s.

Or, for that matter, Bunker Hunt's.

V.

After slowly recovering from the financial and emotional shocks brought on by Bunker's silver play, the remaining children of H. L. Hunt began to go their separate ways during the 1980s. Some flourished; some didn't. Lamar, who lost a good deal of his money in silver, folded his soccer team, the Tornado, in 1981, then, after years of squabbling with a competing tennis curcuit, surrendered to a merger and dissolution of World Championship Tennis. He retained control of the Kansas City Chiefs, but his best days were now behind him. His sister Caroline, after years as a relatively anonymous Dallas housewife, rebounded from a divorce to open a luxury Dallas hotel, the Mansion at Turtle Creek, and soon began buying and building hotels around the world. Ray Hunt, finally free of the first family's intrigues, continued to build Hunt Oil. After finding oil in his first international venture, in the North Sea, he signed a drilling accord with the government of the remote Middle Eastern nation of Yemen in 1981. Over the next twenty years Hunt Oil would draw billions of barrels of oil from Yemen's sandy soil.

What drove the second-generation Hunts apart—emotionally as well as financially—was the massive $1.1 billion loan Bunker and Herbert had taken to escape the silver debacle. The collateral was Placid Oil, still owned by the six siblings' various trusts. That arrangement was fine so long as oil prices continued to rise; in the early 1980s, Placid was valued at $2.2 billion, twice the size of the loan. But even before oil prices began to drop in 1983, Margaret and the other siblings realized that if the worst happened—if Bunker and Herbert defaulted on the loan—the banks would go after Placid, which could cost the "uninvolved" siblings millions of dollars.

Margaret moved to sever all her financial ties to Placid, and to Bunker and Herbert. All through 1982 and 1983 family attorneys worked to disentangle Margaret, Caroline, and Hassie's holdings from those of Bunker and Herbert's. In the end, the three walked away with some of Placid's most promising oil fields. Bunker and Herbert, shrugging off several years of awful publicity—both were obliged to testify about their silver dealings before Congress in 1980—struggled to break free from the confines of their $1.1 billion debt. The oil fields Margaret and Caroline took from Placid left it a smaller company, with less cash to pay the banks each month. At first the two brothers had some success reducing their debt load, raising $410 million by selling off oil properties to the Petro-Lewis Corp. in 1982; they later raised another $161 million by selling stock in a number of other companies, including Louisiana Land & Exploration and Gulf Resources. Bunker wanted to sell even more assets, but the bank consortium's managing partner, Morgan Guaranty of New York, resisted, insisting it would need new collateral to replace anything that was sold. Bunker lobbied the other banks to remove Morgan and succeeded, replacing it with two banks, Bankers Trust and Republic of Dallas, who seemed more reasonable.

No sooner had Bunker achieved some elbow room, however, than oil prices began to drop. It was Penrod that first felt the shock waves. Penrod leased drilling rigs, but the drop in prices meant every major oil company froze or scaled back its hunt for new reserves until prices stabilized. As the months wore on, more and more Penrod rigs fell idle. The company had four hundred million dollars in debt, and in late 1983 Bunker asked its banks for better terms, along with fifty-seven million dollars in new loans. The banks agreed, but asked for more collateral. Bunker and Herbert reluctantly went along, handing over 476 acres of land along the North Central Expressway in suburban Richardson. The first payments on the new debt were delayed until May 1985. For the moment, Penrod seemed secure.

The bad news, however, just kept coming. While the Hunts' oil companies at least still functioned, Bunker and Herbert's sugar empire was falling apart. A venture that began with the takeover of Great Western Sugar in 1974 had expanded over the years, with the acquisitions of several other sugar companies, eventually rolled into a single holding company named Hunt International Resources, known as HIRCO. The company had been piling up annual losses for years when it was blindsided by the soft-drink industry's decision to begin flavoring Coca-Cola, Pepsi, and other drinks

with corn and artificial sweeteners instead of sugar. All through the early
1980s, sugar prices fell. When HIRCO's suppliers demanded new contracts
to reflect the lower prices, the Hunts refused, then sued. Their business
evaporated. Bunker tried to sell the sugar factories, but no one wanted them
anymore.

In March 1985 HIRCO's sugar subsidiaries filed for bankruptcy protection, citing $200 million in liabilities on barely $175 million in assets. A
month later, HIRCO itself followed. Six months after that, a federal judge
wrenched all the business from Hunt control, placing it under the supervision of a Dallas trustee. By a conservative estimate, the sugar business had
cost Bunker and Herbert more than $1 billion.

Through it all, the Hunts managed to keep up appearances. Placid and
Penrod were still large and functioning; even at their darkest moments,
Bunker and Herbert never faced anything like what had happened to Clint
Murchison. In 1983 they moved all their various offices into the gleaming
new Thanksgiving Tower in downtown Dallas. Bunker was still able to fly
the world to watch his horses run races in California and Europe; in 1985
Forbes magazine estimated he remained the nineteenth-wealthiest man in
the world, though his net worth, nine hundred million dollars, was a fraction of what it had been a decade before. He still made headlines from time
to time, as when he donated money to the Nicaraguan Contra guerrillas
secretly funded by a White House aide named Oliver North. By and large,
the Hunts remained the Hunts. If not for their massive debts, much of it dating to the silver bailout, they might have weathered the storm. But at every
turn, silver continued to haunt them. In late 1984 the IRS sued Bunker and
Herbert, seeking two hundred million dollars in back taxes related to silver.
In March 1985, following a five-year federal investigation, the CFTC finally
filed a complaint against the brothers, formally charging them with unlaw-
fully conspiring to corner the silver market. It took years to sell all their
silver. When the last of it was finally gone, Bunker and Herbert had lost
billions.

All through 1984 and 1985, meanwhile, oil prices continued to fall. By
mid-1985 both Placid and Penrod, the twin pillars supporting the Hunt
empire, were on the verge of defaulting on their bank debts. Barely half of

Penrod's rigs were working; those that were fetched rates less than half what they demanded at the height of the drilling boom a decade earlier. Its loss in 1984 alone topped one hundred million dollars. Bunker and Herbert continued selling off pieces of Placid, but half its cash flow was still going to pay off the remaining silver debt. It no longer had sufficient cash to replace the reserves it was selling; at this rate, Placid would wither and die. Its last chance was a risky play Bunker and Herbert were eyeing in the Green Canyon area of the Gulf of Mexico, which analysts believed could contain billions of barrels of new oil. If Placid could find an elephant in the deep waters of the Gulf, Bunker believed, it might yet find a way to survive.

By mid-1985, however, both Placid and Penrod had told the banks they might be forced to miss debt payments. Bunker and Herbert pleaded for time. The banks refused. They had seen how the brothers stiffed their sugar lenders, and were in no mood for assurances. They demanded to be repaid, and didn't especially care where the money came from. "They made no bones about it," Herbert recalled. "They said, 'If Placid can't pay, get it from some other member of the Hunt family.'"

Negotations between the brothers and their banks were already tense when oil prices suddenly went into free fall in late 1985. All through 1986 they continued to drop like a stone, eventually hitting $12.88 a barrel that summer. In a panic, the bank consortium demanded more and more collateral. By and large, Bunker and Herbert refused. What they needed was more cash, not less, they insisted, if they were to have any hopes of hitting marketable amounts of oil in Green Canyon.

With oil prices plummeting, Placid missed its first debt payment on March 27; two months later, Penrod missed a payment as well. Five days later the banks called the loans. Bunker and Herbert had run out of time. Three weeks later, pushed to the brink of bankruptcy, they filed a massive $3.6 billion lawsuit against twenty-three different banks, charging them with fraudulent scheming to "dismantle and destroy" Placid and Penrod. A month later, the banks countersued Bunker, Herbert, and Lamar, demanding immediate repayment on total debts of more than $1.3 billion.

In late August 1986 the whole mess was thrown into a federal court when Placid filed for Chapter 11 bankruptcy. Hordes of attorneys from across the country descended on Dallas for the fireworks. In legal arguments that stretched for more than a year, the banks demanded immediate repayment of their loans. Hunt attorneys argued for the payments to be delayed, to give Placid a chance of finding oil in Green Canyon. The banks won many of the motions; the Hunts, after replacing their Boston attorneys in December 1986, hired a Dallas attorney named Stephen Susman, who began winning some motions of his own.

By mid-1987 Susman had entered into talks aimed at settling all the litigation and constructing a schedule to repay the banks. But for the Hunt brothers themselves, it was too late. Oil had fallen too far; there was simply too much debt. Herbert was the first to fall, filing for bankrupcy protection on July 22, 1987. Bunker held out as long as he could, but with the banks demanding shares of his remaining fortune to satisfy Placid's debts, he had no choice. Finally, on December 1, 1987, like hundreds of derelict oilmen across the state, Bunker followed Herbert into Chapter 11 bankruptcy protection. Lamar filed at the same time; he sold his North Dallas mansion and moved his family into a smaller home in Highland Park, not far from Bunker's. All the brothers' main assets—the sugar companies, Penrod, Placid, now the three brothers themselves—were operating in bankrupcy.

Fighting between the Hunts and the banks would stretch on for years longer, but it would all be anticlimactic. The story was over, at least for Bunker and Herbert. Soon Herbert would begin selling his beloved antique coins. Bunker would be forced to begin selling his horses. In barely ten years, thanks to one of the most harebrained financial schemes of the twentieth century, the two brothers had gone from being among the world's wealthiest men to laughingstocks.

VI.

Like the end of Texas Oil's golden age in the late 1950s, there was no formal announcement that the era of the Big Rich had reached its end. Texas would go on, of course, but something, no one was quite sure what, had died.

When did it happen? Some pointed to the Hunt bankruptcy, some Clint Murchison's, some to the awful auction John Connally underwent after his own bankruptcy. In truth the end came not with a single event but scores of small deaths, of the people and the places and the mood that made Texas what it had been. If forced to pick a date, many would point to the steamy afternoon of June 1, 1987, when the first wrecking balls slammed into the walls of the Shamrock Hotel.

It had been coming for years. In its day the hotel had hosted six American presidents, from Eisenhower to Reagan. By the early 1980s it was rarely full, and the grand soirees that once showcased Frank Sinatra and Milton Berle and Liberace had long since given way to pimply bands playing the proms of teenagers from Pearland and Cypress and Houston's gritty south side. The pool was still there, just as Glenn McCarthy had left it, but the Cork Club was long gone, the tuxedoed "oilionaires" and starlets replaced by winsome retirees. In December 1985 Hilton Hotels announced it was selling the hotel to the expanding Texas Medical Center, whose half-dozen facilities clustered around the hotel, all but begging for space. The Shamrock was to be closed, they said, then torn down, replaced by, of all things, a parking lot.

A neighborhood group sprang up to rescue it. They called themselves "Save the Shamrock." On Sunday, March 16, 1986, the day before St. Patrick's Day, some nine hundred of them paraded down South Main, turned onto Holcombe, and spread out to form a human ring around the old hotel. One man wore a T-shirt that read TEARING DOWN THE SHAMROCK WOULD BE LIKE TEARING DOWN THE ALAMO. Television cameras whirred; local reporters scurried through the crowd, scribbling in their notebooks. At one point there was a stir. A reporter was talking to an elderly man in a dark suit. Word trilled through the throng:

It was Glenn McCarthy.

He was almost eighty years old that day, unrecognizable as the dashing figure of yore, stooped and worn, his face sunken, ravaged by decades of Wild Turkey. He and his wife, Faustine, had been living in suburban La Porte for years by then, if not forgotten by Houston, discarded. "It's a silly thing to tear it down," McCarthy rasped as the reporters wrote down his words. "The people talking about knocking it down have lost their brains. It's a landmark known all over the world. Houston was a small city when it was built."⁵

McCarthy didn't come to the big Irish wake they threw that June the night the Shamrock closed. It was a sad evening. An older woman cried in the lobby. Middle-aged men led their children past the pool, pointing out where they swam as boys. Up in the penthouse a bagpiper serenaded the last visitors as they took the elevators down after midnight. Nor was McCarthy there the day they auctioned off thirty-five thousand bits of his memories, hundreds of Shamrock towels and robes, and thousands of pieces of Shamrock dinnerware. One man bought the entire breakfast room and opened it as The Shamrock Café in a Houston strip mall.

The building remained empty after that, padlocked and musty, until the wrecking crews arrived at the end of May 1987. They didn't rig explosive charges and take it down in one deafening bang, deciding instead to use the wrecking balls to knock it down piece by piece, blow by blow, like the fists of a frustrated Atlas pummeling its great granite facade. It took months for it all to crumble away, until the crews left in December, leaving behind nothing but heaps of rubble.

Glenn McCarthy's kidneys gave out a month later. Doctors wheeled him out of surgery at St. Luke's Hospital, and Faustine placed him in a nursing home. He lingered for nearly a year, finally dying, a day after his eighty-first birthday, on December 26, 1988. The newspapers hailed him as a "Texas giant" even as they tried to explain to younger readers who he had been. A thousand people attended his memorial service, where a bar singer from the Shamrock's glory days led renditions of "You'll Never Walk Alone" and "Londonderry Air." They buried him in Houston's Glenwood Cemetery, steps from the grave of Howard Hughes. "He was a tough man, and the drinking and fight stories are true enough, but he wasn't anything like that character in *Giant*," one old friend told a reporter. "He was a man, and he took a good deal of pride in that. I don't think there's ever been anyone like him, before or since."

Twenty years—twenty years now since the worst of it was over, Clint Murchison Jr.'s death, the Hunt bankruptcies, those awful days when another Texas banking chain seemed to fail every Monday, when the office towers around Houston and Dallas were so empty they were called "see-through" buildings—the days when the era of the Big Rich finally came to its crashing end. Texas lives on, of course, as do many of the Hunts and Basses and Cullens and Murchisons, but it's not the same today, not really. None of the Big Four families remain true players on the nation's business or political stage. When one of their own, George W. Bush, entered the White House in 2001, some of the old names began popping up at presidential dinners. Ray Hunt, still thriving, was one; in 2007 Hunt Oil received a lucrative concession to drill in northern Iraq. Sid Bass, whose family, along with the Hunts, ranked among Bush's largest financial backers, was photographed alongside the president, Laura Bush, and the queen of England.

Today, though, the Hunts and the Basses and all the rest are mere rich people, and Texas, despite its colorful cultural history, is—dare I say it—just another state. Was it ever anything else? Were they? To a Texan, the answer is yes. To many native Texans, it must feel as if something has been lost in the transition from family to corporate empires. A bit of bravado, maybe, a bit of pride. It's as if the knighted nobles have been replaced by a faceless plutocracy, all very clean and efficient, yet somehow lacking some of that boisterous Texas joie de vivre. Call it the state's maturation, as many have;

call it the days when Texas finally entered the modern age. Maybe this is just an adopted son's nostalgia, but I call it just a little sad.

This book is an epitaph for a Texas era I was too young to experience myself, except for its tumultuous end. But for those who lived through it, the age of the Big Rich, especially the golden years of the 1950s, was a time when Texas seemed poised to rival New York or California as a center of economic, political, and cultural influence. It's certainly had an impact, especially in political circles, where Texas politicians and even a few oil-men still make their voices heard, often loudly; witness the group of Lone Star businessmen led by oilman T. Boone Pickens who created the Swift Boat campaign against John Kerry in 2004. George W. Bush rode into the White House on a river of oil money, much of it from Texas; by one count, he received fifteen times as much money from energy companies and their executives as Al Gore. "Oil and gas money," the *New York Times* noted in 2000, "has been the essential lubricant of George Bush's political career."

But as a cultural or intellectual bellwether, Texas never quite fulfilled the promise of those first honeymoon years after 1948 when, whether you loved them or hated them, the Big Four oilmen seemed poised to trigger real change in America. Had they risen to prominence fifteen or twenty years earlier, when they first became rich, maybe they would have. But by the 1950s America was changing, and the most prominent oilmen, many born during the nineteenth century, weren't. In time, as conservative poli-tics moved into the mainstream, so did most Texas oilmen, quietly becom-ing just one more reliable source of campaign money for politicians from Barry Goldwater to John McCain.

Oil still matters in Texas, but it no longer dominates the zeitgeist. Working rigs, those actually looking for oil and gas, fell from a high of 1,318 in 1981 to 311 a decade later; even today, with world oil prices rising to record highs, the Texas rig count has only rebounded into the 900 range. By the mid-1990s taxes on the oil and gas industry provided barely 7 percent of Texas governmental revenues, less than a quarter the level in the late 1970s. Where the state still dominates is in oil refining; those massive plants lining the Gulf Coast still produce more gasoline and oil by-products than any place outside the Arabian Peninsula. It's not so much oil income as oil savvy that fuels much of the state's growth today. Just as Spindletop-era gushers educated the first Texas oilmen, the years since have produced an entire class of modern international oilmen, many of whom can be found today in fields from Nigeria to Indonesia to the deserts of Iraq. Despite a looming national recession, Houston and much of Texas is booming today not because of the oil beneath its dirt but the expertise its engineers and executives have built over the years.

"There's hardly any oil and gas production in a 40-mile radius of Houston," Mayor Bill White told *Slate* in 2008. "It's the knowledge that has concentrated here that is driving things."

As its oil reserves shrank, Texas discovered new sources of wealth, which has gone a long way toward smoothing its rough edges. Texas is far more urban, and far more educated, than it was fifty years ago. More Fortune 500 companies are based there than in any other state. The epicenter of its mid-1990s economic recovery was Austin, where a University of Texas kid named Michael Dell started a computer company that transformed the area into a salsafied Silicon Valley. Dell is the smiling face of the new Texas, at least the one chambers of commerce exalt. There are still plenty of wealthy oilmen in Houston and Dallas, but at charity dinners they sit beside just as many wealthy electronics and engineering CEOs. When Forbes published its list of the richest Americans in 2007, the wealthiest person in Texas wasn't even an oilman. It was Michael Dell. The richest Texas oilman, Dan Duncan, ranked 39th. Of the Big Four families, it's no surprise, the Basses remain the wealthiest; Robert Bass, at $5.5 billion, ranked 57th, Sid and Lee at $3 billion, 117th, and Ed, $2.5 billion, down at 165th. The only Hunt to make the list was Ray, with $4 billion, who ranked 82nd. None of the Murchisons or Cullens even make the list anymore.

It's startling how quickly people forget the men who made so much of this possible. When Glenn McCarthy died in 1988, the *Houston Post* headline actually termed him "happy-go-lucky," about the furthest thing from what he was. Today, outside the oil museums in East Texas and Midland and at Spindletop, you have to look hard to find monuments to the Big Rich. The Shamrock is long gone. The Cowboys will leave Texas Stadium for a new home in Arlington soon; Clint Jr.'s landmark facility will probably be torn down. The home Big Clint built in the 1930s is gone now, as is Clint Jr.'s mansion; both were razed to make way for subdivisions.

The Basses have left far more footprints, beginning with all those Sid Richardson university buildings and youth camps. The family's most visible legacy remains downtown Fort Worth, much of it revitalized with the help of Bass money. Perry Bass, who died in 2006 at the age of ninety-one, also gave millions to his alma mater, Yale; today the university has a Perry R. Bass Center for Molecular and Structural Biology. His son Robert gave $13 million to refurbish the Yale library, now known as The Bass Library. In Dallas, Ray Hunt's Reunion Center still shapes the downtown skyline; his half sister Caroline's Mansion on Turtle Creek remains one of the state's finest restaurants. Down in Houston, the Cullens built a skyscraper, the Cullen Center, on the edge of downtown in the 1960s. The family office is still there, on a high floor. Oscar Wyatt, the cantankerous oilman recently convicted of paying bribes to Saddam Hussein, lived in Roy Cullen's old mansion for forty years. A new oilman has moved in now, a nice fellow, neighbors say.

The Cullens have remained out of the public eye since their last dustup with the Baron Enrico di Portanova, who relaunched his legal assault on the family in the late 1970s. At the time, it was the largest probate case ever brought in the United States, front-page news in the *Wall Street Journal*. The Baron, alas, lost the case, once and for all. He died of lung cancer in Houston in 2000. His 28-bedroom Acapulco mansion, Arabesque, used in the James Bond film *License to Kill*, went on the market a few years back. Asking price: $29 million. Roy Cullen's most prominent son-in-law, Corbin Robertson, died in 1991. His son, Corbin Jr., known as "Corby," runs Quintana today and has branched into private equity, buying a string of energy-related companies, as well as several West Virginia coal companies.

In Fort Worth, Lee Bass still runs the family's oil-related businesses, the heart of which remain Sid Richardson's West Texas and Louisiana oilfields; the Louisiana fields were hit hard during Hurricane Katrina. A well-regarded civic leader, Lee sold one of his great uncle's last assets, a West Texas pipeline company, for $1.6 billion in 2005; ironically, the buyer was Southern Union, the company Clint Murchison founded so many years ago. Ed Bass dabbles in real estate. Robert Bass remains an active investor and philanthropist. According to associates, he also remains estranged from Sid, who splits his time between homes in Fort Worth and on New York's

Fifth Avenue. St. Joe's, now renamed San Jose Island, is still the Bass family's private retreat. Richard Rainwater, now based in South Carolina, is one of America's preeminent investors, as are several other Bass alumni.

Many of the Murchison grandchildren remain in the Dallas area. Clint Jr.'s sons fished several family assets out of the ruins of bankruptcy and quietly run them to this day. They don't have much to do with the other side of the family. John Murchison's widow, Lupe, died a few years back. Jane Murchison Haber remained active in volunteer activities in Dallas and New York for years. She died in 2001. Clint Jr.'s second wife, Anne, is still alive and living in East Texas, where she has a Web site devoted to religious writings and poetry. Not long ago Clint Jr.'s son Burk published a recipe book based on his childhood vacations at Spanish Cay, now long sold.

When I spoke to Burk and his brother Robert, they expressed concern that I might delve into the tawdrier aspects of their father's life, especially his drug use, which they strongly deny. Both speak lovingly of Clint Jr. and want readers to know that whatever his flaws, he was a brilliant and sensitive man, a caring father and an inspiration to them all. "At the end of his life he made some bad decisions, and listened to people with bad advice," Robert told me. "But he was always a man of honor, a man of his word."

Many of H. L. Hunt's heirs remain prominent in Dallas; between children, grandchildren, and great-grandchildren, there are more than one hundred Hunts in all. After years as a beloved civic volunteer, Margaret Hunt Hill died in 2007; one of her grandsons is now suing his father and uncles for a greater share of his inheritance. Hassie Hunt died in 2005. Lamar remained active in sports circles for years, cofounding Major League Soccer in 1996 and owning two of its franchises; he died in 2006. His sister Caroline owns a string of chic hotels; her net worth has been estimated at $600 million. One of H.L.'s daughters by Ruth, Swanee Hunt, was President Clinton's ambassador to Austria, then founded the Women and Public Policy Program at Harvard's Kennedy School of Government. Her sister June remains an evangelist with her own radio program broadcast from Dallas. Frania Tye Lee died in 2002 at the age of ninety-eight. All her children, after belatedly accepting their settlement with the Hunts in the early 1980s, have passed away.

Both Placid Oil and Penrod Drilling were sold off during the 1990s; Bunker sued in vain to block the Placid sale, saying it wasn't what his father would have wanted. All told, Bunker and Herbert's combined losses from the crash in oil prices and their boneheaded silver play have been widely estimated at about $5 billion. After years of legal wrangling, the two brothers finally sorted out their troubles with the government in 1994. Together they agreed to turn over about $140 million in fines and back taxes; part of that went to the State of New York, where a jury found them guilty on civil charges of conspiring to corner the silver market. While embroiled in corporate and personal bankruptcies, Bunker was obliged to sell off many of his assets, including his Circle T ranch and all his thoroughbred racehorses. In a single day in 1988, he sold 580 horses for almost $47 million.

Not that either brother will be accepting charity anytime soon. Bunker emerged from his legal travails with a net worth estimated at $175 million, thanks to a private trust that lay outside his creditors' reach; Herbert's wealth was believed to be in the same range. Both plunged back into business, this time separately. Herbert called his new company Petro-Hunt, which his family runs today. Bunker has a small oil company, too, and by 2000 he was doing well enough to begin buying racehorses again. By 2004 he owned almost eighty. He and Herbert, now both in their mid-eighties, continue to live quietly in Dallas.

It's hard to know what H. L. Hunt or Clint Murchison or Roy Cullen would think of America today. They would probably hate it; Lord knows it wouldn't think much of them. No doubt their hearts would have warmed to see the White House run for eight years by a true conservative, and a Texas oilman at that. All the trickles of conservative sentiment they nurtured going back to the 1930s joined a rushing river of conservatism that put first Ronald Reagan in office, then George H. W. Bush, then his son, populating the halls of Congress with thorny troublemakers like Tom DeLay and Dick Armey and Phil Gramm. Texas oil money didn't put any of them in office, not by itself, but all that cash, along with a brash Texas spirit, long ago empowered a state that might have gone down as just another Mississippi or Alabama. Texas oil, and to a degree the Big Four families, brought true national power to the state, and it's a power America grapples with to this day.

THANK YOUS

There're a lot of folks to thank on a book like this, beginning with Scott Moyers, who first suggested I write something about Texas oil. Scott began this project as my editor and ended up as one of my agents, which is a story in itself. Thanks also to Scott's new colleagues, my longtime agents, Andrew Wylie and Jeff Posternak, and to Brian Siberell at the Creative Artists Agency in Los Angeles.

Much of the on-site work for *The Big Rich* was done in Texas libraries and courthouses. I'd like to thank the very helpful staff at the Special Collections room at the University of Houston, which houses, among its many valuable archives, the papers of George Fuermann; the Center for American History and the Lyndon Baines Johnson Presidential Library at the University of Texas at Austin; the Rice University library in Houston; the Permian Basin Oil Museum in Midland; the Angelo State University Library in San Angelo; the public library in Houston, which holds certain papers of John Henry Kirby; the Sam Houston Regional Library and Research Center in Liberty, which holds the papers of Martin Dies; the Special Collections room at the University of Texas Arlington Library, which holds the papers of both George Armstrong and Ida Darden; the Texas Tech University library in Lubbock; the Henderson County Historical Society in Athens; the Rosenburg Library in Galveston; as well as the public libraries in Dallas, Austin, Beaumont, Fort Worth, Galveston, Midland, Monahans, Odessa, Kermit, Athens, Mexia, Temple, Waco, and Wichita Falls.

At the Cullen Foundation in Houston, Alan Stewart was a gracious guide to the family's history. Thanks to Carl Freund, Joseph Pratt at the University of Houston, and Stephen Fox at Rice University for invaluable guidance. At the Equitable Insurance Company, Jon Cross uncovered scads of information in the corporate archives on the company's unfortunate dealings with Glenn McCarthy. A special thanks to Sid Richardson's "old family friend," who insisted on remaining anonymous; I could not have assembled so complete a picture of Richardson's life without this person's help. In New York, John Ortved and Chris Bateman were indefatigable researchers who helped track down hard-to-find newspaper and magazine articles. Melissa Goldstein collected all the photos. Thanks to Doug Stumpf and Marla Burrough for reading the manuscript, and to my new editor, Laura Stickney, for whipping it into fighting shape. Thanks as well to Graydon Carter, the editor of *Vanity Fair*, the best boss a writer could possibly have.

Many thanks to my parents, Mac and Mary Burrough of Temple, Texas, for inspiration, shelter, and sympathetic ears, and to Paul Kerr, for being a great friend to our family. And last, but never least, all my love to Marla, Griffin, and Dane, for everything.

NOTES

CHAPTER 1: "THERE'S SOMETHING DOWN THERE..."
1. Daniel Yergin. *The Prize: The Epic Quest for Oil, Money & Power* (New York: Free Press, 1991), 86–89.

CHAPTER 2: THE CREEKOLOGIST
The principal source for this chapter is Roy Cullen's authorized biography, *Hugh Roy Cullen: A Story of American Opportunity* by Kilman and Wright. Additional material was drawn from an unpublished manuscript by Ed Kilman, held in the Cullen family archives in Houston, as well as various newspaper articles that accompanied Cullen's gifts to the University of Houston in 1937 and at his death.
1. Cited in Diana Davids Olien and Roger M. Olien. *Oil in Texas: The Gusher Age, 1895–1945* (Austin: University of Texas Press, 2002), 9.
2. Kilman and Wright, 130.
3. Cited in "Lillie and Me," unpublished manuscript, held in Cullen family storeroom.
4. Unpublished Kilman manuscript.
5. Kilman and Wright, 141–42.

CHAPTER 3: SID AND CLINT
The principal sources of information about Clint Murchison's early years are his authorized biography, *Clint*, by his longtime secretary, Ernestine Orrick Van Buren, and *The Murchisons* by Jane Wolfe. The story of Sid Richardson's early years is taken from several magazine articles published in the 1950s, especially "The Billionaire Bachelor" (*Collier's*, 1954); a biographical essay published by Richardson's longtime secretary's son, contained in archives at the Henderson County Historical Society; land and deed records in Ward, Winkler, and Henderson Counties; plus the recollections of a longtime family friend.
1. Wolfe, 31.
2. Van Buren, 40.

CHAPTER 4: THE BIGAMIST AND THE BOOM
The principal source for information about H. L. Hunt's early years is Hunt's self-published *Hunt Heritage*. Additional information is drawn from *Texas Rich* by Harry Hurt and *H.L. and Lyda* by

Hunt's oldest daughter, Margaret Hunt Hill. Information about Hunt's relationship with Frania Tye is derived from Hurt, with several scenes derived from Hill. Background on Dad Joiner and the East Texas field is taken from many sources, notably *The Last Boom* by Michel Halbouty and James Clark. The best summary of Hunt's dealings with Joiner is Hurt, augmented by contemporary newspaper reports.

1. Hunt, *Hunt Heritage*, 88.

CHAPTER 5: THE WORST OF TIMES, THE BEST OF TIMES

Information in this chapter was derived from a variety of sources. The story of Murchison's work in the early 1930s is taken from Van Buren, *Clint*; Wolfe, *The Murchisons*; and *Southern Union* by N. P. Chesnutt. Information about Cullen and West's sale to Humble was found in *History of Humble Oil & Refining Company* by Henrietta M. Larson and Kenneth Wiggins Porter. The story of the Tom O'Connor field is told in Kilman and Wright. Information about George Strake and the Conroe field is derived from a 1956 commemorative booklet written by Patrick O'Bryan, as well as interviews with George Strake Jr. The story of Sid Richardson's work during the Depression is derived from a variety of sources, including Van Buren and land and lease records filed in Winkler County. Information about Richardson's various loans is taken from Winkler County records. Information about Richardson's relationship with Charles E. Marsh is derived from unpublished correspondence between the two men contained in the Marsh papers at the Lyndon B. Johnson Presidential Library.

1. Van Buren, 100.
2. Ibid.
3. James Presley, *Saga of Wealth: The Rise of the Texas Oilman* (New York: G. P. Putnam's Sons, 1978), 116.

CHAPTER 6: THE BIG RICH

Information in this chapter was derived from a variety of sources. The story of Murchison's life during the 1930s is taken from Wolfe and Van Buren. The story of Richardson's life during the 1930s is taken from the recollections of "The Old Friend," as well as Van Buren. Perry Bass's recollections of St. Joe's are taken from *Aransas; The Life of a Texas Coastal County* by William Allen and Sue Hastings Taylor. Richardson's home is described in *The Architecture of O'Neil Ford* by David Dillon. The story of Cullen's life during the 1930s is taken from newspaper clippings contained in the Cullen archives; interviews with his grandson, Roy Cullen; as well as Kilman and Wright. Information about Hunt is principally derived from Hurt and Hill.

1. Wolfe, 101.
2. Van Buren, 283.
3. David Dillon, *The Architecture of O'Neil Ford* (Austin: University of Texas Press), 1999.
4. Fuermann, *Reluctant Empire*, 38.
5. Hill, 115–17.
6. Ibid., 110.
7. Hurt, 134.
8. Hill, 211–13.
9. Ibid., 214–16.
10. Hurt, 128–30.
11. Ibid., 133.
12. Ibid.

CHAPTER 7: BIRTH OF THE ULTRACONSERVATIVES

Information in this chapter is derived from a variety of books and Texas archives, including the George W. Armstrong papers at the University of Texas Arlington; the Maco Stewart papers at The Rosenberg Library in Galveston; and the John Henry Kirby papers at the Houston Public

Library. Biographical information on Kirby is taken from contemporary accounts as well as *John Henry Kirby* by Mary Lasswell. Kirby's activities with Vance Muse are described in *The Establishment in Texas Politics* by George Norris Green, as well as numerous magazine and newspaper articles. The story of Richardson and Murchison's dealings with the Roosevelts is derived from IRS depositions contained in the Westbrook Pegler papers at the Herbert Hoover Presidential Library.

1. Cited in William Anderson, *The Wild Man from Sugar Creek.*
2. Cited in correspondence, George Armstrong papers, University of Texas Arlington.
3. Letter, Lewis Valentine Ulrey to J. Frank Norris, April 25, 1938, contained in Maco Stewart papers, The Rosenberg Library, Galveston, Texas.
4. Cited in Barney Farley, *Fishing Yesterday's Gulf Coast* (College Station: Texas A&M Press, 2002).

CHAPTER 8: WAR AND PEACE

Information in this chapter is drawn from Hill, Hurt, Van Buren, Wolfe, Kilman and Wright, and other books, as well as contemporary newspaper articles. Work on the Inch lines is described in Don Carleton, *A Breed So Rare: The Life of J. R. Parten, Liberal Texas Oil Man, 1896–1992* (Austin: Texas State Historical Association, 1998) as well as in Christopher J. Castaneda and Joseph A. Pratt, *From Texas to the East: A Strategic History of the Texas Eastern Corporation* (College Station: Texas A&M Press, 1993). Everette DeGolyer's Middle Eastern travels are described in Yergin and in Lon Tinkle, *Mr. De: A Biography of Everette Lee DeGolyer* (Boston: Little Brown, 1970).

1. Cited in Richard Goodwin, *Texas Oil, American Dreams: A Study of the Texas Independent Producers and Royalty Owners Association* (Austin: Texas State Historical Association, 1996).
2. Hurt, 219.

CHAPTER 9: THE NEW WORLD

Information in this chapter is derived from contemporary newspaper and magazine articles, as well as Wallace Davis, *Corduroy Road: The Story of Glenn H. McCarthy* (Houston: Anson Jones Press, 1951). The story of McCarthy's dealings with Equitable are taken from reports in the Equitable archives in New York. Biographical information about Edna Ferber came from Julie Goldsmith Gilbert, *Ferber: A Biography of Edna Ferber and Her Circle* (Garden City, N.Y.: Doubleday, 1978).

1. Cited in Leon Jaworski, *Confession and Avoidance: A Memoir* (Garden City, N.Y.: Doubleday, 1979), 54–55.
2. *Houston Chronicle*, Sept. 26, 1950.
3. *New York Times*, Sept. 28, 1952.

CHAPTER 10: "A CLUMSY AND IMMEASURABLE POWER"

Information in chapters 10 and 11 is taken from a variety of contemporary magazine and newspaper articles, as well as books cited below.

1. "Texas Business and McCarthy," *Fortune*, May 1954.
2. *Houston Chronicle*, Oct. 29, 1948.
3. Kilman and Wright, 270–71.
4. Kilman and Wright, 281.
5. *The Nation*, Nov. 3, 1951.
6. Hurt, 151.
7. Jerome Tuccille, *Kingdom: The Story of the Hunt Family of Texas* (Washington: Beard Books, 1984), 229.
8. Ibid., 231.
9. Caro, 274.

10. Ibid.
11. James Reston, Jr., *Lone Star: The Life of John Connally* (New York: Harper & Row, 1989), 162.
12. Ibid., 163.
13. Caro, *The Means of Ascent*.
14. Kilman and Wright, 293.
15. Cited in George Norris Green, *The Establishment in Texas Politics: The Primitive Years, 1938–1957* (Norman: University of Oklahoma Press, 1979), 147.
16. Drew Pearson papers, LBJ Presidential Library.
17. Jack Anderson, *Washington Expose* (Washington DC: Public Affairs Press, 1967), 214–16.
18. Cited in Louise Galambos, and Daun Van Ee, ed. *The Papers of Dwight David Eisenhower: The Presidency: Keeping the Peace* (Baltimore: Johns Hopkins University Press, 2001).
19. "Texas Business and McCarthy," *Fortune*, May 1954.
20. Hurt, 158–62.
21. *Fortune*, May 1954.
22. Wolfe, *The Murchisons*, 186.
23. Reston, 166.
24. Anthony Summers, *The Arrogance of Power: The Secret World of Richard Nixon* (New York: Penguin Books, 2000), 85.
25. Reston, 167.
26. Summers, *Official and Confidential: The Secret Life of J. Edgar Hoover* (New York: G.P. Putnam's Sons, 1993).
27. Ibid., 188.
28. "Texas Business and McCarthy," *Fortune*, May 1954.

CHAPTER 11: "TROGLODYTE, GENUS TEXANA"
1. Wolfe, *The Murchinsons*, 200.
2. Ibid., 201.
3. *New York Times*, April 23, May 2, and May 9, 1954.
4. Wolfe, 214–15.
5. *Time*, May 22, 1954.
6. *The Nation*, March 22, 1954.
7. *Houston Post*, May 23, 1954.
8. *New York Times*, Aug. 18, 1952.
9. Ibid., March 20, 1955.
10. *Dallas Morning News*, Nov. 3, 1961.
11. *Houston Post*, May 9, 1954.
12. Thanks to the Margaret Chase Smith library for this correspondence.
13. Fuermann, *Reluctant Empire*, 50.
14. Robert Sherrill, *The Accidental President* (New York: Grossman Publishers, 1967), 271; *The Washington Post*, September 22, 1970.
15. Caro, *Master of the Senate*, 663.
16. The payment would eventually be linked to a lobbyist representing Superior Oil of Houston.
17. John B. Judis, *William F. Buckley, Jr: Patron Saint of the Conservatives* (New York: Simon & Schuster, 1988), 120.

CHAPTER 12: THE GOLDEN YEARS
Information in this chapter is derived from a variety of sources, including Hurt, Hill, Van Buren, Wolfe's, *The Murchisons*, Kilman and Wright, as well as John Bainbridge, *The Super Americans* (New York: Holt, Rinehart and Winston, 1961).

1. Undated article by Stanley Walker in *The New Yorker*.
2. *Dallas Morning News*, Oct. 26, 1956.
3. Wolfe, *The Murchisons*, 150.
4. Kai Bird, *The Chairman: John L. McCloy. The Making of the American Establishment* (New York: Simon & Schuster, 1992), 431.
5. *Houston Post*, Dec. 5, 1964.
6. Cited in James Presley, *Saga of Wealth: The Rise of the Texas Oilman* (New York: G.P. Putnam's Sons, 1978), 219.

CHAPTER 13: RISING SONS

Information in this chapter is derived from Hill; Hurt; Van Buren; Wolfe, *The Murchisons;* and contemporary newspaper and magazine articles, especially those in the *New York Times, Life,* and *Time.*

1. Hill, 261.
2. Ibid., 251.
3. Hurt, 182.
4. *Time*, June 16, 1961.
5. Wolfe, *The Murchisons*, 168.
6. Dick Hitt, *Classic Clint: The Laughs and Times of Clint Murchison Jr.* (Plano, Tex.: Wordware Publishing, 1992); Hitt, 42.
7. Wolfe, *The Murchisons*, 270–71.
8. Hurt, 184.
9. Hitt, 82.
10. Ibid., 85.

CHAPTER 14: SUN, SEX, SPAGHETTI—AND MURDER

Information in this chapter is derived from Hurt and Wolfe's *The Murchisons*, as well as contemporary newspaper and magazine articles. The story of the Di Portanova-Cullen feud is derived from depositions and filings at the Harris County Courthouse and especially John Davidson, "The Very Rich Life of Enrico Di Portanova," *Texas Monthly*, March 1982.

1. Hurt, 238–39.
2. Ibid., 240–42.
3. Cited in Don Graham, *Cowboys and Cadillacs: How Hollywood Looks at Texas* (Austin: Texas Monthly Press, 1983).
4. *Houston Chronicle*, Jan. 28, 1961.
5. Cited in Warren Leslie, *Dallas Public and Private* (Dallas; SMU Press, 1998).
6. Wolfe, *The Murchisons*, 344.
7. Ibid., 345.
8. Ibid., 286.
9. This scene is elegantly described in James Conaway, *The Texans* (New York: Popular Library, 1978).
10. *Wall Street Journal*, Sept. 6, 1983.

CHAPTER 15: WATERGATE, TEXAS-STYLE

The principal source for this chapter is Hurt, augmented by contemporary newspaper accounts, especially in the *Dallas Morning News*.

1. Hurt, 277–78.
2. Ibid., 281–82.
3. Ibid., 284.
4. Ibid., 293.
5. Hurt, 296.

6. Fay, 23.
7. Hurt, 424.

CHAPTER 16: THE LAST BOOM
The primary source materials for this chapter are Hurt; Wolfe, *The Murchisons*; and personal interviews.
1. Wolfe, *The Murchisons*, 353.
2. Ibid., 360.
3. Ibid., 371.
4. Ibid., 373.
5. Ibid., 378.
6. Hurt, 354.
7. *Dallas Morning News*, Sept. 14, 1975.
8. Hurt, 391.

CHAPTER 17: THE GREAT SILVER CAPER
Information in this chapter is derived from Hurt, contemporary newspaper and magazine articles, and especially Stephen Fay, *Beyond Greed: The Hunt Family's Bold Attempt to Corner the Silver Market* (London: Penguin Books, 1982).
1. Cited in Bruce McNall, with Michael D'Antonio, *Fun While It Lasted* (New York: Hyperion, 2003).
2. Cited in Fay, 110–11.

CHAPTER 18: THE BUST
Information on Clint Murchison Jr. in this chapter is derived from Wolfe, *The Murchisons*, and from contemporary newspaper accounts, especially those in the *Dallas Morning News*. Sections on the Bass family were taken from personal interviews augmented with press accounts. The story of the Hunt family's travails and the Shamrock's demise were taken from personal interviews and contemporary newspaper and magazine accounts.
1. Wolfe, *The Murchisons*, 389.
2. Ibid., 410.
3. Ibid., 411.
4. *Dallas Morning News*, Oct. 25, 1986.
5. *Houston Chronicle*, March 17, 1986.

BIBLIOGRAPHICAL NOTES

Research for this book was drawn from a variety of sources, including interviews, documents from two dozen archives in Texas and elsewhere, plus more than two hundred books and thousands of newspaper and magazine articles on subjects from the families themselves to the Texas oil industry to the rise of modern American conservatism.

Much of my research is original, but much is not. It's important to differentiate between the two, mostly to give credit to the writers from whose work I've borrowed. Chief among these are Harry Hurt, author of the definitive Hunt-family biography, Texas Rich, and Jane Wolfe, author of The Murchisons. Both authors' work informed just about every chapter in this one, especially in later sections. Chapter 15, which deals with the wiretapping travails of Bunker and Herbert Hunt, was drawn largely from Hurt. Likewise, the portrait of Clint Murchison Jr. contained here is based mostly on Wolfe. In both cases I found I simply couldn't improve significantly on what these two talented authors had done more than two decades ago.

Three others books are important to note as well. Clint, an authorized biography of Clint Murchison by his longtime secretary, Ernestine Orrick Van Buren, was invaluable, as was Hugh Roy Cullen's authorized biography, 1954's Hugh Roy Cullen: A Story of American Opportunity. The third book I would mention is H.L. & Lyda by Margaret Hunt Hill.

What follows is a list of books I cite or drew from for The Big Rich:

Abramson, Howard S. Hero in Disgrace: The True Discoverer of the North Pole, Frederick A. Cook. San Jose: toExcel, 1991.

Alexander, Herbert E. Financing Politics: Money, Elections and Political Reform. Washington, D.C.: Congressional Quarterly, 1992.

———. Money in Politics. Washington, D.C.: Public Affairs, 1972.

Allen, George E. Presidents Who Have Known Me. New York: Simon & Schuster, 1950.

———. Presidents Who Have Known Me, 1960 Edition. New York: Simon & Schuster, 1960.

Allen, William, and Sue Hastings Taylor. Aransas: The Life of a Texas Coastal County. Austin: Eakin, 1997.

Ambrose, Stephen E. Eisenhower: Soldier and President. New York: Simon & Schuster, 1990.

Anderson, Jack. Washington Exposé. Washington, D.C.: Public Affairs, 1967.

Anderson, William. *The Wild Man from Sugar Creek: The Political Career of Eugene Talmadge.*
Baton Rouge: Louisiana State University Press, 1975.

Armstrong, George. *Memoirs of George W. Armstrong.* Austin: Steck, 1958.

————. *The Zionists.* Privately published, 1950.

Ashman, Charles. *Connally: The Adventures of Big Bad John.* New York: William Morrow & Co.,
1974.

Bainbridge, John. *The Super Americans.* New York: Holt, Rinehart and Winston, 1961.

Baker, Bobby, with Larry L. King. *Wheeling and Dealing: Confessions of a Capitol Hill Operator.*
New York: W.W. Norton, 1978.

Ball, Douglas, and Dan S. Turner. *This Fascinating Oil Business.* Indianapolis: Bobbs-Merrill,
1965.

Barragy, T. J. *Gathering Texas Gold: J. Frank Dobie and the Men Who Saved the Longhorns.* Cayo
Del Grullo Press, n.d.

Bell, Douglas, ed. *The Radical Right.* Garden City, N.Y.: Doubleday, 1964.

Bird, Kai. *The Chairman: John J. McCloy and the Making of the American Establishment.* New
York: Simon & Schuster, 1992.

Bjerre-Poulsen, Niels. *Right Face: Organizing the American Conservative Movement 1945–65.*
Copenhagen: Museum Tusculanum Press, 2002.

Boatright, Mody. *Folklore of the Oil Industry.* Dallas: SMU Press, 1963.

Boatright, Mody C., and William A. Owens. *Tales from the Derrick Floor: A People's History of the
Oil Industry.* Garden City, N.Y.: Doubleday, 1970.

Bredeson, Carmen. *The Spindletop Gusher: The Story of the Texas Oil Boom.* Brookfield, Conn.:
Millbrook Press, 1997.

Brown, Stanley H. *H.L Hunt.* Chicago: Playboy Press, 1976.

Bryce, Robert. *Cronies: Oil, the Bushes and the Rise of Texas, America's Superstate.* New York:
Public Affairs, 2004.

Burst, Ardis. *The Three Families of H. L. Hunt.* New York: Weidenfield & Nicholson, 1988.

Byrd, David Harold. *I'm an Endangered Species.* Houston: Pacesetter Press, 1978.

Carleton, Don E. *A Breed So Rare: The Life of J. R. Parten, Liberal Texas Oil Man, 1896–1992.*
Austin: Texas State Historical Association, 1998.

————. *Red Scare! Right-wing Hysteria, Fifties Fanaticism and Their Legacy in Texas.* Austin: Texas
Monthly, 1985.

Caro, Robert A. *Master of the Senate.* New York: Knopf, 2002.

————. *Means of Ascent.* New York: Random House, 1990.

————. *Path to Power: The Years of Lyndon Johnson.* New York: Random House, 1981.

Cartwright, Gary. *Blood Will Tell: The Murder Trials of T. Cullen Davis.* New York: Pocket,
1979.

Castaneda, Christopher James. *Invisible Fuel: Manufactured and Natural Gas in America,
1800–2000.* New York: Twayne, 1999.

————. *Regulated Enterprise: Natural Gas Pipelines and Northeastern Markets, 1938–1954.* Columbus: Ohio State, 1993.

Castaneda, Christopher J., and Joseph A. Pratt. *From Texas to the East: A Strategy History of the Texas Eastern Corporation.* College Station: Texas A&M Press, 1993.

Castaneda, Christopher J., and Clarance M. Smith. *Gas Pipelines and the Emergency of America's Regulatory State.* Cambridge: Cambridge Press, 1996.

Chestnutt, N. P. *Southern Union.* El Paso: Mangan Books, 1979.

Clark, James A., and Michel T. Halbouty. *The Last Boom: The Exciting Saga of the Discovery of the Greatest Oil Field in America.* Shearer, 1972.

————. *Spindletop: The True Story of the Oil Discovery that Changed the World.* Houston: Gulf Publishing, 1952.

Coleman, Loren. *Tom Slick and the Search for the Yeti.* Boston: Faber & Faber, 1989.

Conaway, James. *The Texans.* New York: Popular Library, 1978.

Connolly, John, with Mickey Herskowitz. *In History's Shadow.* New York: Hyperion, 1993.

Conway, Flo, and Jim Siegelman. *Holy Terror: The Fundamentalist War on America's Freedoms in Religion, Politics and Our Private Lives.* Garden City, N.Y.: Doubleday, 1982.

Cook, Blanche Wiesen. *Eleanor Roosevelt: Volume 2, The Defining Years.* New York: Penguin, 1999.

Cook, Fred J. *The Nightmare Decade.* New York: Random House, 1971.

Crawford, Anne Fears, and Jack Keever. *John B. Connolly: Portrait in Power.* Austin: Jenkins, 1973.

Dallek, Robert. *Lone Star Rising: Lyndon Johnson and His Times 1908–1960.* New York: Oxford Press, 1991.

Davidson, Chandler. *Race and Class in Texas Politics.* Princeton: Princeton University, 1990.

Davis, Wallace. *Corduroy Road: The Story of Glenn H. McCarthy.* Houston: Anson Jones Press, 1951.

Day, James M. *The Black Giant: A History of the East Texas Oil Field and Oil Industry Skulduggery & Trivia.* Austin: Eakin Press, 2003.

Diamond, Sara. *Roads to Dominion: Right-Wing Movements and Political Power in the United States.* New York: Guilford, 1995.

Dies, Martin. *Martin Dies' Story.* New York: Bookmailer, 1963.

Dillon, David. *The Architecture of O'Neil Ford.* Austin: University of Texas Press, 1999.

Donahue, Jack. *Wildcatter: The Story of Michel T. Halbouty and the Search for Oil.* New York: McGraw Hill, 1979.

Dugger, Ronnie. *The Politician: The Life and Times of Lyndon Johnson.* Old Saybrook: Konecky & Konecky, 1982.

Edwards, Lee. *The Conservative Revolution: The Movement That Remade America.* New York: Free Press, 1999.

Eisenhower, Dwight D. *Mandate for Change: The White House Years.* Garden City, N.Y.: Doubleday, 1963.

Engler, Robert. *The Politics of Oil*. Chicago: University of Chicago Press, 1961.

Epstein, Benjamin R., and Arnold Forster. *The Radical Right*. New York: Random House, 1966.

Farley, Barney. *Fishing Yesterday's Gulf Coast*. College Station: Texas A&M Press, 2002.

Fay, Stephen. *Beyond Greed: The Hunt Family's Bold Attempt to Corner the Silver Market*. London: Penguin Books, 1982.

Fehrenbach, T. F. *Lone Star: A History of Texas and the Texans*. New York: Da Capo, 2000.

Ferber, Edna. *Giant*. Garden City, N.Y.: Doubleday, 1952.

Ferrell, Robert H., ed. *The Eisenhower Diaries*. New York: W.W. Norton & Co., 1981.

Flemmons, Jerry. *Amon: The Life of Amon Carter Sr. of Texas*. Austin: Jenkins Press, 1978.

Flynn, John T. *The Road Ahead: America's Creeping Revolution*. New York: Devin-Adair, 1950.

———. *The Roosevelt Myth*. New York: Devin-Adair, 1948.

Forster, Arnold, and Benjamin R. Epstein. *Danger on the Right*. New York: Random House, 1964.

Frederickson, Kari. *The Dixiecrat Revolt and the End of the Solid South*. Chapel Hill: North Carolina Press, 2001.

Fried, Albert. *McCarthyism: The Great American Scare*. New York: Oxford, 1997.

Fuermann, George. *Houston: The Feast Years*. Houston: Premier. 1962.

———. *Houston: Land of the Big Rich*. Garden City, N.Y.: Doubleday, 1951.

———. *Reluctant Empire: The Mind of Texas*. Garden City, N.Y.: Doubleday, 1957.

Galambos, Louise, and Daun Van Ee, eds. *The Papers of Dwight David Eisenhower: The Presidency: Keeping the Peace*. Baltimore: Johns Hopkins University Press, 2001.

Garay, Ronald. *Gordon McLendon: The Maverick of Radio*. New York: Greenwood Press, 1992.

Gellermann, William. *Martin Dies*. New York: John Day, 1944.

Gentry, Curt. J. *Edgar Hoover: The Man and the Secrets*. New York: W.W. Norton & Co., 1991.

Gilbert, Julie Goldsmith. *Ferber: A Biography of Edna Ferber and Her Circle*. Garden City, N.Y.: Doubleday, 1978.

Goodman, Walter. *The Committee: The Extraordinary Career of the House Committee on Un-American Activities*. New York: Farrar, Straus and Giroux, 1964.

Goodwyn, Frank. *Lone-Star Land: 20th Century Texas in Perspective*. New York: Knopf, 1955.

Goodwyn, Lawrence. *Texas Oil, American Dreams: A Study of the Texas Independent Producers and Royalty Owners Association*. Austin: Texas State Historical Association, 1996.

Graham, Don. *Cowboys and Cadillacs: How Hollywood Looks at Texas*. Austin: Texas Monthly, 1983.

———. *Kings of Texas: The 150-Year Saga of an American Ranching Empire*. Hoboken, N.J.: John Wiley & Sons, 2003.

———. *Lone Star Literature*. New York: W.W. Norton, 2003.

———. *Texas: A Literary Portrait*. San Antonio: Corona, 1985.

Grant, Joseph M. *The Great Texas Banking Crash.* Austin: University of Texas Press, 1996.

Green, George Norris. *The Establishment in Texas Politics: The Primitive Years, 1938–1957.* Norman: University of Oklahoma Press, 1979.

Griffith, Robert. *The Politics of Fear: Joseph R. McCarthy and the Senate.* Amherst: University of Massachusetts Press, 1970.

Gunther, John. *Inside U.S.A.* New York: New Press, 1997.

Hardeman, D. B., and Donald C. Bacon. *Rayburn: A Biography.* New York: Madison Books, 1987.

Heale, M. J. *McCarthy's Americans: Red Scare Politics in State and Nation, 1935–1965.* Athens: University of Georgia Press, 1998.

Heard, Alexander. *The Costs of Democracy.* Chapel Hill: University of North Carolina Press, 1960.

Helgesen, Sally. *Wildcatters: A Story of Texans, Oil and Money.* Garden City, N.Y.: Doubleday, 1981.

Hill, Margaret Hunt, with Burt Boyer and Jane Boyer. *H.L. and Lyda: Growing up in the H. L. Hunt and Lyda Bunker Hunt Family as Told by Their Eldest Daughter.* Little Rock: August House, 1994.

Himmelstein, Jerome L. *To the Right: The Transformation of American Conservatism.* Berkeley: University of California Press, 1990.

Hitt, Dick. *Classic Clint: The Laughs and Times of Clint Murchison Jr.* Plano, Tex.: Wordware Publishing, 1992.

Hunt, H. L. *HLH Columns.* Dallas: HLH Products, 1965.

———. *Hunt Heritage.* Dallas: HLH Products, 1973.

———. *Weekly Strength.* Dallas: HLH Products, n.d.

———. *Why Not Speak.* Dallas: HLH Products, 1964.

Hurt, Harry, III. *Texas Rich: The Hunt Dynasty from the Early Oil Days through the Silver Crash.* New York: Norton, 1982.

Hyne, Norman J. *Nontechnical Guide to Petroleum Geology, Exploration, Drilling, and Production.* Tulsa: Penn Well, 2001.

Ickes, Harold L. *Fightin' Oil.* New York: Knopf, 1943.

Isser, Steve. *Texas Oil and the New Deal: Populist Corruption.* Lewiston, N.Y.: Edwin Mellen, 2001.

Jaworski, Leon, with Mickey Herskowitz. *Confession and Avoidance: A Memoir.* Garden City, N.Y.: Doubleday, 1979.

Johnson, Arthur M. *Petroleum Pipelines and Public Policy, 1906–1959.* Cambridge: Harvard University Press, 1967.

Judis, John. *William F. Buckley, Jr.: Patron Saint of the Conservatives.* New York: Simon & Schuster, 1988.

Kelley, Kitty. *The Family: The Real Story of the Bush Dynasty.* New York: Doubleday, 2004.

Kilman, Ed, and Theon Wright. *Hugh Roy Cullen: A Story of American Opportunity*. New York: Prentice-Hall, 1954.

Knowles, Ruth Sheldon. *The Greatest Gamblers: The Epic of American Oil Exploration*. Norman: University of Oklahoma Press, 1978.

Lamour, Dorothy, as told to Dick McInnes. *Dorothy Lamour: My Side of the Road*. Englewood Cliffs, N.J.: Prentice-Hall, 1980.

Larson, Henrietta M., and Kenneth Wiggins Porter. *History of Humble Oil & Refining Company*. New York: Harper & Brothers, 1959.

Lasswell, Mary. *John Henry Kirby*. Austin: Encino Press, 1967.

Leach, Joseph. *The Typical Texan: Biography of an American Myth*. Dallas: Southern Methodist University Press, 1952.

Leslie, Warren. *Dallas Public and Private: Aspects of an American City*. Dallas: SMU Press, 1998.

Lind, Michael. *Made in Texas: George W. Bush and the Southern Takeover of American Politics*. New York: Basic, 2003.

Linsley, Judith Walker, Ellen Walker Rienstra, and Jo Ann Stiles. *Giant Under the Hill: A History of the Spindletop Oil Discovery at Beaumont, Texas, in 1901*. Austin: Texas State Historical Association, 2002.

Lipset, Seymour Martin, and Earl Raab. *The Politics of Unreason: Right-Wing Extremism in America, 1790–1970*. New York: Harper & Row, 1970.

Livingstone, Harrison E. *The Radical Right and the Murder of John F. Kennedy*. Victoria, B.C.: Trafford, 2004.

Luksa, Frank. *Cowboys Essential*. Chicago: Triumph, 2006.

Lundberg, Ferdinand. *The Rich and the Super-Rich*. New York: Bantam, 1968.

McAlister, Wayne H. *Life on Matagorda Island*. College Station: Texas A&M Press, 2004.

McClellan, Barr. *Blood Money & Power: How LBJ Killed JFK*. New York: Hannover House, 2003.

McDaniel, Robert W., and Henry C. Dethloff. *Patillo Higgins and the Search for Texas Oil*. College Station: Texas A&M Press, 1989.

McEnteer, James. *Deep in the Heart: The Texas Tendency in American Politics*. Westport, Conn.: Praeger, 2004.

McNall, Bruce, with Michael D'Antonio. *Fun While It Lasted*. New York: Hyperion, 2003.

Mallison, Sam T. *The Great Wildcatter*. Charleston: Education Foundation of West Virginia, 1953.

Marcus, Stanley. *Minding the Store*. Boston: Little, Brown, 1974.

Marshall, J. Howard, III. *Done in Oil: An Autobiography*. College Station: Texas A&M Press, 1994.

Martin, William. *A Prophet with Honor: The Billy Graham Story*. New York: William Morrow, 1991.

———. *With God on Our Side: The Rise of the Religious Right in America*. New York: Broadway Books, 1996.

Micklethwait, John, and Adrian Wooldridge. *The Right Nation: Conservative Power in America.* New York: Penguin Press, 2004.

Miles, Ray. *King of the Wildcatters: The Life and Times of Tom Slick, 1883–1930.* College Station: Texas A&M Press, 1996.

Miller, Ann, with Norma Lee Browning. *Miller's High Life.* Garden City, N.Y.: Doubleday, 1972.

Miller, Char, ed. *Fifty Years of the Texas Observer.* San Antonio: Trinity Press, 2004.

Miller, Jeff. *Going Long: The Wild 10-Year Saga of the Renegade American Football League.* Chicago: Contemporary Books, 2003.

Minutaglio, Bill. *City on Fire: The Forgotten Disaster That Devastated a Town and Ignited a Landmark Legal Battle.* New York: HarperCollins, 2003.

———. *First Son: George W. Bush and the Bush Family Dynasty.* New York: Three Rivers Press, 1999.

Moncrief, Charlie. *Wildcatters: The True Story of How Conspiracy, Greed, and the IRS Almost Destroyed a Legendary Texas Oil Family.* Washington, D.C.: Regnery, 2002.

Mooney, Booth. *LBJ: An Irreverent Chronicle.* New York: Thomas Y. Crowell, 1976.

Morehead, Richard. *50 Years in Texas Politics.* Austin: Eakin, 1982.

Morris, Willie. *North Toward Home.* New York: Random House, 1967.

Moser, John E. *Right Turn: John T. Flynn and the Transformation of American Liberalism.* New York: New York University Press, 2005.

Myres, Samuel D. *The Permian Basin: Petroleum Empire of the Southwest: Era of Advancement.* El Paso: Permian Press, 1977.

Nash, George H. *The Conservative Intellectual Movement in America.* Wilmington, Del.: Intercollegiate Studies Institute, 1996.

Nevin, David. *The Texans: What They Are—and Why.* New York: Bonanza, 1968.

Nixon, Richard. *Six Crises.* New York: Simon & Schuster, 1962.

O'Connor, Richard. *The Oil Barons.* Boston: Little, Brown, 1971.

Olien, Diana Davids, and Roger M. Olien. *Oil in Texas: The Gusher Age, 1895–1945.* Austin: University of Texas Press, 2002.

Olien, Roger M. *From Token to Triumph: The Texas Republicans Since 1920.* Dallas: SMU Press, 1982.

Olien, Roger M., and Diana Davids Olien. *Easy Money: Oil Promoters and Investors in the Jazz Age.* Chapel Hill: University of North Carolina Press, 1990.

———. *Life in the Oil Fields.* Austin: Texas Monthly, 1986.

———. *Oil & Ideology: The Cultural Creation of the American Petroleum Industry.* Chapel Hill: University of North Carolina Press, 2000.

———. *Oil Booms: Social Change in Five Texas Towns.* Lincoln: University of Nebraska Press, 1982.

———. *Wildcatters: Texas Independent Oilmen.* Austin: Texas Monthly, 1984.

Oshinsky, David M. *A Conspiracy So Immense: The World of Joe McCarthy.* New York: Free Press, 1983.

Owen, Edgar Wesley. *Trek of the Oil Finders: A History of Exploration for Petroleum.* Tulsa: American Association of Petroleum Geologists, 1975.

Parmet, Herbert S. *Eisenhower and the American Crusades.* New York: Macmillan, 1972.

Pavlik, Gregory P. *Forgotten Lessons: Selected Essays of John T. Flynn.* Irvington-on-Hudson, N.Y.: Foundation for Economic Education, 1996.

Perlstein, Rick. *Before the Storm: Barry Goldwater and the Unmaking of the American Consensus.* New York: Hill & Wang, 2001.

Perry, George Sessions. *Texas: A World in Itself.* Gretna, La.: Pelican, 1975.

Pittman, H. C. *Inside the Third House: A Veteran Lobbyist Takes a 50-Year Frolic Through Texas Politics.* Austin: Eakin, 1992.

Polito, Robert. *Savage Art: A Biography of Jim Thompson.* New York: Random House, 1995.

Pope, Clarence. *An Oil Scout in the Permian Basin.* El Paso: Permian Press, 1972.

Posner, Gerald. *Case Closed: Lee Harvey Oswald and the Assassination of JFK.* New York: Random House, 1993.

Pratt, Joseph A., Tyler Priest, and Christopher J. Castaneda. *Offshore Pioneers: Brown & Root and the History of Offshore Oil and Gas.* Houston: Gulf Publishing, 1997.

Presley, James. *Saga of Wealth: The Rise of the Texas Oilman.* New York: G.P. Putnam's Sons, 1978.

Pressler, Paul. *The Texas Regulars.* Garland, Tex.: Hannibal Books, 2001.

Reaves, William E., Jr. *Texas Art and a Wildcatter's Dream: Edgar B. Davis and the San Antonio Art League.* College Station: Texas A&M Press, 1998.

Reeves, Thomas C. *The Life and Times of Joe McCarthy.* New York: Stein & Day, 1982.

Reston, James, Jr. *The Lone Star: The Life of John Connolly.* New York: Harper & Row, 1989.

Rister, Carl Coke. *Oil! The Titan of the Southwest.* Norman: University of Oklahoma Press, 1949.

Rogers, John William. *The Lusty Texans of Dallas.* New York: E.P. Dutton & Co., 1951.

Roosevelt, Elliott. *As He Saw It.* New York: Duell, Sloan & Pearce, 1946.

Roosevelt, Elliott, and James Brough. *A Rendezvous with Destiny: The Roosevelts of the White House.* New York: Dell, 1973.

———. *The Roosevelts of Hyde Park: An Untold Story.* New York: G.P. Putnam's Sons, 1973.

Rundell, Walter, Jr. *Early Texas Oil: A Photographic History, 1866–1936.* College Station: Texas A&M Press, 1977.

Rusher, William A. *The Rise of the Right.* New York: William Morrow, 1984.

Russell, Dick. *The Man Who Knew Too Much.* New York: Carroll & Graf, 1992.

Sanders, M. Elizabeth. *The Regulation of Natural Gas: Policy and Politics, 1938–1978.* Philadelphia: Temple University Press, 1981.

Schoenwald, Jonathan M. *A Time for Choosing: The Rise of Modern American Conservatism.* Oxford: Oxford University Press, 2001.

Schweizer, Peter, and Rochelle Schweizer. *The Bushes: Portrait of a Dynasty.* New York: Random House, 2004.

Scott, John David. *True Legacy: The Biography of a Unique Texas Oilman . . . W. A. 'Monty' Moncrief*. Privately published, 1982.

Scott, Peter Dale. *Deep Politics and the Death of JFK*. Berkeley: University of California Press, 1993.

Shaffer, Roger E. *Spindletop Unwound: A True Story of Greed, Ambition and Murder in the First Degree*. Plano: Republic of Texas Press, 1997.

Sheedy, Sandy. *Texas Big Rich: Exploits, Eccentricities and Fabulous Fortunes Won and Lost*. New York: William Morrow & Co., 1990.

Sheil, Mark, and Tony Fitzmaurice, eds. *Cinema and the City: Film and Urban Societies in a Global Context*. Malden, Mass.: Blackwell, 2001.

Sherill, Robert. *The Accidental President*. New York: Grossman, 1967.

———. *The Oil Follies of 1970–1980: How the Petroleum Industry Stole the Show (And Much More Besides)*. New York: Anchor Press, 1983.

Singerman, Philip. *Red Adair: An American Hero*. London: Bloomsbury, 1989.

Smith, Jean Edward. *Lucius D. Clay: An American Life*. New York: Henry Holt, 1990.

Smoot, Dan. *People Along the Way*. Tyler, Tex.: Tyler Press, 1993.

Spratt, John S. *The Road to Spindletop: Economic Change in Texas, 1875–1901*. Dallas: Southern Methodist University Press, 1955.

Stenehjem, Michele Flynn. *An American First: John T. Flynn and the America First Committee*. New Rochelle, N.Y.: Arlington House, 1976.

Summers, Anthony. *The Arrogance of Power: The Secret World of Richard Nixon*. New York: Penguin, 2000.

———. *Official and Confidential: The Secret Life of J. Edgar Hoover*. New York: G.P. Putnam's Sons, 1993.

Tair, Samuel W., Jr. *The Wildcatters: An Informal History of Oil-Hunting in America*. Princeton: Princeton University Press, 1946.

Tananbaum, Duane. *The Bricker Amendment Controversy: A Test of Eisenhower's Political Leadership*. Ithaca: Cornell University Press, 1988.

Texas Permian Historical Society, eds. *Water, Oil, Sand and Sky: A History of Ward County, Texas*. Pamphlet, 1962.

Thompson, Craig. *Since Spindletop: A Human Story of Gulf's First Half-Century*. Gulf, 1951.

Thompson, Jim. *South of Heaven*. New York: Fawcett, 1967.

Thompson, Thomas. *Blood and Money*. New York: Carroll & Graf, 1976.

Tinkle, Lon. *Mr. De: A Biography of Everette Lee DeGolyer*. Boston: Little, Brown, 1970.

Troyan, Michael. *A Rose for Mrs. Miniver: The Life of Greer Garson*. Lexington: University Press of Kentucky, 1999.

Tuccille, Jerome. *Kingdom: The Story of the Hunt Family of Texas*. Washington: Beard Books, 1984.

Van Buren, Ernestine Orrick. *Clint: Clinton Williams Murchison, a Biography*. Austin: Eakin, 1986.

Walker, Stanley. *Texas*. New York: Viking, 1962.

White, Theodore. *America in Search of Itself: The Making of the President, 1956–1980*. New York: Harper & Row, 1982.

Wolfe, Jane. *Blood Rich: When Oil Billions, High Fashion and Royal Intimacies Are Not Enough*. Boston: Little, Brown, 1993.

———. *The Murchisons: The Rise and Fall of a Texas Dynasty*. New York: St. Martin's, 1989.

Wolfskill, George. *The Revolt of the Conservatives: A History of the American Liberty League, 1914–1940*. Boston: Houghton Mifflin, 1962.

Wolfskill, George, and John A. Hudson. *All but the People: Franklin D. Roosevelt and His Critics, 1933–39*. London: Macmillan, 1969.

Yergin, Daniel. *The Prize: The Epic Quest for Oil, Money & Power*. New York: Free Press, 1991.

Zirbel, Craig I. *The Texas Connection: The Assassination of John F. Kennedy*. Scottsdale, Ariz.: Texas Connection, 1991.

INDEX